Studies in Advanced Mathematics

Harmonic Analysis and Applications

Studies in Advanced Mathematics

Titles Included in the Series

JOHN J. BENEDETTO
University of Maryland

Harmonic Analysis and Applications

CRC PRESS

Boca Raton   New York   London   Tokyo

Acquiring Editor: Tim Pletscher
Cover Designer: Denise Craig
Associate Marketing Manager, Direct Marketing: Becky McEldowney
Associate Marketing Manager: Susie Carlisle
Senior Project Editor: Carol Whitehead
Manufacturing Assistant: Sheri Schwartz

Library of Congress Cataloging-in-Publication Data

Benedetto, John.
    Harmonic analysis and applications / John J. Benedetto.
      p. cm.--(Studies in advanced mathematics)
    Includes bibliographical references and index.
    ISBN 0-8493-7879-6 (alk. paper)
    1. Harmonic analysis. I. Title. II. Series.
QA403.B377 1996
515'.2433--dc20
                                        96-20445
                                                CIP

No claim to original U. S. Government works
International Standard Book Number 0-8493-7879-6
Library of Congress Card Number 96-20445
Printed in the United States of America 1 2 3 4 5 6 7 8 9 0
Printed on acid-free paper

About the Author

John J. Benedetto is a harmonic analyst.

He has published over 100 papers, including 9 monographs, in wavelet and frame theory, in spectral analysis, estimation, and synthesis, and on the uncertainty principle, speech compression, and noise reduction.

Dr. Benedetto has had 20 Ph.D. students. He is Editor-in-Chief of *The Journal of Fourier Analysis and Applications* published by CRC Press and is Series Editor of Birkhäuser's Applied and Numerical Harmonic Analysis series. The current work as well as his co-edited book, *Wavelets: Mathematics and Applications*, are titles in the Studies in Advanced Mathematics series published by CRC Press.

To my parents, Vienna and John,
and
To my wife, Catherine

Contents

Preface

Among the various points of view in harmonic analysis, we shall define and emphasize the *analysis* and *synthesis* of functions in terms of *harmonics*. My *goal* is to present harmonic analysis at a level that exhibits its vitality, intricacy and simplicity, power, elegance, and usefulness. Despite an array of deep theoretical results and important abstract structures associated with harmonic analysis, I *believe* the subject lives because of its essential role in understanding a host of engineering, mathematical, and scientific ideas. Goals and beliefs are not always realities, but this book is my attempt to quantify the preceding aspiration and sentiment.

Harmonic Analysis and Applications is both a textbook and essay.

Textbook

The listing in Prologue I provides the material for an upper level undergraduate course in harmonic analysis and some of its applications. We refer to this course as Course I, and I have taught such a course for many years to engineering, physics, computer science, and mathematics students. The first 30 exercises in each chapter are appropriate for Course I. The exercises range from elementary to difficult and from theoretical to computational and/or computer oriented. Those that are not assigned might provide extracurricular and titillative reading, without the burden of proof (and grading).

Prologue II describes Fourier transforms for various settings such as the real line $\mathbb{R}$, the integers $\mathbb{Z}$, finite intervals, or finite sets of points. Hopefully, it will be useful and provide perspective.

Chapter 1 presents the Fourier analysis of integrable and square integrable (finite energy) functions on $\mathbb{R}$. Chapter 3 presents the Fourier analysis of finite and infinite sequences as well as of functions defined on finite intervals. Chapter 2 presents distribution theory. In my opinion, distribution theory provides a useful vantage point for dealing with many ideas from engineering, physics, and mathematics;

and, in my experience, the calculational presentation herein is assimilable at the undergraduate level.

Notwithstanding the importance of Chapter 2, I have sometimes adjusted Course I by going directly to the Course I material in Chapter 3 after finishing the corresponding material in Chapter 1. This has given me time to present elementary wavelet theory and various algorithms and computer exercises associated with wavelets. The blend of Fourier transforms and Fourier series from Chapters 1 and 3 is an ideal background for discussing wavelets.

Besides wavelet theory, the topics from Chapters 1 and 3 have also served as a natural background for presenting results from the following areas:

The theory of frames and their applications;

Regular and irregular sampling theory, and applications;

Uncertainty principle inequalities;

Fourier analysis and signal processing.

Essay

It is natural to seek the fundamental components of complex phenomena, and then to describe such phenomena in terms of these components. With regard to our opening sentence about the *analysis* and *synthesis* of functions in terms of *harmonics*, the *analysis* is the determination of the harmonics or *components* for a given function, and the *synthesis* is the reconstruction or *decomposition* of this function in terms of its components.

Historically, there are the *atoms* (components) and *atomic decompositions* of the ancient Greek philosophers as well as of physicists and modern harmonic analysts [FJW91]. There are Leibniz' ultimate components or elements of the universe, the so-called *monads* or individual centers of force; and there are Nobel Laureate Dennis Gabor's elementary quanta or components of information, the so-called *logons* [Gabo46]. There are the *basic primary forms* of cubists such as Gris, Braque, and Picasso, whose paintings were created (a veritable inspirational synthesis) in terms of simple, often geometrical, components, e.g., [Gold59]; and there is a comparable musical paradigm in terms of harmonics as originally discovered in antiquity [Pie83]. There is *Riemann's decomposition* of the number theoretic psi function in terms of certain elementary components, which are *nontrigonometric waveforms*; and this decomposition is used to investigate zeros of the Riemann zeta function [Bom92], cf., Example 2.4.6*g* and Exercise 2.32.

This is a book about *classical harmonic analysis*. The fundamental components are *trigonometric functions*; and we shall deal with these components and the corresponding spectral *trigonometric decompositions* inherent in Fourier's original approach [Zyg59]. Classical harmonic analysis arose naturally in eighteenth and nineteenth century mathematical physics, in studies about the propagation of heat and the decomposition of electromagnetic waves. In our own century, classical harmonic analysis has continued to flourish, with a magnificent array of applications and with profound theoretical developments including a fundamental role in wavelet theory [Dau92], [Mey90].

The general area of harmonic analysis has many other personalities and themes and levels of abstraction than that of classical harmonic analysis. Alas, we shall not deal with most of them in this book. For example, although harmonic analysis at the time of Fourier could be described in terms of the spectral theory of a second-order differential operator, that subject has modern-day group-invariant versions of which we have just said our last word. Similarly, we shall *not* deal with representation theory, Banach algebras, or locally compact groups G, even though these subjects have a significant relationship with harmonic analysis. On the other hand, our treatment of classical harmonic analysis does have a Banach algebraic flavor, and is a substantial part of the harmonic analysis of phenomena defined on G in the case G is commutative.

We shall present a systematic treatment of classical harmonic analysis, and give careful proofs of the basic theorems. A feature of our presentation is that we are also providing expositions and perspectives on many topics. Some of these are extensive, such as our treatment of Wiener's Generalized Harmonic Analysis in Section 2.9. Further, there are several lengthy historical comments, e.g., Section 3.2. Because of the proofs and perspectives and exercises, this book can also serve as a textbook for courses more advanced than Course I.

Our mathematical emphasis has been in the direction of real analysis with very little complex analysis; and our point of view has been to deal thoroughly with central spaces such as L^1 and L^2, as opposed to reporting on some of the important results concerning BMO or Triebel-Lizorkin spaces. We have not stressed the group-theoretical underpinnings of harmonic analysis (even in classical terms), since there are other fundamental (nonalgebraic) characteristics of classical harmonic analysis. Our applications are limited by the usual constraints: author prejudice and author limitation. This is compensated to some extent by a serious bibliography, referenced at appropriate junctures in the text. We have also introduced some applications, not only because of their importance, but because of their intrinsic relationship with theoretical developments, e.g., Sections 3.6 and 3.7.

Notation and Idiosyncracies

The excessive number of references to [Ben xx] should not lead a newcomer to false conclusions about this author's contributions. On the other hand, I can attest that these references will contain thorough bibliographies highlighting the real contributors and will serve to keep our own bibliography for this book more manageable.

The sections labeled, *x.y.z* Definition, will always define a term, and that term will be italicized. However, they may also contain some elementary calculations and examples, and some exposition about the term being defined.

We sometimes use the symbol "≡" to define notation. For example, we write "we first evaluate $a \equiv \int_0^\infty e^{-u^2}\,du$" so that we can deal with "a" instead of the more complicated right side in the ensuing calculation. Generally, we use "=" to define notation in displayed items or when the context is clear.

$\mathbb{C}$ is the set of complex numbers, $\mathbb{R}$ is the set of real numbers, $\mathbb{Q}$ is the set of rational numbers, $\mathbb{Z}$ is the set of integers, and $\mathbb{N}$ is the set of natural numbers, i.e., $\mathbb{N} = \{1, 2, 3, \ldots\}$.

Instead of denoting integration of a function f over $\mathbb{R}$ by $\int_{-\infty}^\infty f(t)dt$, we shall often write $\int f(t)dt$, i.e.,

$$\int f(t)dt = \int_{-\infty}^\infty f(t)dt.$$

$\varnothing$ designates the empty set. $\operatorname{Re} c$, resp., $\operatorname{Im} c$, is the real part, resp., imaginary part, of $c \in \mathbb{C}$. $\operatorname{sgn} t$ is 1 or -1 depending on whether $t > 0$ or $t < 0$. If $X \subseteq \mathbb{R}$, then $X^c = \mathbb{R} \backslash X$, the complement of X; and $\mathbf{1}_X$ is the characteristic function of X equal to 1 if $t \in X$ and equal to 0 if $t \in X^c$. Other notation and notions are introduced as needed or in the Appendices, and they are referenced in the list of notation or in the index.

Acknowledgements

First, I wish to express my appreciation to my former graduate students, George Benke, Jean-Pierre Gabardo, Joseph Lakey, David Walnut, and Georg Zimmermann. They carefully read portions of the book and provided valuable comments and invaluable insights.

I have had the good fortune of friendship and learning from some of the modern masters in subjects ranging from spectral synthesis and Banach algebras to Wiener's Generalized Harmonic Analysis to signal processing. I shall not risk an incomplete list except to acknowledge Carl Herz, Stylianos Pichorides, and Harry Pollard, each of whom has passed away in recent years. I want to thank all of my graduate students who have shared some of the intellectual excitement of Fourier analysis with me. Also, my collaborations have been extremely rewarding, and it is a pleasure to express my gratitude to my coauthors, many of whom are referenced in the book.

There was specific assistance, on issues ranging from mathematics to birthplaces of mathematicians to computers to publishing, from Elmar Winkelnkemper, Larry Washington, C. Robert Warner, Sadahiro Saeki, Markus Melenk, Jean Mawhin, Nora Konopka, Raymond Johnson, Andrzej Hulanicki, Christian Houdré, Michael Hernandez, Hans Heinig, Christopher Heil, Melissa Harrison, Israel Gohberg, Ward Evans, David Colella, Christiane Carton-Lebrun, and Gary Benedetto. Finally, I had the extraordinary good fortune to have a great typist; my thanks to Stephanie Smith.

Prologue I

Course I

No proofs of theorems, propositions, or lemmas are required unless indicated.

Definition 1.1.1
Definition 1.1.2
Remark 1.1.3
Remark 1.1.4
Definition 1.1.5
Theorem 1.1.6
Remark 1.1.8
Proposition 1.1.9 and proof
Proposition 1.1.10 and proof
Proposition 1.1.11
Remark 1.1.13

Theorem 1.2.1
Example 1.2.2

Example 1.3.1
Example 1.3.2
Example 1.3.3
Example 1.3.4

Theorem 1.4.1
Remark 1.4.5

Definition 1.5.11
Proposition 1.5.2 and proof
Remark 1.5.3a

Definition 1.6.1
Proposition 1.6.2 and proof
Proposition 1.6.3 and proof
Proposition 1.6.4a
Example 1.6.5a
Example 1.6.6
Theorem 1.6.9 and proofs of parts b,c
Proposition 1.6.11

Remark 1.7.1
Theorem 1.7.6 (= Theorem 1.1.6)
Theorem 1.7.8
Example 1.7.10

Example 1.8.1
Example 1.8.2
Example 1.8.3

Remark 1.9.1
Remark 1.9.2
Remark 1.9.3

Definition 1.10.1
Theorem 1.10.2
Proposition 1.10.3 and proof

Prologue II

Fourier Transforms, Fourier Series, and Discrete Fourier Transforms

Domain of the function f	Definition of the Fourier transform $\hat{f} = F$	Domain of the Fourier transform F
$\mathbb{R}$	$F(\gamma) = \hat{f}(\gamma) = \displaystyle\int f(t) e^{-2\pi i t \gamma}\, dt$	$\hat{\mathbb{R}} = \mathbb{R}$

Inversion Formula: $f(t) = \int F(\gamma) e^{2\pi i t \gamma}\, d\gamma$

Notes. • Integration is over $\mathbb{R}$.
  • The domain of the Fourier transform is denoted by $\hat{\mathbb{R}}$.

$\mathbb{Z}$	$F(\gamma) = \displaystyle\sum f[n] e^{-\pi i n \gamma/\Omega}$	$\mathbb{T}_{2\Omega} = \hat{\mathbb{R}}/2\Omega\mathbb{Z}$

Inversion Formula: $f[n] = \frac{1}{2\Omega} \int_{-\Omega}^{\Omega} F(\gamma) e^{\pi i n \gamma/\Omega}\, d\gamma$

Notes. • $\Omega > 0$ is fixed and summation is over $\mathbb{Z}$.
  • F is a *Fourier series* with *Fourier coefficients* $f = \{f[n] : n \in \mathbb{Z}\}$.
  • F is 2Ω-periodic on $\hat{\mathbb{R}}$, and $\mathbb{T}_{2\Omega} = \hat{\mathbb{R}}/2\Omega\mathbb{Z}$ denotes this domain.

$\mathbb{T}_{2\Omega}$	$F[n] = \dfrac{1}{2\Omega} \displaystyle\int_{-\Omega}^{\Omega} f(\gamma) e^{-\pi i n \gamma/\Omega}\, d\gamma$	$\mathbb{Z}$

Inversion Formula: $f(\gamma) = \sum F[n] e^{\pi i n \gamma/\Omega}$

$\mathbb{Z}_N$	$F[n] = \displaystyle\sum_{m=0}^{N-1} f[m] e^{-2\pi i m n/N}$	$\mathbb{Z}_N$

Inversion Formula: $f[m] = \frac{1}{N} \sum_{n=0}^{N-1} F[n] e^{2\pi i m n/N}$

Notes. • F is a *Discrete Fourier Transform* (DFT).
  • F is N-periodic on $\mathbb{Z}$, and $\mathbb{Z}_N$ denotes this domain.

1

Fourier Transforms

1.1. Definitions and Formal Calculations

1.1.1 Definition. Integrable Functions

Set

$$L^1_{\text{loc}}(\mathbb{R}) = \left\{ f : \mathbb{R} \to \mathbb{C} : \forall a < b, \quad \int_a^b |f(t)|\, dt < \infty \right\}$$

and

$$L^1(\mathbb{R}) = \left\{ f : \mathbb{R} \to \mathbb{C} : \|f\|_{L^1(\mathbb{R})} = \int_{-\infty}^\infty |f(t)|\, dt < \infty \right\}.$$

$L^1_{\text{loc}}(\mathbb{R})$ is the space of *locally integrable functions* on $\mathbb{R}$ and $L^1(\mathbb{R})$ is the space of *integrable functions* on $\mathbb{R}$.

Let $g \in L^1_{\text{loc}}(\mathbb{R})$. $\int_{-\infty}^\infty g(t)\, dt$, which we designate frequently by $\int g(t)\, dt$, is

$$\lim_{S \to \infty, T \to \infty} \int_{-S}^T g(t)\, dt.$$

The *Cauchy principal value, pv $\int g(t)\, dt$,* is

$$\lim_{T \to \infty} \int_{-T}^T g(t)\, dt.$$

If $\int g(t)\, dt$ exists, then $pv \int g(t)\, dt$ exists and the two integrals have the same value. In the opposite direction, $pv \int g(t)\, dt$ may exist while $\int g(t)\, dt$ does not, e.g., $g(t) = t$. On the other hand, if $pv \int g(t)\, dt$ exists and either g is *even*, i.e., $g(t) = g(-t)$, or $g \geq 0$ on $\mathbb{R}$, then

$$pv \int g(t)\, dt = \int g(t)\, dt,$$

e.g., Exercise 1.1.

We have been purposely vague about the definition of $\int_a^b g(t)\, dt$. If you know the Lebesgue integral, then fine! If not, the Riemann integral works for most of our calculations and most of the functions we shall consider.

1.1.2 Definition. Fourier Transform

The *Fourier transform* of $f \in L^1(\mathbb{R})$ is the function F defined as

$$F(\gamma) = \int_{-\infty}^{\infty} f(t)e^{-2\pi it\gamma}\, dt, \qquad \gamma \in \hat{\mathbb{R}}\,(=\mathbb{R}),$$

cf., Exercise 1.15. Notationally, we write the pairing between the function f and F in one of the following ways:

$$f \longleftrightarrow F, \qquad \hat{f} = F, \qquad f = \check{F}.$$

The space of Fourier transforms of L^1-functions is denoted by $A(\hat{\mathbb{R}})$, i.e.,

$$A(\hat{\mathbb{R}}) = \{F : \hat{\mathbb{R}} \to \mathbb{C} : \exists f \in L^1(\mathbb{R}) \text{ such that } \hat{f} = F\}.$$

1.1.3 Remark. Inversion Formula

Let $f \in L^1(\mathbb{R})$ and let $\hat{f} = F$. The *Fourier transform inversion formula* is

$$f(t) = \int F(\gamma)e^{2\pi it\gamma}\, d\gamma. \tag{1.1.1}$$

We shall prove (1.1.1) in Section 1.7, but let us formally derive it now.

1.1.4 Remark. Formal Calculation of the Inversion Formula

The Dirac δ "function" (actually, it is a probability measure) can be thought of in terms of the "formula"

$$\forall f, \qquad f(t) = \int f(u)\,\delta(t - u)\, du; \tag{δ}$$

and a model of the uncertainty principle is the "formula"

$$\delta(t) = \int e^{2\pi it\gamma}\, d\gamma. \tag{UP}$$

If the Fourier pair $f \longleftrightarrow F$ is given, then (δ) and (UP) allow us to make the following formal calculation:

$$\int F(\gamma)e^{2\pi it\gamma}\,d\gamma = \iint f(u)e^{-2\pi iu\gamma}e^{2\pi it\gamma}\,du\,d\gamma$$

$$= \iint f(u)e^{2\pi i(t-u)\gamma}\,d\gamma\,du \qquad (1.1.2)$$

$$= \int f(u)\,\delta(t-u)\,du = f(t).$$

(δ) and (UP) are nonsense as they stand but do have a sense intuitively. (δ) is easily motivated, e.g., Section 1.6 and Section 2.1. The rationale for (UP) involves the fact $f'(t) \longleftrightarrow (2\pi i\gamma)F(\gamma)$ (which we shall verify shortly), which in turn allows us to use $\hat{\mathbb{R}}$ as the domain of the momentum and thereby to invoke the usual interpretation from physics of the uncertainty principle, e.g., [vN55, Chapter III.4, esp. page 235], [Wey50a, page 77 and Appendix I]. Mathematically, (UP) can be given a precise meaning in terms of the notion of "oscillatory integral" [Hör83, Volume I, Section 7.8, especially (7.8.5)].

In a more elementary, but also fundamental way, we can think of (UP) in terms of the following idealized piano experiment. The standard for concert pitch is that the A above middle C should have 440 vibrations per second. Thus, the A four octaves down (and the last key on the piano) should have 27.5 vibrations per second. Suppose we could strike this last key for a time interval of 1/30 seconds, i.e., the hammer strikes the string and 1/30 seconds later the damper returns to the string, thereby stopping the sound. In particular, a complete vibration for this key does not occur. We then have very precise time information, represented by $\delta(t)$ on the left side of (UP), but correspondingly imprecise frequency information since the emitted sound is anything but the desired pure periodic pitch of this low A. In fact, this imprecise frequency information can be thought of as the "noisy sum" of many pure tones, represented by the integral on the right side of (UP).

Even assuming the validity of (something like) (δ) and (UP), the calculation (1.1.2) leaves something to be desired because of the casual switching of order of integration we have made, cf., Section 1.7.

1.1.5 Definition. Bounded Variation

A function $f : \mathbb{R} \to \mathbb{C}$ has *bounded variation* on an interval $I \subseteq \mathbb{R}$ if there is a constant M such that for every finite set $t_0 < t_1 < \cdots < t_n,\ t_j \in I$, we have

$$\sum_{j=1}^{n} |f(t_j) - f(t_{j-1})| \leq M.$$

In this case we write $f \in BV(I)$, and note that we could have $I = \mathbb{R}$. If $f : \mathbb{R} \to \mathbb{C}$ has bounded variation on each interval of finite length, then f *locally has bounded variation*, and we write $f \in BV_{\text{loc}}(\mathbb{R})$.

Functions having bounded variation on bounded intervals I have graphs of finite length; and such functions are a natural generalization of continuously differentiable functions, e.g., [Ben76, Chapter 4].

One form of the inversion formula that we shall verify in Section 1.7 is the *Jordan pointwise inversion formula*.

1.1.6 Theorem. Jordan Theorem

Let $f \in L^1(\mathbb{R})$ and assume $f \in BV[t - \epsilon, t + \epsilon]$ for some $t \in \mathbb{R}$ and $\epsilon > 0$. Then

$$\frac{f(t+) + f(t-)}{2} = \lim_{\Omega \to \infty} \int_{-\Omega}^{\Omega} F(\gamma)e^{2\pi i t\gamma} \, d\gamma, \qquad (1.1.3)$$

where $f \longleftrightarrow F$. If f is continuous at t, then the left side of (1.1.3) can be replaced by $f(t)$.

Another important pointwise inversion formula, but one which we shall not verify, is the following result due to Pringsheim, e.g., [RL55].

1.1.7 Theorem. Pringsheim Theorem

Let $f \in BV(\mathbb{R})$ and assume

$$\lim_{|t| \to \infty} f(t) = 0.$$

Then

$$F(\gamma) = \int f(t)e^{-2\pi i t\gamma} \, dt$$

exists for all γ except possibly $\gamma = 0$, and

$$\forall t \in \mathbb{R}, \qquad \frac{f(t+) + f(t-)}{2} = \lim_{\Omega \to \infty, \epsilon \to 0} \int_{\epsilon < |\gamma| < \Omega} F(\gamma)e^{2\pi i t\gamma} \, d\gamma.$$

For the remainder of Section 1.1 our statements and proofs are formal, in the sense that we have not been concerned with correct mathematical hypotheses for asserting the existence of the Fourier transform or using the inversion formula.

1.1.8 Remark. Formal calculations for $f \longleftrightarrow F$

Consider the formal pairing $f \longleftrightarrow F$, where $f = f_1 + i f_2$ and $F = F_1 + i F_2$. Then, by the definition of F, we obtain

$$F_1(\gamma) + i F_2(\gamma) = \int (f_1(t) + i f_2(t))(\cos 2\pi t\gamma - i \sin 2\pi t\gamma) \, dt,$$

and so

$$F_1(\gamma) = \int (f_1(t) \cos 2\pi t\gamma + f_2(t) \sin 2\pi t\gamma) \, dt$$

and

$$F_2(\gamma) = \int (f_2(t) \cos 2\pi t\gamma - f_1(t) \sin 2\pi t\gamma) \, dt.$$

Similarly, by (1.1.1), we have

$$f_1(t) + i f_2(t) = \int (F_1(\gamma) + i F_2(\gamma))(\cos 2\pi t\gamma + i \sin 2\pi t\gamma) \, d\gamma$$

and, hence,

$$f_1(t) = \int (F_1(\gamma) \cos 2\pi t\gamma - F_2(\gamma) \sin 2\pi t\gamma) \, d\gamma$$

and

$$f_2(t) = \int (F_2(\gamma) \cos 2\pi t\gamma + F_1(\gamma) \sin 2\pi t\gamma) \, d\gamma.$$

1.1.9 Proposition.

Consider the formal pairing $f \longleftrightarrow F$. f is real and even if and only if F is real and even. In this case

$$F(\gamma) = 2 \int_0^\infty f(t) \cos 2\pi t\gamma \, dt$$

and

$$f(t) = 2 \int_0^\infty F(\gamma) \cos 2\pi t\gamma \, d\gamma.$$

PROOF.   Suppose f is real and even. Then $f_2 = 0$ and $f = f_1$; and we apply the calculations of Remark 1.1.8 to obtain

$$F(\gamma) = \int f(t)(\cos 2\pi t\gamma - i \sin 2\pi t\gamma) \, dt$$

$$= \int f(t) \cos 2\pi t\gamma \, dt = 2 \int_0^\infty f(t) \cos 2\pi t\gamma \, dt.$$

In particular, F is real and even. A similar calculation works for the opposite direction. ∎

1.1.10 Proposition.

Consider the formal pairing $f \longleftrightarrow F$. f is real if and only if $F(\gamma) = \overline{F(-\gamma)}$.

PROOF.   If f is real, then $f = f_1$; and we apply the calculations of Remark 1.1.8 to obtain

$$F_1(\gamma) = \int f_1(t) \cos 2\pi t \gamma \, dt$$

and

$$F_2(\gamma) = -\int f_1(t) \sin 2\pi t \gamma \, dt.$$

Thus, we have $F_1(-\gamma) = F_1(\gamma)$ and $F_2(-\gamma) = -F_2(\gamma)$, and, hence,

$$\overline{F(-\gamma)} = \overline{F_1(-\gamma) + i F_2(-\gamma)} = F_1(-\gamma) - i F_2(-\gamma)$$
$$= F_1(\gamma) + i F_2(\gamma) = F(\gamma).$$

For the converse, suppose $\overline{F(\gamma)} = F(-\gamma)$. Therefore,

$$F_1(\gamma) - i F_2(\gamma) = F_1(-\gamma) + i F_2(-\gamma),$$

i.e., $F_1(\gamma) = F_1(-\gamma)$ and $F_2(\gamma) = -F_2(-\gamma)$; hence, F_1 is even and F_2 is odd. Using these facts and the calculations of Remark 1.1.8, we calculate

$$f_2(t) = \int (F_1(\gamma) \sin 2\pi t \gamma + F_2(\gamma) \cos 2\pi t \gamma) \, dt = 0,$$

from which we conclude that f is real.   ∎

1.1.11 Proposition.

Consider the formal pairing $f \longleftrightarrow F$ and write

$$F(\gamma) = A(\gamma) e^{i\phi(\gamma)},$$

where $A(\gamma) \geq 0$ and $\phi(\gamma) \in \mathbb{R}$. If f is real, then

$$f(t) = \int A(\gamma) \cos(2\pi t \gamma + \phi(\gamma)) \, d\gamma.$$

PROOF. We calculate

$$\cos(2\pi t\gamma + \phi(\gamma)) = \cos 2\pi t\gamma \cos \phi(\gamma) - \sin 2\pi t\gamma \sin \phi(\gamma)$$

and so

$$\begin{aligned}
A(\gamma)\cos(2\pi t\gamma + \phi(\gamma)) \\
&= A(\gamma)\cos \phi(\gamma)\cos 2\pi t\gamma - A(\gamma)\sin \phi(\gamma)\sin 2\pi t\gamma \\
&= F_1(\gamma)\cos 2\pi t\gamma - F_2(\gamma)\sin 2\pi t\gamma
\end{aligned}$$

since

$$F(\gamma) = F_1(\gamma) + i F_2(\gamma) = A(\gamma)(\cos \phi(\gamma) + i \sin \phi(\gamma)).$$

Using this fact and the hypothesis that f is real, we see from the calculations of Remark 1.1.8 that

$$f(t) = \int A(\gamma)\cos(2\pi t\gamma + \phi(\gamma))\, d\gamma. \quad \blacksquare$$

1.1.12 Remark. Amplitude and Phase

Consider the formal pairing $f \longleftrightarrow F$ and write $F(\gamma) = A(\gamma)e^{i\phi(\gamma)}$. $A(\gamma)$ is the *amplitude* and $\phi(\gamma)$ is the *phase angle* of $F(\gamma)$.

The inversion formula (1.1.1) allows us to think of a signal f as a "sum" (integral)

$$``f(t) = \sum_{\gamma} \left(A(\gamma)e^{i\varphi(\gamma)}\right) e^{2\pi i t\gamma}\text{"}$$

of exponentials $e^{2\pi i t\gamma}$ with complex coefficients $A(\gamma)\,e^{i\phi(\gamma)}$. Different phase angles φ can produce quite different-looking signals f even if $A(\gamma)$ remains the same. As an elementary example, note that if $\varphi(\gamma)$ is replaced by $\varphi(\gamma) - 2\pi u\gamma$ for some fixed $u \in \mathbb{R}$, then $f(t)$ is replaced by the translate $f(t - u)$, cf., Theorem 1.2.1d.

The amplitude squared $|A(\gamma)|^2$ can be thought of as the amount of energy of f in the frequency band about a small neighborhood of γ; and $|A|^2$ is often a measurable quantity in, for example, signal processing, spectroscopy, and fluid mechanics. The physical measurement is based on ideas about correlations and translations of f, e.g., Exercise 1.33, Example 2.8.10, and Michelson's invention of the interferometer [Loe66], [Mic62]. The resulting 0-phase information or some windowed form of it is called the *spectrogram* or *periodogram* or *power spectrum* of f, e.g., Section 2.8 and Definition 2.9.5. A basic methodology in engineering and the sciences is to approximate a reconstruction of f from its spectrogram. This is a first step in harmonic analysis signal reconstruction technology.

1.1.13 Remark. A Table

Consider the formal pairing $f \longleftrightarrow F$, where $F = F_1 + i F_2$.

$$f \text{ real if and only if } \overline{F(\gamma)} = F(-\gamma),$$

$$F(\gamma) = \int f(t) \cos 2\pi t \gamma \, dt - i \int f(t) \sin 2\pi t \gamma \, dt,$$

$$f(t) = 2 \int_0^\infty (F_1(\gamma) \cos 2\pi t \gamma - F_2(\gamma) \sin 2\pi t \gamma) \, d\gamma,$$

$$= 2\mathrm{Re} \int_0^\infty F(\gamma) e^{2\pi i t \gamma} \, d\gamma.$$

$$f \text{ real and even if and only if } F \text{ real and even,}$$

$$F(\gamma) = 2 \int_0^\infty f(t) \cos 2\pi t \gamma \, dt,$$

$$f(t) = 2 \int_0^\infty F(\gamma) \cos 2\pi t \gamma \, d\gamma.$$

$$f \text{ real and odd if and only if } F \text{ odd and } F \text{ imaginary,}$$

$$F(\gamma) = -2i \int_0^\infty f(t) \sin 2\pi t \gamma \, dt,$$

$$f(t) = 2i \int_0^\infty F(\gamma) \sin 2\pi t \gamma \, d\gamma.$$

$$f \text{ imaginary if and only if } \overline{F(\gamma)} = -F(-\gamma),$$

e.g., Exercise 1.2.

In light of Remark 1.1.13 we ask:

Question. Can we characterize the case when f is not only real but is also nonnegative?

Answer. Yes, but the answer is not simple, and it involves the notion of *positive definite functions*, which play a key role in theoretical considerations associated with spectral estimation, e.g., Example 2.7.9*b*, Theorem 2.7.10, and Sections 3.6 and 3.7.

1.2. Algebraic Properties of Fourier Transforms

Notationally, for a fixed γ, we set

$$e_\gamma(t) = e^{2\pi i t \gamma};$$

and, for a fixed u and a given function f, we set

$$(\tau_u f)(t) = f(t - u).$$

$\tau_u f$ is the *translation* of f by u.

1.2.1 Theorem. Algebraic Properties of Fourier Transforms

a. *Linearity. Consider* $f_j \longleftrightarrow F_j$ *and let* $c_j \in \mathbb{C}$, $j = 1, 2$, *where* $f_j \in L^1(\mathbb{R})$. *Then* $(c_1 f_1 + c_2 f_2)\hat{}(\gamma) = (c_1 \hat{f}_1 + c_2 \hat{f}_2)(\gamma)$, *i.e.,*

$$c_1 f_1 + c_2 f_2 \longleftrightarrow c_1 F_1 + c_2 F_2.$$

b. *Symmetry. Consider* $f \longleftrightarrow F$, *where* $f \in L^1(\mathbb{R})$ *and* $F \in L^1(\hat{\mathbb{R}})$. *Then* $\hat{F}(\gamma) = f(-\gamma)$, *i.e.,*

$$F(t) \longleftrightarrow f(-\gamma).$$

c. *Conjugation. Consider* $f \longleftrightarrow F$, *where* $f \in L^1(\mathbb{R})$. *Then* $(\bar{f})\hat{}(\gamma) = \bar{\hat{f}}(-\gamma)$, *i.e.,*

$$\overline{f(t)} \longleftrightarrow \overline{F(-\gamma)},$$

cf., Proposition 1.1.10.

d. *Translation (time shifting). Consider* $f \longleftrightarrow F$, *where* $f \in L^1(\mathbb{R})$, *and take* $u \in \mathbb{R}$. *Then* $(\tau_u f)\hat{}(\gamma) = e^{-2\pi i u \gamma} \hat{f}(\gamma)$, *i.e.,*

$$(\tau_u f)(t) \longleftrightarrow e_{-u}(\gamma) F(\gamma).$$

e. *Modulation (frequency shifting). Consider* $f \longleftrightarrow F$, *where* $f \in L^1(\mathbb{R})$, *and take* $\lambda \in \hat{\mathbb{R}}$. *Then* $(e^{2\pi i t \lambda} f(t))\hat{}(\gamma) = \hat{f}(\gamma - \lambda)$, *i.e.,*

$$e_\lambda(t) f(t) \longleftrightarrow F(\gamma - \lambda) = \tau_\lambda F(\gamma).$$

f. *Time dilation (time scaling). Consider* $f \longleftrightarrow F$, *where* $f \in L^1(\mathbb{R})$, *and take* $\lambda \in \mathbb{R} \backslash \{0\}$. *Define the* λ-*dilation of* f *by*

$$f_\lambda(t) = \lambda f(\lambda t).$$

Then $\hat{f}_\lambda(\gamma) = (\lambda/|\lambda|) \hat{f}(\gamma/\lambda)$, *i.e.,*

$$f_\lambda(t) \longleftrightarrow \frac{\lambda}{|\lambda|} F\left(\frac{\gamma}{\lambda}\right).$$

The formal proof of this result is easy, but we should comment on the hypotheses of part *b*. It turns out that the integrability of f and F is sufficient for the validity of the pointwise inversion formula (1.1.1), which is used in the proof of part *b*. The verification of this sufficiency requires some work, and we shall deal with it in Theorem 1.7.8.

1.2.2 Example. Dilation and Modulation

a.   *Dilation.* Let $\mathbf{1}_{[-T,T)}$ be the *rectangular pulse* on $\mathbb{R}$ defined as

$$\mathbf{1}_{[-T,T)}(t) = \begin{cases} 1, & -T \leq t < T, \\ 0, & \text{otherwise.} \end{cases}$$

If $T = 1$ we write $\mathbf{1} = \mathbf{1}_{[-1,1)}$. Now define the function $f = h\mathbf{1}_{[-T,T)}, h > 0$. To fix ideas, take $\lambda \in (0, 1)$. Then the graphs of f and f_λ are as shown in Figure 1.1.

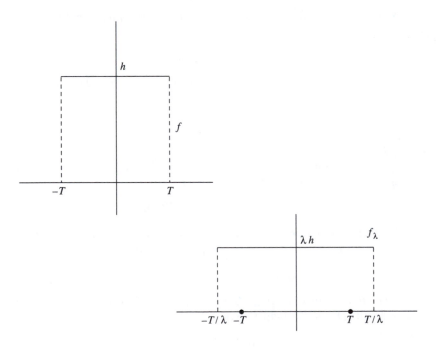

FIGURE 1.1.

b.   *Modulation.* Consider a *carrier wave* $\cos 2\pi t \gamma_0$, where $\gamma_0 > 0$ is the *carrier frequency*. If $f \in L^1(\mathbb{R})$, then $f(t) \cos 2\pi t \gamma_0$ is the resulting *modulated signal*, e.g., Figure 1.2.

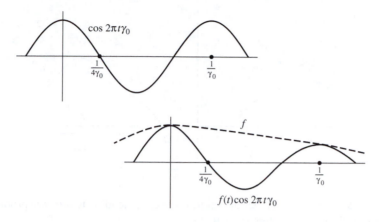

FIGURE 1.2.

We shall compute the Fourier transform of $f(t) \cos 2\pi t \gamma_0$ in terms of $\hat{f}(\gamma) = F(\gamma) = A(\gamma)e^{i\phi(\gamma)}$:

$$f(t) \cos 2\pi t \gamma_0 = f(t) \frac{e^{2\pi i t \gamma_0} + e^{-2\pi i t \gamma_0}}{2}$$

$$\longleftrightarrow \frac{1}{2}(F(\gamma - \gamma_0) + F(\gamma + \gamma_0)) = A_c(\gamma)e^{i\phi_c(\gamma)},$$

where $A_c(\gamma)$ is the amplitude of $\frac{1}{2}(F(\gamma - \gamma_0) + F(\gamma + \gamma_0))$ and $\phi_c(\gamma)$ is its phase angle.

Suppose $f \in L^1(\mathbb{R})$ has the property that $\hat{f}$ vanishes off the interval $[-\Omega, \Omega]$. In this case, we say that f is Ω-*bandlimited*, and we use the notation $\Omega-BL$ to designate this property of a function. To fix ideas, consider the amplitude A as in Figure 1.3.

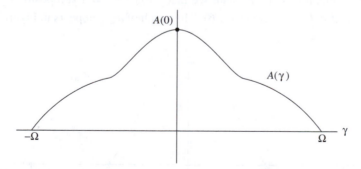

FIGURE 1.3.

If $\gamma_0 \geq \Omega$, then A_c has the graph of Figure 1.4, indicating that A splits into two parts, each having half the amplitude of the original.

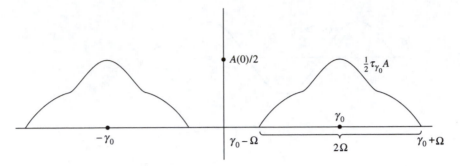

FIGURE 1.4.

If $\gamma_0 < \Omega$, then the two lobes in Figure 1.4 overlap, leading to *aliasing* problems, e.g., Example 1.9.4 and Remark 3.10.12.

1.3. Examples

1.3.1 Example. The Sinc or Dirichlet Function

Let $f(t) = \mathbf{1}_{[-T,T)}(t)$. Clearly, we have $\hat{\mathbf{1}}_{[-T,T)}(\gamma) = (\sin 2\pi T\gamma)/(\pi\gamma)$. Notationally, we write

$$d(\gamma) = \frac{\sin \gamma}{\pi \gamma}$$

and

$$\mathrm{sinc}\, \gamma = \frac{\sin \pi \gamma}{\pi \gamma},$$

so that $\hat{\mathbf{1}}_{[-T,T)}(\gamma) = d_{2\pi T}(\gamma)$. If $\lambda > 0$, then $\hat{\mathbf{1}}_{[-\lambda/2\pi,\lambda/2\pi]} = d_\lambda$. We refer to d as the *Dirichlet function*, and shall see that $\int d(\gamma)\, d\gamma = 1$ (Proposition 1.6.3), noting that $d \notin L^1(\hat{\mathbb{R}})$ (Exercise 1.6). The graph of $d_{2\pi T}$ appears in Figure 1.5.

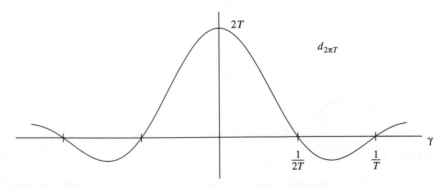

FIGURE 1.5.

1.3.2 Example. The Poisson Function

Let $f(t) = e^{-2\pi r|t|}$, $r > 0$. To compute $\hat{f}$ we calculate

$$\int_{-T}^{T} e^{-2\pi r|t|} e^{-2\pi i t \gamma} \, dt$$

$$= \int_{-T}^{0} e^{2\pi r t} e^{-2\pi i t \gamma} \, dt + \int_{0}^{T} e^{-2\pi r t} e^{-2\pi i t \gamma} \, dt$$

$$= \frac{1}{2\pi (r - i\gamma)} \left(1 - e^{-2\pi T (r - i\gamma)}\right) - \frac{1}{2\pi (r + i\gamma)} \left(e^{-2\pi T (r + i\gamma)} - 1\right)$$

$$= \frac{1}{2\pi} \left(\frac{1}{r - i\gamma} + \frac{1}{r + i\gamma}\right) + \epsilon(T).$$

Clearly, $\lim_{T \to \infty} \epsilon(T) = 0$ since $r > 0$; and, hence,

$$\hat{f}(\gamma) = \frac{1}{\pi} \frac{r}{r^2 + \gamma^2}.$$

We write

$$p(\gamma) = \frac{1}{\pi(1 + \gamma^2)}$$

so that $p_{(1/r)}(\gamma) = \hat{f}(\gamma)$. Thus, if $\lambda = 1/r > 0$, then

$$e^{-2\pi |t|/\lambda} \longleftrightarrow p_\lambda(\gamma).$$

We refer to p as the *Poisson function*, and observe that $\int p(\gamma) \, d\gamma = 1$, e.g., Example 1.6.5.

Note that the exponential decrease of f is transformed into the polynomial decrease of $\hat{f}$, cf., Exercise 1.5 where the point of nondifferentiability in this example is replaced by a discontinuity.

1.3.3 Example. The Gaussian

Let $f(t) = e^{-\pi r t^2}$, $r > 0$. We could calculate $\hat{f} = F$ by means of contour integrals, but we choose a real and, by now, classical approach [Fel66, page 476]. By definition of F, which is real and even, we have

$$F'(\gamma) = -2\pi i \int t e^{-\pi r t^2} e^{-2\pi i t \gamma} \, dt. \tag{1.3.1}$$

Noting that

$$\frac{d}{dt}\left(e^{-\pi r t^2}\right) = -2\pi r t\, e^{-\pi r t^2},$$

we rewrite (1.3.1) as

$$F'(\gamma) = -2\pi i \int \frac{-1}{2\pi r}\left(e^{-\pi r t^2}\right)' e^{-2\pi i t \gamma}\, dt$$

$$= \frac{i}{r}\left[e^{-\pi r t^2} e^{-2\pi i t \gamma}\,\Big|_{-\infty}^{\infty} - \int e^{-\pi r t^2}(-2\pi i \gamma)e^{-2\pi i t \gamma}\, dt\right]$$

$$= \frac{-2\pi \gamma}{r} F(\gamma).$$

Thus, F is a solution of the differential equation

$$F'(\gamma) = -\frac{2\pi \gamma}{r} F(\gamma); \tag{1.3.2}$$

and (1.3.2) is solved by elementary means with solution

$$F(\gamma) = C e^{-\pi \gamma^2 / r},$$

e.g., Exercise 1.9.

Taking $\gamma = 0$ and using the definition of the Fourier transform, we see that

$$C = \int e^{-\pi r t^2}\, dt.$$

In order to calculate C we first evaluate $a \equiv \int_0^\infty e^{-u^2}\, du$:

$$a^2 = \int_0^\infty e^{-x^2}\, dx \int_0^\infty e^{-y^2}\, dy$$

$$= \int_0^\infty \int_0^\infty e^{-(x^2+y^2)}\, dx\, dy = \int_0^{\pi/2}\int_0^\infty e^{-r^2} r\, dr\, d\theta$$

$$= \frac{\pi}{4}\int_0^\infty e^{-u}\, du = \frac{\pi}{4}.$$

Thus, $a = \sqrt{\pi}/2$ and so

$$\int e^{-u^2}\, du = \sqrt{\pi}.$$

Consequently,

$$C = \int e^{-\pi r t^2}\, dt = \frac{1}{\sqrt{\pi r}} \int e^{-u^2}\, du = \frac{1}{\sqrt{r}}.$$

Therefore, we have shown that

$$e^{-\pi r t^2} \longleftrightarrow \frac{1}{\sqrt{r}} e^{-\pi \gamma^2 / r}.$$

We write

$$g(t) = \frac{1}{\sqrt{\pi}} e^{-t^2}$$

so that if $\lambda > 0$, then

$$g_\lambda(t) \longleftrightarrow e^{-(\pi \gamma / \lambda)^2}.$$

In particular,

$$\frac{1}{\sqrt{r}} g_{\sqrt{\pi r}} \longleftrightarrow g_{\sqrt{\pi/r}}$$

and hence $g_{\sqrt{\pi}} \longleftrightarrow g_{\sqrt{\pi}}$. We refer to g as the *Gauss function* or *Gaussian*, and note that $\int g(\gamma)\, d\gamma = 1$.

1.3.4 Example. The Fejér Function

Let $f(t) = \max(1 - |t|, 0)$. On $[-1, 1]$, the graph of f consists of the equal legs of an isosceles triangle of height 1; f vanishes outside $[-1, 1]$, e.g., Figure 1.6. The Fourier transform of f is

$$F(\gamma) = \int_0^1 (1 - t)e^{-2\pi i t \gamma}\, dt + \int_{-1}^0 (1 + t)e^{-2\pi i t \gamma}\, dt$$

$$= \left[\frac{1}{2\pi i \gamma} + \frac{1}{(2\pi i \gamma)^2}\left(e^{-2\pi i \gamma} - 1\right)\right] + \left[-\frac{1}{2\pi i \gamma} - \frac{1}{(2\pi i \gamma)^2}\left(1 - e^{2\pi i \gamma}\right)\right]$$

$$= \frac{-2 + 2\cos 2\pi \gamma}{(2\pi i \gamma)^2} = \frac{2(1 - \cos 2\pi \gamma)}{(2\pi \gamma)^2}$$

$$= \frac{2(1 - [\cos \pi \gamma \cos \pi \gamma - \sin \pi \gamma \sin \pi \gamma])}{(2\pi \gamma)^2} = \frac{\sin^2 \pi \gamma}{(\pi \gamma)^2},$$

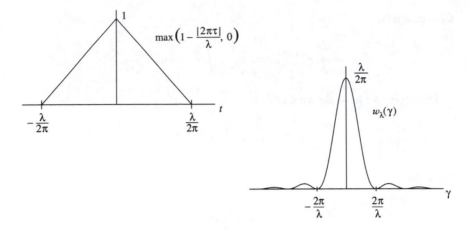

FIGURE 1.6.

i.e.,

$$\hat{f}(\gamma) = \left(\frac{\sin \pi \gamma}{\pi \gamma}\right)^2.$$

Notationally, we set $\Delta = f$ and

$$w(\gamma) = \frac{1}{2\pi}\left(\frac{\sin \gamma/2}{\gamma/2}\right)^2,$$

so that $w_{2\pi}(\gamma) = \hat{\Delta}(\gamma)$, i.e.,

$$\Delta \longleftrightarrow w_{2\pi}. \tag{1.3.3}$$

Thus, if $\lambda > 0$, then

$$\max\left(1 - \frac{|2\pi t|}{\lambda}, 0\right) = \frac{\lambda}{2\pi}\Delta_{2\pi/\lambda}(t) \longleftrightarrow w_\lambda(\gamma). \tag{1.3.4}$$

We refer to w as the *Fejér function*, and note that

$$\int w(\gamma)\,d\gamma = 1$$

(Proposition 1.6.3). Fejér's surname at birth was Weiss.

   Besides the integrability of w_λ, a key difference between d_λ and w_λ is the fact that $w_\lambda \geq 0$.

1.4. Analytic Properties of Fourier Transforms

1.4.1 Theorem. Analytic Properties of Fourier Transforms

Let $f \in L^1(\mathbb{R})$, $f \longleftrightarrow F$.

a. *Boundedness. For each $\gamma \in \hat{\mathbb{R}}$, $|F(\gamma)| \leq \|f\|_{L^1(\mathbb{R})}$.*

b. *Continuity. F is uniformly continuous on $\hat{\mathbb{R}}$, i.e., for all $\epsilon > 0$, there is $\delta > 0$ such that for each γ and each λ for which $|\lambda| < \delta$, we have $|F(\gamma+\lambda)-F(\gamma)| < \epsilon$. In particular, F is continuous on $\hat{\mathbb{R}}$.*

c. *Riemann–Lebesgue Lemma. $\lim_{|\gamma|\to\infty} F(\gamma) = 0$.*

d. *Time differentiation. Suppose that $f^{(m)}$, $m \geq 1$, exists everywhere and that $f^{(m)} \in L^1(\mathbb{R})$. Assume*

$$f(\pm\infty) = \cdots = f^{(m-1)}(\pm\infty) = 0, \qquad (1.4.1)$$

where $f(\pm\infty) = 0$ indicates that $\lim_{t\to+\infty} f(t) = 0$ and $\lim_{t\to-\infty} f(t) = 0$. Then

$$f^{(m)}(t) \longleftrightarrow (2\pi i\gamma)^m F(\gamma).$$

e. *Frequency differentiation. Suppose $t^m f(t) \in L^1(\mathbb{R})$. Then $tf(t), \ldots, t^{m-1}f(t) \in L^1(\mathbb{R})$, $F^{(1)}, \ldots, F^{(m)}$ exist everywhere, and*

$$\forall j = 0, 1, \ldots, m, \qquad (-2\pi it)^j f(t) \longleftrightarrow F^{(j)}(\gamma).$$

PROOF. **a.** $|F(\gamma)| \leq \int |f(t)| |e^{-2\pi it\gamma}| \, dt = \|f\|_{L^1(\mathbb{R})}$.

b. We begin with the estimate

$$|F(\gamma + \lambda) - F(\gamma)|$$

$$\leq \int |f(t)| |e^{-2\pi it\gamma}(e^{-2\pi it\lambda} - 1)| \, dt = \int |f(t)| |e^{-2\pi it\lambda} - 1| \, dt. \qquad (1.4.2)$$

Let $f_{[\lambda]}(t) = |f(t)||e^{-2\pi it\lambda} - 1|$ so that $\lim_{\lambda\to 0} f_{[\lambda]}(t) = 0$ for all t and $|f_{[\lambda]}(t)| \leq 2|f(t)|$. Thus, LDC (the Lebesgue Dominated Convergence Theorem, Theorem A.9 from Appendix A) applies to the right side of (1.4.2), which is independent of $\gamma \in \hat{\mathbb{R}}$. Consequently, we have

$$\forall \epsilon > 0, \quad \exists \lambda_0 > 0 \quad \text{such that} \quad \forall \lambda \in (0, \lambda_0) \quad \text{and} \quad \forall \gamma \in \hat{\mathbb{R}},$$
$$|F(\gamma + \lambda) - F(\gamma)| < \epsilon.$$

This is the desired uniform continuity.

c. Suppose $f = \mathbf{1}_{[a,b]}$ and $\gamma \neq 0$. Then

$$|\hat{f}(\gamma)| = \left| \int_a^b e^{-2\pi i t \gamma} \, dt \right| = \frac{1}{2\pi |\gamma|} \left| e^{-2\pi i b \gamma} - e^{-2\pi i a \gamma} \right| \leq \frac{1}{\pi |\gamma|},$$

and the right side tends to 0 as $|\gamma|$ tends to infinity.

Therefore, $\lim_{|\gamma| \to \infty} |\hat{f}(\gamma)| = 0$ if $f = \sum_{j=1}^n c_j \mathbf{1}_{[a_j, b_j]}$, where $b_j \leq a_{j+1}$.

For arbitrary $f \in L^1(\mathbb{R})$, $f \longleftrightarrow F$, we take $\epsilon > 0$; and we shall find $\gamma_\epsilon > 0$ such that if $|\gamma| > \gamma_\epsilon$ then $|\hat{f}(\gamma)| < \epsilon$. To this end we invoke Theorem A.5 and choose

$$g = \sum_{j=1}^n c_j \mathbf{1}_{[a_j, b_j]}, \qquad g \longleftrightarrow G,$$

where $b_j \leq a_{j+1}$, for which $\|f - g\|_{L^1(\mathbb{R})} < \epsilon/2$. Consequently, we have

$$\forall \gamma \in \hat{\mathbb{R}}, \quad |F(\gamma)| \leq |F(\gamma) - G(\gamma)| + |G(\gamma)|$$

$$\leq \|f - g\|_{L^1(\mathbb{R})} + |G(\gamma)| < \frac{\epsilon}{2} + |G(\gamma)|.$$

From the previous step we can take $\gamma_\epsilon > 0$ such that $|\gamma| > \gamma_\epsilon$ implies $|G(\gamma)| < \epsilon/2$. This completes the proof.

d. By integration by parts (Theorem A.22), we compute

$$\int_{-S}^T f^{(m)}(t) e^{-2\pi i t \gamma} \, dt = f^{(m-1)}(t) e^{-2\pi i t \gamma} \Big|_{-S}^T$$

$$+ 2\pi i \gamma \int_{-S}^T f^{(m-1)}(t) e^{-2\pi i t \gamma} \, dt = f^{(m-1)}(t) e^{-2\pi i t \gamma} \Big|_{-S}^T$$

$$+ 2\pi i \gamma \left(f^{(m-2)}(t) e^{-2\pi i t \gamma} \Big|_{-S}^T + 2\pi i \gamma \int_{-S}^T f^{(m-2)}(t) e^{-2\pi i t \gamma} \, dt \right)$$

$$= \cdots = \sum_{j=0}^{m-1} (2\pi i \gamma)^j$$

$$\times \left(f^{(m-(j+1))}(T) e^{-2\pi i T \gamma} - f^{(m-(j+1))}(-S) e^{2\pi i S \gamma} \right)$$

$$+ (2\pi i \gamma)^m \int_{-S}^T f(t) e^{-2\pi i t \gamma} \, dt.$$

Letting $S, T \to \infty$, the right side converges to $(2\pi i \gamma)^m F(\gamma)$ and the result is proved.

e.   Without loss of generality let $m = 1$ and fix $\gamma \in \hat{\mathbb{R}}$. Then

$$\frac{F(\gamma + \lambda) - F(\gamma)}{\lambda} = \int f(t) e^{-2\pi i t \gamma} \left(\frac{e^{-2\pi i t \lambda} - 1}{\lambda} \right) dt,$$

and we designate the integrand by $f(t, \lambda)$ (γ is fixed).

By the mean value theorem we have the estimate

$$\left| \frac{e^{-2\pi i t \lambda} - 1}{\lambda} \right| = \left| \frac{\cos 2\pi t \lambda - 1}{2\pi t \lambda} 2\pi t - i \frac{\sin 2\pi t \lambda}{2\pi t \lambda} 2\pi t \right|$$

$$\leq 2\pi |t| \left| \frac{\cos 2\pi t \lambda - 1}{2\pi t \lambda} \right| + 2\pi |t|$$

$$\leq 2\pi |t| \frac{|\sin \xi| |2\pi t \lambda|}{|2\pi t \lambda|} + 2\pi |t| \leq 4\pi |t|.$$

Consequently,

$$|f(t, \lambda)| \leq 4\pi |t f(t)| \quad \text{a.e.,} \tag{1.4.3}$$

and we also know that

$$\lim_{\lambda \to 0} f(t, \lambda) = -2\pi i t f(t) e^{-2\pi i t \gamma} \quad \text{a.e.} \tag{1.4.4}$$

since

$$\lim_{\lambda \to 0} \frac{\cos 2\pi t \lambda - 1}{\lambda} = 2\pi t \lim_{\alpha \to 0} \frac{\cos \alpha - 1}{\alpha} = 0$$

and

$$\lim_{\lambda \to 0} \frac{-i \sin 2\pi t \lambda}{\lambda} = -2\pi i t.$$

By (1.4.3) and (1.4.4) we can invoke LDC and assert that

$$\exists \lim_{\lambda \to 0} \frac{F(\gamma + \lambda) - F(\gamma)}{\lambda} = \int (-2\pi i t) f(t) e^{-2\pi i t \gamma} \, dt. \quad \blacksquare$$

1.4.2 Remark. The Role of Absolute Continuity

a.   Suppose $f^{(m)}$ exists a.e. and $f^{(m)} \in L^1(\mathbb{R})$. If $a, b \in \mathbb{R}$ and $c \in \mathbb{C}$, then

$$F(t) = c + \int_a^t f^{(m)}(u) \, du$$

is absolutely continuous on $[a, b]$ and $F(a) = c$. FTCI (Theorem A.20) implies $F' = f^{(m)}$ a.e. on $[a, b]$. This does not imply that $f^{(m-1)} \in AC_{\text{loc}}(\mathbb{R})$; and closely related to this phenomenon is the fact that it is *not* necessarily true that

$$(f')\hat{}(\gamma) = 2\pi i \gamma \hat{f}(\gamma) \quad \text{on } \hat{\mathbb{R}} \tag{1.4.5}$$

when $f, f' \in L^1(\mathbb{R})$. For example, if f_C is the Cantor function for the usual 1/3-Cantor set C on $[0, 1]$, e.g., [Ben76, page 22], then $f = \tau_{-1} f_C + (1 - f_C) \mathbf{1}_{[0,1]}$ defines a continuous compactly supported function of bounded variation on $\mathbb{R}$ for which $f' = 0$ a.e. In particular, $f, f' \in L^1(\mathbb{R})$ and (1.4.5) fails.

In this regard, note that if $f, f' \in L^1(\mathbb{R})$ and $f \in AC_{\text{loc}}(\mathbb{R})$, then

$$\int f'(t)\, dt = 0$$

[Ben76, Theorem 4.16].

  b.   The formula,

$$\int f^{(m)}(t) e^{-2\pi i t \gamma}\, dt = (2\pi i \gamma)^m \hat{f}(\gamma), \tag{1.4.6}$$

is true by the proof of Theorem 1.4.1d if we replace the hypothesis, $f^{(m)} \in L^1(\mathbb{R})$, by $f^{(m)} \in L^1_{\text{loc}}(\mathbb{R})$. In either case, the application of integration by parts in the proof, "going from m to $m - 1$", is subtle since the everywhere differentiability of $f^{(m)}$ allows us to conclude that $f^{(m-1)} \in AC_{\text{loc}}(\mathbb{R})$ [Ben76, Theorem 4.15], and this smoothness allows us to integrate by parts.

  c.   Equation (1.4.6) is also valid, without the aformentioned subtlety, if the hypotheses, that $f^{(m)}$ is everywhere differentiable and $f^{(m)} \in L^1(\mathbb{R})$, are replaced by the hypothesis that $f^{(m)}$ be piecewise continuous, cf., the example of part a and the delicate issues that can arise in [Ben76, Section 4.6].

  d.   The hypothesis (1.4.1) of Theorem 1.4.1d is not required. For simplicity, let $m = 1$ and assume $f, f' \in L^1(\mathbb{R})$ and $f \in AC_{\text{loc}}(\mathbb{R})$, cf., part b. For fixed $a \in \mathbb{R}$ and $c \in \mathbb{C}$, set $F(t) = c + \int_a^t f'(u)\, du$. By FTCI, $F \in AC_{\text{loc}}(\mathbb{R})$ and $F' = f'$ a.e. Since $f \in AC_{\text{loc}}(\mathbb{R})$, we have $f = F + C$ on $[a, \infty)$ and so

$$\forall t \in [a, \infty), \qquad f(t) = F(a) + C + \int_a^t f'(u)\, du.$$

Therefore, $f(a) = F(a) + C$ and

$$f(t) - f(a) = \int_a^t f'(u)\, du.$$

Thus, $f' \in L^1(\mathbb{R})$ implies $\lim_{t \to \pm\infty} f(t) = L_\pm$, and we have $L_+ = L_- = 0$ since $f \in L^1(\mathbb{R})$. This is (1.4.1). Also, this calculation shows that

$$\forall a \in \mathbb{R}, \qquad f(a) = -\int_a^\infty f'(u)\, du.$$

e. In our proof of Theorem 1.4.1*d* we did not require that $f^{(j)} \in L^1(\mathbb{R})$ for $0 < j < m$. We only used the fact that each such $f^{(j)} \in AC_{\text{loc}}(\mathbb{R})$. It is true, however, that if f, $f^{(m)} \in L^1(\mathbb{R})$, then $f^{(j)} \in L^1(\mathbb{R})$ for $0 < j < m$, e.g., [BC49, pages 29–30].

Proposition 1.4.3 below is an extension for $m < 0$ of Theorem 1.4.1*d*.

1.4.3 Proposition.

Let $f \in L^1(\mathbb{R})$, $f \longleftrightarrow F$. *Define* $g(t) = \int_{-\infty}^t f(u)\, du$ *and assume* $g \in L^1(\mathbb{R})$. *(Note that* $\int f(t)\, dt = 0$ *since* $g \in L^1(\mathbb{R})$, *e.g., Exercise 1.10.) Then* $F(\gamma) = 2\pi i \gamma \hat{g}(\gamma)$ *for* $\gamma \in \hat{\mathbb{R}}$, *and so*

$$\forall \gamma \in \hat{\mathbb{R}} \backslash \{0\}, \qquad \hat{g}(\gamma) = \frac{1}{2\pi i \gamma} F(\gamma),$$

i.e.,

$$g(t) = \int_{-\infty}^t f(u)\, du \longleftrightarrow \frac{1}{2\pi i \gamma} F(\gamma),$$

where $\lim_{\gamma \to 0} F(\gamma)/(2\pi i \gamma) = \hat{g}(0)$.

PROOF. We calculate

$$\int_{-S}^T g'(t) e^{-2\pi i t \gamma}\, dt$$

$$= g(t)(-2\pi i \gamma) e^{-2\pi i t \gamma}\bigg|_{-S}^T + \int_{-S}^T g(t)(2\pi i \gamma) e^{-2\pi i t \gamma}\, dt.$$

Since $g(\pm\infty) = 0$ and $g'(t) = f(t)$ a.e. by FTCI, we can conclude that $2\pi i \gamma \hat{g}(\gamma) = F(\gamma)$. ∎

1.4.4 Example. $C_0(\hat{\mathbb{R}}) \backslash A(\hat{\mathbb{R}}) \neq \varnothing$

Theorem 1.4.1*b,c* allow us to conclude that $A(\hat{\mathbb{R}}) \subseteq C_0(\hat{\mathbb{R}})$, where $C_0(\hat{\mathbb{R}})$ is the space of continuous functions F on $\hat{\mathbb{R}}$ for which

$$\lim_{|\gamma| \to \infty} F(\gamma) = 0.$$

It is relatively easy to check that this inclusion is proper. For example, if F is defined as

$$F(\gamma) = \begin{cases} \dfrac{1}{\log \gamma}, & \text{if } \gamma > e, \\[2mm] \dfrac{\gamma}{e}, & \text{if } 0 \le \gamma \le e, \end{cases} \qquad (1.4.7)$$

on $[0, \infty)$ and as $-F(-\gamma)$ on $(-\infty, 0]$, then $F \in C_0(\hat{\mathbb{R}})$. The fact that $F \notin A(\hat{\mathbb{R}})$ depends on the divergence of $\int_e^\infty d\gamma/(\gamma \log \gamma)$. Instead of providing the details we refer ahead to Example 3.3.4a where the analogous calculation for Fourier series coefficients is not only verified but motivated, cf., [Gol61, pages 8–9].

The function in (1.4.7) is not an isolated example. In fact, $A(\hat{\mathbb{R}})$ is only a set of first category in $C_0(\hat{\mathbb{R}})$, Exercise 1.40. Even more, a Baire category argument can also be used to show the existence of $F \in C_c(\hat{\mathbb{R}})$ for which $F \notin A(\hat{\mathbb{R}})$. Explicit examples of such functions are more difficult to construct, but it is possible to do so, e.g., define

$$B(\gamma) = \begin{cases} \dfrac{1}{n}\sin(2\pi 4^n \gamma), & \text{if } \dfrac{1}{2^{n+1}} \le |\gamma| \le \dfrac{1}{2^n}, \\[3mm] 0, & \text{if } \gamma = 0 \text{ or } |\gamma| > \dfrac{1}{2}, \end{cases}$$

[Her85]. ("B" is for "butterfly".)

It is natural to ask for an *intrinsic characterization* of $A(\hat{\mathbb{R}})$, i.e., to seek a theorem of the form "$F \in C_0(\hat{\mathbb{R}})$ is an element of $A(\hat{\mathbb{R}})$ if and only if . . .", where ". . ." is a statement about the behavior of F on $\hat{\mathbb{R}}$. This is an open problem, e.g., [Kah70].

1.4.5 Remark. Perspective on the Operational Calculus

Theorem 1.4.1 is a major component of the operational calculus used in classical electrical engineering and in solving various differential equations. Typically, a calculus problem, e.g., a differential equation, is transformed into an algebra problem by Theorem 1.4.1d; the algebra problem is solved, and the solution is transformed by an inversion formula into the solution of the original problem. A feature of this formalism is the notion of *convolution*.

1.5. Convolution

1.5.1 Definition. Convolution

Let $f, g \in L^1(\mathbb{R})$. The *convolution* of f and g, denoted by $f * g$, is

$$f * g(t) = \int f(t-u)g(u)\, du = \int f(u)g(t-u)\, du.$$

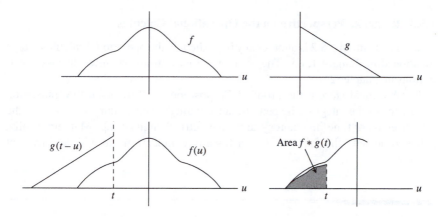

FIGURE 1.7.

It is not difficult to prove that $f * g \in L^1(\mathbb{R})$ (Exercise 1.31). Later, we shall demonstrate the role of convolution in the method alluded to in Remark 1.4.5. The algebraic properties of convolution are the subject of Exercise 1.32.

Convolution is illustrated in Figure 1.7.

1.5.2 Proposition.

Let $f, g \in L^1(\mathbb{R})$, with corresponding Fourier pairs $f \longleftrightarrow F$ and $g \longleftrightarrow G$. Then $f * g \in L^1(\mathbb{R})$ and $(f * g)\hat{} = \hat{f}\hat{g}$, i.e.,

$$f * g \longleftrightarrow FG.$$

PROOF. As indicated in Definition 1.5.1, the assertion that $f * g \in L^1(\mathbb{R})$ is the task of Exercise 1.31. Assuming this fact, we use the Fubini–Tonelli Theorem (Theorem A.14) to compute

$$(f * g)\hat{}(\gamma) = \iint f(t - u)g(u)e^{-2\pi i t\gamma}\, du\, dt$$

$$= \iint f(t - u)g(u)e^{-2\pi i(t-u)\gamma}\, e^{-2\pi i u\gamma}\, du\, dt$$

$$= \int \left(\int f(t - u)e^{-2\pi i(t-u)\gamma}\, dt \right) g(u)e^{-2\pi i u\gamma}\, du$$

$$= \int \hat{f}(\gamma)g(u)e^{-2\pi i u\gamma}\, du = \hat{f}(\gamma)\hat{g}(\gamma). \qquad \blacksquare$$

1.5.3 Remark. Perspective on the Operational Calculus

a. Proposition 1.5.2 is another key ingredient in the operational calculus recipe mentioned in Remark 1.4.5. The complete story unfolds when we discuss distributions in Chapter 2.

b. A critical step in the proof of Proposition 1.5.2 involved the translation invariance of the integral (the penultimate equality). This feature accounts for the effectiveness of time-invariant systems in electrical engineering. Mathematically, it has to do with the fact that $\mathbb{R}$ is a (locally compact) group with an invariant measure.

1.6. Approximate Identities and Examples

The following notion is critical in approximating the unit impulse and for providing examples in applications including signal processing and spectral estimation.

1.6.1 Definition. Approximate Identity

An *approximate identity* is a family $\{k_{(\lambda)} : \lambda > 0\} \subseteq L^1(\mathbb{R})$ of functions with the properties:

a) $\forall \lambda > 0, \int k_{(\lambda)}(t)\,dt = 1$;

b) $\exists K$ such that $\forall \lambda > 0, \|k_{(\lambda)}\|_{L^1(\mathbb{R})} \leq K$;

c) $\forall \eta > 0, \lim_{\lambda \to \infty} \int_{|t| \geq \eta} |k_{(\lambda)}(t)|\,dt = 0$.

Caveat. The subscript "(λ)" in Definition 1.6.1 does not necessarily denote a dilation. The following result, however, shows that dilations yield a large class of approximate identities.

1.6.2 Proposition.

Let $k \in L^1(\mathbb{R})$ have the property that $\int k(t)\,dt = 1$. The family $\{k_\lambda : k_\lambda(t) = \lambda k(\lambda t), \lambda > 0\} \subseteq L^1(\mathbb{R})$ of dilations is an approximate identity.

PROOF. To verify the condition of Definition 1.6.1*a*, we compute

$$\int k_\lambda(t)\,dt = \lambda \int k(\lambda t)\,dt = \int k(t)\,dt = 1.$$

For part *b* we compute

$$\int |k_\lambda(t)|\,dt = \lambda \int |k(\lambda t)|\,dt = \int |k(u)|\,du = K < \infty,$$

where K is finite since $k \in L^1(\mathbb{R})$.

For part *c*, take $\eta > 0$ and compute

$$\int_{|t|\geq\eta} |k_\lambda(t)|\, dt = \lambda \int_{|t|\geq\eta} |k(\lambda t)|\, dt = \int_{|u|\geq\lambda\eta} |k(u)|\, du;$$

this last term tends to 0 as λ tends to ∞ since $\eta > 0$ and because of the definition of the integral. ∎

1.6.3 Proposition.

$$\int_{-\infty}^{\infty} \frac{\sin t}{t}\, dt = \int_{-\infty}^{\infty} \frac{\sin^2 t}{t^2}\, dt = \pi.$$

(Contrary to our consistent notation, we have written the limits of integration $\pm\infty$ since the first integral is an improper Riemann integral.)

PROOF. *a.* To prove that these integrals are equal, let $u = \sin^2 t$ and $dv = dt/t^2$ in the second integral so that

$$\int_0^\infty \frac{\sin^2 t}{t^2}\, dt = -\frac{\sin^2 t}{t} \Big|_0^\infty + \int_0^\infty \frac{2}{t} \sin t \cos t\, dt$$

$$= \int_0^\infty \frac{\sin 2t}{t}\, dt = \int_0^\infty \frac{\sin t}{t}\, dt.$$

b. We now show that

$$\int_0^\infty \frac{\sin t}{t}\, dt = \frac{\pi}{2}$$

to complete the result. There are several ways to accomplish this computation including a contour integral calculation. We choose the following method.

Let $F(\sigma)$ be the Laplace transform,

$$F(\sigma) = \int_0^\infty e^{-\sigma t} \frac{\sin t}{t}\, dt = \mathcal{L}\left(\frac{\sin t}{t}\right)(\sigma).$$

We assert that $F(\sigma)$ is a continuous function on $[0, \infty)$ and that the formal calculation,

$$\forall \sigma > 0, \qquad \exists F'(\sigma) = -\int_0^\infty e^{-\sigma t} \sin t\, dt,$$

is in fact true. (The verification of these claims involves uniform convergence.) The convergence of $F(0)$ is clear by an alternating series argument.

It is easy to see that $\mathcal{L}(\sin t)(\sigma) = 1/(1+\sigma^2)$, $\sigma > 0$, either by direct calculation using integration by parts or by using the general Laplace transform formula,

$\mathcal{L}(g^{(2)})(\sigma) = \sigma^2 \mathcal{L}(g)(\sigma) - \sigma g(0) - \sigma g'(0)$, for the special function $g(t) = \sin t$. Thus, using FTC, we compute

$$F(\sigma) - F(0) = \int_0^\sigma F'(\eta) d\eta = -\int_0^\sigma \mathcal{L}(\sin t)(\eta) \, d\eta$$

$$= -\int_0^\sigma \frac{d\eta}{1 + \eta^2} = -\tan^{-1}\sigma, \quad \sigma > 0.$$

We know $F(\infty) = 0$ by LDC, and so

$$F(0) = \lim_{\sigma \to \infty} \tan^{-1}\sigma = \frac{\pi}{2}. \quad \blacksquare$$

1.6.4 Remark. The Dirichlet and Fejér Kernels

a. The family $\{d_\lambda\}$ of dilations of $d(t)$ is the *Dirichlet kernel*, and the family $\{w_\lambda\}$ of dilations of $w(t)$ is the *Fejér kernel*. The Fejér kernel is an approximate identity by Propositions 1.6.2 and 1.6.3. The Dirichlet kernel is not an approximate identity since $d_\lambda \notin L^1(\mathbb{R})$, whereas Proposition 1.6.3 highlights a similarity between d_λ and w_λ, cf., Exercises 1.17 and 1.46. Although $\{d_\lambda\}$ is not an approximate identity, it does possess the property that its "mass" accumulates at the origin, while its Fourier transform tends to the function identically 1 on $\hat{\mathbb{R}}$ as $\lambda \to \infty$.

b. By definition of convolution,

$$\frac{1}{T}\mathbf{1}_{[-T/2,T/2]} * \mathbf{1}_{[-T/2,T/2]}(t) = \max\left(1 - \frac{|t|}{T}, 0\right). \tag{1.6.1}$$

Thus, by Example 1.3.1 and Proposition 1.5.2, we have another proof of the Fourier transform pairing (1.3.4), viz.,

$$\max\left(1 - \frac{|t|}{T}, 0\right) \longleftrightarrow w_{2\pi T}(\gamma),$$

since

$$\frac{1}{T}d_{\pi T}(\gamma)^2 = w_{2\pi T}(\gamma).$$

Equation (1.6.1) asserts that the "convolution of rectangles" is a "triangle". Further steps are carried out in Exercise 1.17.

Yet another way to prove (1.3.4) is to introduce the following translation and dilation of the *Haar wavelet*, viz.,

$$f(t) = \begin{cases} \dfrac{1}{T}, & \text{if } t \in [-T, 0), \\[2mm] -\dfrac{1}{T}, & \text{if } t \in [0, T), \\[2mm] 0, & \text{otherwise.} \end{cases}$$

An elementary computation gives

$$g(t) = \int_{-\infty}^{t} f(u)\, du = \max\left(1 - \frac{|t|}{T}, 0\right),$$

and, in particular, $f, g \in L^1(\mathbb{R})$. From Proposition 1.4.3 we obtain the Fourier transform pairing,

$$\max\left(1 - \frac{|t|}{T}, 0\right) \longleftrightarrow \frac{1}{2\pi i \gamma} \hat{f}(\gamma). \tag{1.6.2}$$

Since $f = \frac{1}{T}\big(\tau_{-T/2}\mathbf{1}_{[-T/2,T/2)} - \tau_{T/2}\mathbf{1}_{[-T/2,T/2)}\big)$, we can easily compute

$$\frac{1}{2\pi i \gamma} \hat{f}(\gamma) = w_{2\pi T}(\gamma),$$

and so (1.3.4) is obtained from (1.6.2).

1.6.5 Example. The Poisson Kernel

   a.   The family $\{p_\lambda\}$ of dilations of $p(t)$, e.g., Example 1.3.2, is an approximate identity by Proposition 1.6.2 and the fact that

$$\int \frac{dt}{\pi(1 + t^2)} = 1. \tag{1.6.3}$$

$\{p_\lambda\}$ is the *Poisson kernel*, and p_λ and $\hat{p}_\lambda$ are graphed in Figure 1.8.
   b.   We can do the elementary computation (1.6.3) by means of direct integration, cf., [Rud66, page 4], or in the following more complicated way. In Example 1.3.2 we verified that $e^{-2\pi|t|} \longleftrightarrow p(\gamma)$ so that, once we have the inversion theory of Section 1.7, we shall have the pairing $p(t) \longleftrightarrow e^{-2\pi|\gamma|} \equiv P(\gamma)$. Thus, $\hat{p}(0) = P(0) = 1$ by Theorem 1.1.6 (Theorem 1.7.6), and this is (1.6.3).
   c.   The *Paley–Wiener Logarithmic Integral Theorem* is the following assertion. *Let ϕ be a nonnegative function for which $\int \phi^2(\gamma)\, d\gamma < \infty$. There is a function f vanishing on $(-\infty, 0)$ for which $\int |f(t)|^2\, dt < \infty$ and $|\hat{f}| = \varphi$ a.e. if and only if*

$$\int \frac{|\log \phi(\gamma)|}{1 + \gamma^2}\, d\gamma < \infty,$$

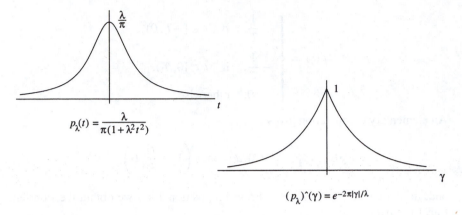

$$p_\lambda(t) = \frac{\lambda}{\pi(1+\lambda^2 t^2)}$$

$$(p_\lambda)^\wedge(\gamma) = e^{-2\pi|\gamma|/\lambda}$$

FIGURE 1.8.

cf., Section 1.10 for the definition of $\hat{f}$ in this case. A function f is referred to as a *causal signal* in the case it vanishes on $(-\infty, 0)$; and a function f is a *signal of finite energy* in the case $\int |f(t)|^2 \, dt < \infty$.

The Paley–Wiener Logarithmic Integral Theorem is used to characterize causal signals, e.g., [OS75], [Pap77], cf., Remark 3.7.10. We have stated it now to highlight the appearance of the Poisson function, to mention finite energy signals which we shall study in Section 1.10, and to give an explicit result which alludes to the profound uniqueness and uncertainty principle properties of Fourier analysis, cf., Example 1.10.6.

1.6.6 Example. The Gauss Kernel

The family $\{g_\lambda\}$ of dilations of $g(t)$, e.g., Example 1.3.3, is an approximate identity by Proposition 1.6.2 and the fact that $\int g(t) \, dt = (1/\sqrt{\pi}) \int e^{-t^2} \, dt = 1$. $\{g_\lambda\}$ is the *Gauss kernel*, and g_λ and $(g_\lambda)^\wedge$ are graphed in Figure 1.9.

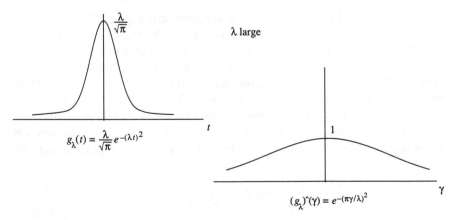

$$g_\lambda(t) = \frac{\lambda}{\sqrt{\pi}} e^{-(\lambda t)^2}$$

$$(g_\lambda)^\wedge(\gamma) = e^{-(\pi\gamma/\lambda)^2}$$

FIGURE 1.9.

1.6.7 Example. Properties of the Poisson Function

a. If $\{p_\lambda : \lambda > 0\}$ is the Poisson kernel, then $\hat{p}_\lambda$ is not differentiable at 0 even though $p_\lambda \in L^1(\mathbb{R})$ and decreases like $1/t^2$ as $|t| \to \infty$, cf., Theorem 1.4.1e. The verification of nondifferentiability is elementary from Example 1.3.2; in fact, expanding $\hat{p}_\lambda$ in a Taylor series, we have

$$\lim_{\gamma \to 0\pm} \frac{\hat{p}_\lambda(\gamma) - \hat{p}_\lambda(0)}{\gamma - 0} = \mp \frac{2\pi}{\lambda}.$$

Further, $\hat{p}_\lambda$ is even, convex, and decreasing to 0 on $(0, \infty)$.

b. Because of Example 1.6.5*a,c*, we next note that

$$I = \int \frac{\log|\gamma|}{1 + \gamma^2}\, d\gamma = 0.$$

In fact, integrating over $(0, \infty)$ and letting $\lambda = 1/\gamma$ we see that $\frac{I}{2} = -\frac{I}{2}$, cf., [AGR88] for a unifying principle to calculate such integrals by real methods.

c. The linear fractional transformation $w = (z - i)/(z + i)$ maps the upper half-plane onto the unit disk with $\mathbb{R}$ mapped onto the unit circle. As such, the Poisson function p assumes the role of the Jacobian:

$$\frac{1}{2\pi}\int_0^{2\pi} g\left(e^{i\theta}\right) d\theta = \int f(t) p(t)\, dt,$$

where $g(e^{i\theta})$ is defined on the unit circle and

$$f(t) = g\left(e^{i\theta}\right) = g\left(\frac{t-i}{t+i}\right).$$

d. Note that if $a, b > 0$, then

$$p_{1/a} * p_{1/b} = p_{1/(a+b)}$$

because of Example 1.3.2 and Proposition 1.5.2.

For a host of other examples we refer to [Harr78].

1.6.8 Example. Central Limit Theorem

a. We can compute

$$\forall t \in \mathbb{R}, \qquad \lim_{n \to \infty} \left(\cos \frac{t}{\sqrt{n}}\right)^n = e^{-t^2/2}, \tag{1.6.4}$$

[Hint. We approximate the cosine by the first two terms of its Taylor series, and have the approximation,

$$\left(\cos\frac{t}{\sqrt{n}}\right)^n \approx \left(1 + \frac{1}{-2n/t^2}\right)^{(-2n/t^2)(-t^2/2)}.$$

The right side converges to the right side of (1.6.4) by the definition of e. The "error terms" can be shown to tend to zero by a variety of methods.]

   b.   The *Central Limit Theorem* in probability theory is *equivalent* to the following result. *Let $f \in L^1(\mathbb{R}) \cap L^\infty(\mathbb{R})$ be nonnegative and continuous, and assume that*

$$\int f(t)\, dt = 1, \qquad \int t f(t)\, dt = 0, \quad and$$

$$\int t^2 f(t)\, dt = 1. \tag{1.6.5}$$

Define $f_1 = f$, $f_n = f * f_{n-1}$ for each $n \geq 2$, and $g_n(t) = n^{1/2} f(n^{1/2}t)$. Then

$$\forall a < b, \qquad \lim_{n\to\infty} \int_a^b g_n(t)\, dt = \frac{1}{\sqrt{2\pi}} \int_a^b e^{-t^2/2}\, dt.$$

There is a dazzling treatment of this material in [Kör88, Chapter 70]. The hypotheses (1.6.5) are equivalent to the statement that f *is the probability density function of a random variable X having mean 0 and variance 1.* We shall explain this terminology in Section 2.8. The Central Limit Theorem deals with the asymptotic behavior of sample means as the sample size increases, and it quantifies the remarkable fact that the sum of a large number of independent random variables approximates a normal, i.e., Gaussian, distribution, e.g., [Kac59], [Lam66, Chapter 3], [Pri81]. It will not come as a surprise that (1.6.4) is related to the Central Limit Theorem.

   The major *elementary* property of approximate identities is given in Theorem 1.6.9a. Theorem 1.6.9b is the special case for the Fejér kernel, and part c is the Fourier transform uniqueness theorem. We prove the uniqueness theorem as a corollary of part b.

1.6.9   Theorem. Approximation and Uniqueness

Let $f \in L^1(\mathbb{R})$.

   a.   *If $\{k_{(\lambda)} : \lambda > 0\} \subseteq L^1(\mathbb{R})$ is an approximate identity, then*

$$\lim_{\lambda\to\infty} \| f - f * k_{(\lambda)} \|_{L^1(\mathbb{R})} = 0.$$

b. *We have*

$$\lim_{\lambda \to \infty} \int \left| f(t) - \int_{-\lambda/2\pi}^{\lambda/2\pi} \left(1 - \frac{2\pi|\gamma|}{\lambda} \right) \hat{f}(\gamma) e^{2\pi i t \gamma} \, d\gamma \right| \, dt = 0.$$

c. *Uniqueness. If $\hat{f} = 0$ on $\hat{\mathbb{R}}$, then f is the 0-function.*

PROOF. **a.** We use the fact that $\int k_{(\lambda)}(t) \, dt = 1$ to compute

$$\| f - f * k_{(\lambda)} \|_{L^1(\mathbb{R})}$$

$$= \int \left| \int k_{(\lambda)}(u) f(t) \, du - \int k_{(\lambda)}(u) f(t-u) \, du \right| dt$$

$$\leq \int |k_{(\lambda)}(u)| \left(\int |f(t) - \tau_u f(t)| \, dt \right) du.$$

Let $\epsilon > 0$. By Theorem A.5, there is $\eta > 0$ with the property that

$$\forall |u| < \eta, \quad \| f - \tau_u f \|_{L^1(\mathbb{R})} < \epsilon/K, \qquad (1.6.6)$$

where $\| k_{(\lambda)} \|_{L^1(\mathbb{R})} \leq K$. Therefore, we have the estimate

$$\| f - f * k_{(\lambda)} \|_{L^1(\mathbb{R})} \leq 2\| f \|_{L^1(\mathbb{R})} \int_{|u| \geq \eta} |k_{(\lambda)}(u)| \, du$$

$$+ \frac{\epsilon}{K} \int_{|u| \leq \eta} |k_{(\lambda)}(u)| \, du$$

$$\leq \epsilon + 2\| f \|_{L^1(\mathbb{R})} \int_{|u| \geq \eta} |k_{(\lambda)}(u)| \, du.$$

Consequently, by the definition of an approximate identity, we have

$$\overline{\lim_{\lambda \to \infty}} \, \| f - f * k_{(\lambda)} \|_{L^1(\mathbb{R})} \leq \epsilon;$$

and so we obtain part *a* since $\epsilon > 0$ can be chosen as small as we like.

b. To begin with, the calculation in Example 1.3.4 shows that

$$w_\lambda(t) = \int_{-\lambda/2\pi}^{\lambda/2\pi} \left(1 - \frac{2\pi|\gamma|}{\lambda} \right) e^{2\pi i t \gamma} \, d\gamma.$$

Then, by the definition of convolution and an application of the Fubini–Tonelli Theorem, we compute

$$f * w_\lambda(t) = \int_{-\lambda/2\pi}^{\lambda/2\pi} \left(1 - \frac{2\pi|\gamma|}{\lambda}\right) \hat{f}(\gamma) e^{2\pi i t \gamma} \, d\gamma.$$

Since $\{w_\lambda\}$ is an approximate identity, part b follows from part a.

Part c follows from part b. In fact, the hypothesis and part b imply $\|f\|_{L^1(\mathbb{R})} = 0$; and so f is the 0-function by Theorem A.5.   ∎

1.6.10 Remark. Inversion Formula for L^1-Norm

Theorem 1.6.9b has the flavor of the inversion results discussed in Section 1.1. For example we could compare Theorem 1.6.9b with Theorem 1.1.6. There are two differences:

i. Theorem 1.1.6 is a pointwise result, whereas we deal with L^1-convergence in Theorem 1.6.9b;

ii. The Dirichlet kernel is used in the statement of Theorem 1.1.6, whereas the Fejér kernel is used in Theorem 1.6.9b.

1.6.11 Proposition.

Let $f \in L^\infty(\mathbb{R})$ be continuous on $\mathbb{R}$. If $\{k_{(\lambda)} : \lambda > 0\} \subseteq L^1(\mathbb{R})$ is an approximate identity, then

$$\forall t \in \mathbb{R}, \qquad \lim_{\lambda \to \infty} f * k_{(\lambda)}(t) = f(t).$$

($L^\infty(\mathbb{R})$ is defined in Definition A.10.)

PROOF. We first compute

$$|f(t) - f * k_{(\lambda)}(t)| = \left| \int k_{(\lambda)}(u)(f(t) - f(t - u)) \, du \right|$$

$$\leq \int |k_{(\lambda)}(u)| \, |f(t) - f(t - u)| \, du$$

for a fixed $t \in \mathbb{R}$.

Let $\epsilon > 0$. Since f is continuous, there is $\eta > 0$ such that if $0 \leq |u| < \eta$, then $|f(t) - f(t - u)| < \epsilon/K$, where $\|k_{(\lambda)}\|_{L^1(\mathbb{R})} \leq K$. This yields the estimate

$$|f(t) - f * k_{(\lambda)}(t)| \leq \epsilon + 2\|f\|_{L^\infty(\mathbb{R})} \int_{|u| \geq \eta} |k_{(\lambda)}(u)| \, du.$$

Consequently, by the definition of an approximate identity, we have

$$\overline{\lim_{\lambda \to \infty}} \, |f(t) - f * k_{(\lambda)}(t)| \leq \epsilon;$$

and so we obtain our result since ϵ can be chosen as small as we like.   ∎

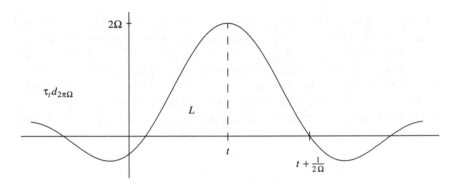

FIGURE 1.10.

1.7. Pointwise Inversion of the Fourier Transform

1.7.1 Remark. Motivation for the Inversion Theorem

The inversion formula in Theorem 1.1.6 for f continuous is

$$f(t) = \lim_{\Omega \to \infty} \int_{-\Omega}^{\Omega} F(\gamma) e^{2\pi i t \gamma} \, d\gamma, \quad f \longleftrightarrow F. \tag{1.7.1}$$

To see that this formula is reasonable we begin with the formal calculation:

$$
\begin{aligned}
\int F(\gamma) e^{2\pi i t \gamma} \, d\gamma &= \iint f(u) e^{2\pi i (t-u)\gamma} \, du \, d\gamma \\
&= \int f(u) \left[\lim_{\Omega \to \infty} \int_{-\Omega}^{\Omega} e^{2\pi i (t-u)\gamma} \, d\gamma \right] du \\
&= \lim_{\Omega \to \infty} \int f(u) \frac{\sin 2\pi (t-u)\Omega}{\pi (t-u)} \, du \\
&= \lim_{\Omega \to \infty} f * d_{2\pi\Omega}(t).
\end{aligned}
\tag{1.7.2}
$$

Observe that the area L of the major lobe of $\tau_t d_{2\pi\Omega}$ in Figure 1.10 remains constant for all Ω:

$$
\begin{aligned}
\int_{t-1/2\Omega}^{t+1/2\Omega} \tau_t d_{2\pi\Omega}(u) \, du &= \int_{-1/2\Omega}^{1/2\Omega} d_{2\pi\Omega}(u) \, du \\
&= \int_{-1/2\Omega}^{1/2\Omega} \frac{\sin 2\pi u\Omega}{\pi u} \, du = \frac{1}{\pi} \int_{-\pi}^{\pi} \frac{\sin u}{u} \, du.
\end{aligned}
\tag{1.7.3}
$$

The calculation (1.7.2) involves changing orders of operations and a principal value argument. These steps must be justified. The basic idea, however, is clear. We expect that $\lim_{\Omega \to \infty} f * d_{2\pi\Omega}(t) = f(t)$ (and this is (1.7.1)) since L remains constant as the major lobes converge to t and since the oscillations of the minor lobes on either side of t become increasingly rapid as $\Omega \to \infty$. The intuition is that the total contribution of the minor lobes will be negligible for large Ω since the Dirichlet kernel take positive and negative values. This intuition is not quite correct since

$$\frac{1}{\pi} \int_{-\pi}^{\pi} \frac{\sin u}{u} \, du > 1$$

(why?)—see Exercise 1.4; but the cancellation is such that we can verify (1.7.1) under the conditions given in Theorem 1.1.6.

We shall have more to say about the phenomenon associated with the integral in (1.7.3) in Section 1.9. Another issue that arises in the calculation (1.7.2) is the fact that $\hat{f}$ need not be in $L^1(\hat{\mathbb{R}})$ for $f \in L^1(\mathbb{R})$.

1.7.2 Example. $f \in L^1(\mathbb{R})$ Does Not Imply $\hat{f} \in L^1(\hat{\mathbb{R}})$

Let $f(t) = H(t)e^{-2\pi rt}$ where $r > 0$ and H is the *Heaviside function* defined as $H = \mathbf{1}_{[0,\infty)}$. Then $\hat{f}(\gamma) = 1/(2\pi(r + i\gamma)) \notin L^1(\hat{\mathbb{R}})$, e.g., Exercise 1.5. This should be compared with the fact that if $f(t) = e^{-2\pi r|t|}$, then $\hat{f} = p_{1/r} \in L^1(\hat{\mathbb{R}})$, cf., the end of Example 1.3.2.

Having made these introductory remarks, let us begin the proof of Theorem 1.1.6, which is Jordan's pointwise inversion formula for the Fourier transform. Theorem 1.1.6 and Theorem 1.7.6 are the same.

1.7.3 Lemma. Second Mean Value Theorem for Integrals or Bonnet Theorem

Let g be continuous on $[a, b]$, and let f be increasing on $[a, b]$. There is $\xi \in [a, b]$ such that

$$\int_a^b f(t)g(t) \, dt = f(a+) \int_a^\xi g(t) \, dt + f(b-) \int_\xi^b g(t) \, dt,$$

e.g., [Apo57, page 217] and Exercise 1.48.

1.7.4 Remark. Jordan Decomposition

The classical form of the *Jordan Decomposition Theorem* for a function $f : [a, b] \longrightarrow \mathbb{C}$ asserts that $f \in BV[a, b]$, e.g., Definition 1.1.5, *if and only if f can be expressed as the difference $f_1 - f_2$ of two increasing functions on $[a, b]$.*

The proof is not difficult. We set $f_1(a) = 0$ and define

$$\forall t \in (a, b], \qquad f_1(t) = \sup \left\{ \sum |f(t_j) - f(t_{j-1})| \right\},$$

where the supremum is taken over every finite set $a = t_0 < t_1 < \cdots < t_n = t$. It is easy to see that f_1 is increasing, and straightforward to check that $f_2 \equiv f_1 - f$ is increasing. This completes the proof.

There are standard measure-theoretic generalizations of this result, e.g., [AB66], [Ben76, Section 5.1], [Rud66].

1.7.5 Lemma.

Let $g \in BV[0, \epsilon]$, $\epsilon > 0$. Then

$$\lim_{\Omega \to \infty} \int_0^{\epsilon} g(t) d_{2\pi\Omega}(t) \, dt = \frac{1}{2} g(0+). \tag{1.7.4}$$

PROOF. **a.** By the Jordan Decomposition Theorem stated in Remark 1.7.4, we assume that g is increasing on $[0, \epsilon]$; and, in particular, from the definition of bounded variation, g is bounded on $[0, \epsilon]$.

b.i. Assume $g(0+) = 0$ and let $\eta > 0$. Since $\int d(t) \, dt = 1$ for the Dirichlet function d (Proposition 1.6.3), there is $C > 0$ such that

$$\forall a, b \in \mathbb{R}, \qquad \left| \int_a^b \frac{\sin t}{\pi t} \, dt \right| \le C.$$

We shall verify that

$$\overline{\lim_{\Omega \to \infty}} \left| \int_0^{\epsilon} g(t) d_{2\pi\Omega}(t) \, dt \right| \le \eta C; \tag{1.7.5}$$

and this will complete the proof of (1.7.4) for the $g(0+) = 0$ case since $\eta > 0$ is arbitrary.

b.ii. Since $g(0+) = 0$, there is $\nu = \nu(\eta) \in (0, \epsilon)$ such that $|g(t)| \le \eta$ for all $t \in (0, \nu)$. Also, by Lemma 1.7.3, using the continuity of $d_{2\pi\Omega}$ and the boundedness and monotonicity of g, there is $\xi \in [0, \nu]$ for which

$$\int_0^{\nu} g(t) d_{2\pi\Omega}(t) \, dt$$

$$= g(0+) \int_0^{\xi} d_{2\pi\Omega}(t) \, dt + g(\nu-) \int_{\xi}^{\nu} d_{2\pi\Omega}(t) \, dt$$

$$= g(\nu-) \int_{2\pi\Omega\xi}^{2\pi\Omega\nu} \frac{\sin u}{\pi u} \, du.$$

Consequently,

$$\left| \int_0^\nu g(t) d_{2\pi\Omega}(t)\, dt \right| \le C|g(\nu-)| \le \eta C. \tag{1.7.6}$$

b.iii.   Note that $(g(t)/t)\mathbf{1}_{[\nu,\epsilon]}(t) \in L^1(\mathbb{R})$ since we don't have to deal with the origin. Therefore, by the Riemann–Lebesgue Lemma,

$$\lim_{\Omega \to \infty} \int_\nu^\epsilon g(t) d_{2\pi\Omega}(t)\, dt = 0.$$

Using this fact, (1.7.6), and the inequality

$$\left| \int_0^\epsilon g(t) d_{2\pi\Omega}(t)\, dt \right|$$
$$\le \left| \int_0^\nu g(t) d_{2\pi\Omega}(t)\, dt \right| + \left| \int_\nu^\epsilon g(t) d_{2\pi\Omega}(t)\, dt \right|,$$

we obtain (1.7.5).

   c.   Finally, suppose $g(0+) \neq 0$. Let $h(t) = g(t) - g(0+)$ so that $h(0+) = 0$ and

$$\lim_{\Omega \to \infty} \int_0^\epsilon h(t) d_{2\pi\Omega}(t)\, dt = 0$$

by part b. Also, we know from Proposition 1.6.3 that

$$\lim_{\Omega \to \infty} \int_0^\epsilon d_{2\pi\Omega}(t)\, dt = \lim_{\Omega \to \infty} \int_0^{2\pi\Omega\epsilon} d(t)\, dt = \frac{1}{2}.$$

Combining these facts, we compute

$$\lim_{\Omega \to \infty} \int_0^\epsilon g(t) d_{2\pi\Omega}(t)\, dt$$
$$= \lim_{\Omega \to \infty} \left(\int_0^\epsilon \big(h(t) + g(0+)\big) d_{2\pi\Omega}(t)\, dt \right) = \frac{g(0+)}{2},$$

since both limits exist when we expand the middle term.   ∎

We can now complete the proof of Jordan's theorem.

1.7.6 Theorem. Jordan Theorem

Let $f \in L^1(\mathbb{R})$ and assume $f \in BV[t - \epsilon, t + \epsilon]$ for some $t \in \mathbb{R}$ and $\epsilon > 0$. Then

$$\frac{f(t+) + f(t-)}{2} = \lim_{\Omega \to \infty} \int_{-\Omega}^{\Omega} F(\gamma) e^{2\pi i t \gamma} \, d\gamma,$$

where $f \longleftrightarrow F$. If f is continuous at t, then the left side can be replaced by $f(t)$.

PROOF. For each $\Omega > 0$, define the "partial sums"

$$S_\Omega(t) = \int_{-\Omega}^{\Omega} e^{2\pi i t \gamma} F(\gamma) \, d\gamma$$

$$= \int f(u) \left(\int_{-\Omega}^{\Omega} e^{2\pi i (t-u)\gamma} \, d\gamma \right) du = f * d_{2\pi\Omega}(t).$$

The calculation is justified by the Fubini–Tonelli Theorem since the double integral on $\mathbb{R} \times [-\Omega, \Omega]$ is absolutely convergent. We write $S_\Omega(t)$ as

$$S_\Omega(t) = \int f(t - u) d_{2\pi\Omega}(u) \, du$$

$$= \int_0^\infty (f(t + u) + f(t - u)) d_{2\pi\Omega}(u) \, du.$$

Let $g(u) = f(t + u) + f(t - u)$, noting that t is fixed, and let $\epsilon > 0$. The result will be proved when we show

$$\lim_{\Omega \to \infty} \int_0^\epsilon g(u) d_{2\pi\Omega}(u) \, du = \frac{f(t+) + f(t-)}{2} \tag{1.7.7}$$

and

$$\lim_{\Omega \to \infty} \int_\epsilon^\infty g(u) d_{2\pi\Omega}(u) \, du = 0. \tag{1.7.8}$$

Equation (1.7.7) is an immediate consequence of Lemma 1.7.5. Equation (1.7.8) follows from the Riemann–Lebesgue Lemma and the fact that $(g(t)/t) \tau_\epsilon H(t) \in L^1(\mathbb{R})$, since $f \in L^1(\mathbb{R})$. ∎

1.7.7 Remark. The Jordan Theorem and Partial Sums

a. We could use other kernels besides the pairing $d_{2\pi\Omega} \longleftrightarrow 1_{[-\Omega,\Omega)}$ to obtain analogues of Jordan's Theorem. The advantage of Jordan's Theorem is that we really are dealing with the "partial sum" S_Ω and not some weighting of F, such as the factor,

$$\left(1 - \frac{2\pi|\gamma|}{\lambda}\right)\mathbf{1}_{[-\lambda/2\pi,\lambda/2\pi]}(\gamma),$$

in Theorem 1.6.9*b*.

b. Jordan's Theorem is the analogue for Fourier transforms of Dirichlet's Theorem for Fourier series. Dirichlet proved his result much earlier in the nineteenth century, and his work contains the first proper definition of the notion of "function". We shall prove Dirichlet's Theorem in Theorem 3.1.6.

If $f \in L^1(\mathbb{R})$ and $\hat{f} \in L^1(\hat{\mathbb{R}})$, we can use Theorem 1.6.9 to obtain the following pointwise inversion theorem, cf., Exercise 1.37.

1.7.8 Theorem. Inversion Formula for $f \in L^1(\mathbb{R}) \cap A(\mathbb{R})$

Let $f \in L^1(\mathbb{R}) \cap A(\mathbb{R})$. Then

$$\forall t \in \mathbb{R}, \qquad f(t) = \int \hat{f}(\gamma)e^{2\pi it\gamma}\, d\gamma. \tag{1.7.9}$$

PROOF.   Since $\hat{f} \in L^1(\hat{\mathbb{R}})$, it follows that g defined by the right side of (1.7.9) is uniformly continuous (Theorem 1.4.1*b*).

Note that

$$\left\| \int \hat{f}(\gamma)e^{2\pi it\gamma}\, d\gamma - \int_{-\lambda/2\pi}^{\lambda/2\pi}\left(1 - \frac{2\pi|\gamma|}{\lambda}\right)\hat{f}(\gamma)e^{2\pi it\gamma}\, d\gamma \right\|_{L^\infty(\mathbb{R})}$$

$$\leq \left\| \int_{|\gamma|>\lambda/2\pi} \hat{f}(\gamma)e^{2\pi it\gamma}\, d\gamma \right\|_{L^\infty(\mathbb{R})}$$

$$+ \left\| \int_{-\lambda/2\pi}^{\lambda/2\pi} \frac{2\pi|\gamma|}{\lambda}\hat{f}(\gamma)e^{2\pi it\gamma}\, d\gamma \right\|_{L^\infty(\mathbb{R})} \tag{1.7.10}$$

$$\leq \int_{|\gamma|>\lambda/2\pi} |\hat{f}(\gamma)|\, d\gamma + \int_{-\lambda/2\pi}^{\lambda/2\pi} \frac{2\pi|\gamma|}{\lambda}|\hat{f}(\gamma)|\, d\gamma.$$

We shall apply LDC to the second integral on the right side of (1.7.10). Let

$$F_\lambda(\gamma) = \frac{2\pi|\gamma|}{\lambda}|\hat{f}(\gamma)|\mathbf{1}_{[-\lambda/2\pi,\lambda/2\pi]}(\gamma),$$

so that $\lim_{\lambda\to\infty} F_\lambda = 0$ a.e. and $|F_\lambda| \leq 2\pi|\hat{f}| \in L^1(\hat{\mathbb{R}})$. Consequently, LDC applies, and hence

$$\lim_{\lambda\to\infty} \int_{-\lambda/2\pi}^{\lambda/2\pi} \frac{2\pi|\gamma|}{\lambda}|\hat{f}(\gamma)|\, d\gamma = 0.$$

From the definition of $L^1(\hat{\mathbb{R}})$, the first integral on the right side of (1.7.10) also tends to 0 as $\lambda \to \infty$. Therefore,

$$\lim_{\lambda \to \infty} \left\| g(t) - \int_{-\lambda/2\pi}^{\lambda/2\pi} \left(1 - \frac{2\pi |\gamma|}{\lambda} \right) \hat{f}(\gamma) e^{2\pi i t \gamma} \right\|_{L^\infty(\mathbb{R})} = 0. \quad (1.7.11)$$

We now invoke Theorem 1.6.9*b*, viz.,

$$\lim_{\lambda \to \infty} \left\| f(t) - \int_{-\lambda/2\pi}^{\lambda/2\pi} \left(1 - \frac{2\pi |\gamma|}{\lambda} \right) \hat{f}(\gamma) e^{2\pi i t \gamma} \right\|_{L^1(\mathbb{R})} = 0, \quad (1.7.12)$$

to obtain a pointwise a.e. result in the following way. Equation (1.7.12) implies that

$$\lim_{\lambda \to \infty} \int_{-\lambda/2\pi}^{\lambda/2\pi} \left(1 - \frac{2\pi |\gamma|}{\lambda} \right) \hat{f}(\gamma) e^{2\pi i t \gamma} \, d\gamma = f(t) \quad \text{in measure.} \quad (1.7.13)$$

(By definition, $\lim_{\lambda \to \infty} f_\lambda = f$ *in measure* if

$$\forall \epsilon > 0, \qquad \lim_{\lambda \to \infty} |\{t : |f_\lambda(t) - f(t)| \geq \epsilon\}| = 0,$$

e.g., [Ben76, page 89].) A basic result due to F. Riesz is that *convergence in measure implies convergence a.e. of a subsequence*, e.g., [Ben76, page 106] and Example A.11, cf., [RN55, page 100] for a motivating footnote from the master. Thus, (1.7.13) can be changed to convergence a.e. for some λ_n instead of λ. Combined with (1.7.11), for λ_n instead of λ, this adjustment of (1.7.13) yields the fact that $f = g$ a.e.; and the result follows since f and g are continuous. ∎

1.7.9 Remark. The Lebesgue Set

As remarked in the proof of Theorem 1.7.8, if $f \in L^1(\mathbb{R})$, then there is $\{\lambda_n\} \subseteq (0, \infty)$ such that

$$\lim_{\lambda_n \to \infty} \int_{-\lambda_n/2\pi}^{\lambda_n/2\pi} \left(1 - \frac{2\pi |\gamma|}{\lambda_n} \right) \hat{f}(\gamma) e^{2\pi i t \gamma} \, d\gamma = f(t) \quad \text{a.e.} \quad (1.7.14)$$

It turns out that λ_n can be replaced by λ in (1.7.14), and that the convergence a.e. can be enlarged to include all t in the *Lebesgue set* for f, e.g., [Gol61, pages 14–16]. The *Lebesgue set* L for $f \in L^1_{\text{loc}}(\mathbb{R})$ is the largest set of points $t \in \mathbb{R}$ for which

$$\lim_{h \to 0} \frac{1}{h} \int_0^h |f(t + u) - f(t)| \, du = 0.$$

Refining FTCI we see that $|\mathbb{R} \setminus L| = 0$ and that L includes all points of continuity of f, e.g., [Ben76, Section 4.4].

1.7.10 Example. Computations with the Inversion Formula

a.   Using the Jordan Theorem we have

$$\frac{1}{\pi} \int_0^\infty \frac{\cos 2\pi t\gamma + \gamma \sin 2\pi t\gamma}{1 + \gamma^2}\, d\gamma$$

$$= \begin{cases} 0, & \text{if } t < 0, \\ \dfrac{1}{2}, & \text{if } t = 0, \\ e^{-2\pi t}, & \text{if } t > 0. \end{cases} \qquad (1.7.15)$$

To see this, let $f(t) = e^{-2\pi t} H(t)$. Note that $f \in L^1(\mathbb{R})$, and that $f \in BV(I)$, not only for bounded intervals I but also for $\mathbb{R}$. Thus, by Theorem 1.7.6,

$$\forall t \in \mathbb{R}, \qquad \frac{f(t+) + f(t-)}{2} = \lim_{\Omega \to \infty} \frac{1}{2\pi} \int_{-\Omega}^{\Omega} \frac{e^{2\pi i t\gamma}}{1 + i\gamma}\, d\gamma, \qquad (1.7.16)$$

since $\hat{f}(\gamma) = 1/(2\pi(1 + i\gamma))$. Clearly, the left side of (1.7.16) is the right side of (1.7.15), so it remains to verify that

$$\frac{1}{2\pi} \int_0^\infty \frac{\cos 2\pi t\gamma + \gamma \sin 2\pi t\gamma}{1 + \gamma^2}\, d\gamma$$

$$= \lim_{\Omega \to \infty} \int_{-\Omega}^{\Omega} \frac{e^{2\pi i t\gamma}}{2\pi(1 + i\gamma)}\, d\gamma. \qquad (1.7.17)$$

We have

$$\frac{e^{2\pi i t\gamma}}{1 + i\gamma} = \frac{\cos 2\pi t\gamma + \gamma \sin 2\pi t\gamma + i(\sin 2\pi t\gamma - \gamma \cos 2\pi t\gamma)}{1 + \gamma^2}.$$

The imaginary part is odd, and so the integral on the right side of (1.7.17) is

$$\frac{1}{2\pi} \int_{-\Omega}^{\Omega} \frac{\cos 2\pi t\gamma + \gamma \sin 2\pi t\gamma}{1 + \gamma^2}\, d\gamma.$$

This yields (1.7.17) since the integrand is even.

b.   Similarly, we can use either Theorem 1.7.6 or Theorem 1.7.8 to verify that

$$\frac{2r}{\pi} \int_0^\infty \frac{\cos 2\pi t\gamma}{r^2 + \gamma^2}\, d\gamma = e^{-2\pi r|t|},$$

e.g., Exercise 1.23a, cf., (1.7.15).

1.8. Partial Differential Equations

Harmonic analysis and partial differential equations (PDEs) have a profound relationship and interaction, e.g., [Hör83]; and the subject of harmonic analysis owes its existence and formative years to PDE, e.g., [Fou1822]. As a result, the literature on harmonic analysis *and* PDE is extraordinarily extensive, e.g., [BMc66], [CB78], [Dav85], [Gus87], [Wei65] for some elementary books. For this reason, and because of our limitations (both space-time and neuronal) and goals, e.g., Preface, we have only selected the following small collection of PDEs. These give a flavor of the aforementioned interaction, but perhaps so little as to be misleading!

1.8.1 Example. A Diffusion Equation

Consider the heat equation with convection,

$$\frac{\partial u}{\partial t} = \frac{k}{4\pi^2}\frac{\partial^2 u}{\partial x^2} + \frac{c}{2\pi i}\frac{\partial u}{\partial x}, \tag{1.8.1}$$

with initial and boundary conditions,

$$u(x,0) = f(x) \tag{1.8.2}$$

and

$$\forall t \geq 0, \qquad u(\pm\infty, t) = u_x(\pm\infty, t) = 0, \tag{1.8.3}$$

respectively. $u(x,t)$ represents the temperature at x when the time is t. The domain of u consists of $\mathbb{R} \times [0, \infty)$ and $f(x)$ is the given temperature on $\mathbb{R}$ when the process begins at time $t = 0$. The term $(c/2\pi i)(\partial u/\partial x)$ is the convection term, and the constants satisfy $k > 0$ and $c \in \mathbb{R}$. We shall compute $u(x,t)$.

Formally, we assume there is a solution $u(\cdot, t) \in L^1(\mathbb{R})$ for each $t \geq 0$ and we let $U(\cdot, t)$ be its Fourier transform, i.e., for each $t \geq 0$, we have the pairing $u(x,t) \longleftrightarrow U(\gamma, t)$. We also assume $f, \hat{f} \in L^1$. Taking Fourier transforms of the functions in (1.8.1) we obtain

$$U_t(\gamma, t) = \frac{k}{4\pi^2}\int u_{xx}(x,t)e^{-2\pi i x\gamma}\,dx + \frac{c}{2\pi i}\int u_x(x,t)e^{-2\pi i x\gamma}\,dx$$

$$= \frac{k}{4\pi^2}\left[u_x(x,t)e^{-2\pi i x\gamma}\Big|_{-\infty}^{\infty} + 2\pi i\gamma \int u_x(x,t)e^{-2\pi i x\gamma}\,dx\right]$$

$$+ \frac{c}{2\pi i}\left[u(x,t)e^{-2\pi i x\gamma}\Big|_{-\infty}^{\infty} + 2\pi i\gamma \int u(x,t)e^{-2\pi i x\gamma}\,dx\right]$$

$$= \frac{2\pi i \gamma k}{4\pi^2} \left[u(x,t)e^{-2\pi i x \gamma} \Big|_{-\infty}^{\infty} +2\pi i \gamma \int u(x,t)e^{-2\pi i x \gamma}\, dx \right]$$

$$+ c\gamma U(\gamma,t) = (-k\gamma^2 + c\gamma)U(\gamma,t),$$

where we have used (1.8.3) and assumed $u_x(\cdot,t), u_{xx}(\cdot,t) \in L^1(\mathbb{R})$ for each $t > 0$. Consequently, for each fixed $\gamma \in \hat{\mathbb{R}}$ we have the ordinary differential equation

$$U_t(\gamma,t) = (-k\gamma^2 + c\gamma)U(\gamma,t), \qquad t > 0;$$

and, as in Example 1.3.3, we can solve it by elementary methods to obtain

$$U(\gamma,t) = C(\gamma)\exp\left([-k\gamma^2 + c\gamma]t\right). \tag{1.8.4}$$

Hence, $U(\gamma,0) = C(\gamma)$; and so, since $U(\gamma,0) = \hat{f}(\gamma)$, we can write

$$U(\gamma,t) = \hat{f}(\gamma)\exp\left([-k\gamma^2 + c\gamma]t\right) \tag{1.8.5}$$

for all $(\gamma,t) \in \hat{\mathbb{R}} \times (0,\infty)$. (At this stage we do not choose to be careful about letting $t = 0$ in (1.8.4).) By completing the square, (1.8.5) becomes

$$U(\gamma,t) = \exp\left(\frac{tc^2}{4k}\right)\hat{f}(\gamma)\exp\left(-tk\left(\gamma - \frac{c}{2k}\right)^2\right) \tag{1.8.6}$$

for all $(\gamma,t) \in \hat{\mathbb{R}} \times (0,\infty)$; and, in particular, $\hat{f}(\gamma)\exp(-tk(\gamma - \frac{c}{2k})^2) \in L^1(\hat{\mathbb{R}})$ for each fixed $t > 0$. Thus, we can apply the inversion theorem, Theorem 1.7.8, to obtain our solution

$$u(x,t) = \exp\left(\frac{tc^2}{4k}\right)\left[f(y) * \left(e^{\pi i y c/k}\, g_{\pi/(tk)^{1/2}}(y)\right)\right](x) \tag{1.8.7}$$

by taking the inverse Fourier transform of (1.8.6) for each $t > 0$.

The calculation of (1.8.7) depends on the convolution formula, Proposition 1.5.2, and the fact that

$$\int e^{-tk(\gamma - c/2k)^2}\, e^{2\pi i y \gamma}\, d\gamma = e^{\pi i y c/k}\int e^{-(tk)\lambda^2}e^{2\pi i y \lambda}\, d\lambda$$

$$= e^{\pi i y c/k}\, g_{\pi/(tk)^{1/2}}(y).$$

We now have to check that $u(x,t)$ in (1.8.7) is really a solution of the system (1.8.1)–(1.8.3). We leave this as Exercise 1.39. Technically, we have no right to begin with a function u, as we did, unless we had available an existence theorem, which, in fact, does exist (sic).

1.8.2   Example. A Dirichlet Problem

a.   Consider Laplace's equation

$$\Delta u = \frac{\partial^2 u}{\partial x^2} + \frac{\partial^2 u}{\partial y^2} = 0 \tag{1.8.8}$$

on the upper half-plane $\mathbb{R} \times [0, \infty)$ with boundary condition,

$$u(x, 0) = f(x), \tag{1.8.9}$$

where $f(x)$ is a given function. The *Dirichlet problem* is to determine whether or not the system (1.8.8) and (1.8.9) has a unique solution in the upper half-plane.

To focus on the problem mathematically, we assume (1.8.8) is valid on $\mathbb{R} \times (0, \infty)$. Other natural assumptions will arise as we proceed with the calculation.

Physically, the system (1.8.8) and (1.8.9) models a steady-state temperature distribution problem. "Steady state" indicates that the average temperature doesn't change with time, i.e., the rate at which heat flows into the upper half-plane is 0. This should be compared with Example 1.8.1, which is not a steady-state problem as reflected by the presence of the term "$\partial u / \partial t$" on the left-hand side of (1.8.1). (Lest there be any confusion, Example 1.8.1 is a one-dimensional temperature distribution problem, and this example is two-dimensional.) Condition (1.8.9) indicates a known temperature distribution along the boundary, and the Dirichlet problem is to determine if a steady-state system with a known temperature distribution on the boundary, e.g., predictable radiators along the walls of a room, characterizes the temperature in the interior.

Formally, we assume there is a solution $u(x, y)$ and that $u(\cdot, y) \in L^1(\mathbb{R})$ for each $y \geq 0$. If $U(\cdot, y)$ is the Fourier transform of $u(\cdot, y)$, then (1.8.8) yields

$$\forall \gamma \in \hat{\mathbb{R}}, \qquad -4\pi^2 \gamma^2 U(\gamma, y) + \frac{d^2}{dy^2} U(\gamma, y) = 0. \tag{1.8.10}$$

For each fixed $\gamma \in \hat{\mathbb{R}}$ we view (1.8.10) as an ordinary differential equation in y. The corresponding characteristic equation is $r^2 - 4\pi^2 \gamma^2 = 0$ so that $r = \pm 2\pi |\gamma|$. Hence, we have

$$U(\gamma, y) = a(\gamma) e^{2\pi y |\gamma|} + b(\gamma) e^{-2\pi y |\gamma|}.$$

Now, we refine the assumption, $u(\cdot, y) \in L^1(\mathbb{R})$, to include the estimate,

$$\exists M \quad \text{such that} \quad \forall y \geq 0, \quad \int |u(x, y)| \, dx \leq M.$$

As such we see that $a(\gamma) = 0$ for $\gamma \neq 0$ since

$$|U(\gamma, y)| \leq \int |u(x, y)|\, dx \leq M.$$

Thus, we obtain

$$U(\gamma, y) = b(\gamma)e^{-2\pi y|\gamma|}.$$

Formally, we have

$$b(\gamma) = \hat{f}(\gamma) \tag{1.8.11}$$

and

$$u(x, y) = \int \frac{y f(u)\, du}{\pi (y^2 + (x - u)^2)}. \tag{1.8.12}$$

In fact, (1.8.11) is clear by (1.8.9) and the equation for U; and (1.8.12) follows by the calculation

$$u(x, y) = \left[f * \left(e^{-2\pi y|\gamma|} \right)^{\vee} \right](x) = f * p_{1/y}(x),$$

which is valid by Theorem 1.7.8 when $f, \hat{f} \in L^1$.

The right-hand side of (1.8.12) is the *Poisson integral formula for the upper half-plane*. It is easy to check that $u(x, y)$ so defined is a solution of Laplace's equation on $\mathbb{R} \times (0, \infty)$. Further, since $f \in L^1(\mathbb{R})$ and $\{p_\lambda\}$ is an approximate identity, we know that

$$\lim_{y \to 0} u(x, y) = f(x)$$

in L^1-norm. In this sense we have solved the Dirichlet problem as far as obtaining a solution.

The uniqueness of this solution is intuitively clear for the following reason. Suppose $f = 0$ on $\mathbb{R}$, the boundary of the upper half-plane. By the definition of a steady-state system, the temperature $u(x, y)$ would also have to be 0 since the temperature of the upper half-plane is not influenced by any other heat flow. Of course, there is a highly developed mathematical theory of uniqueness for solutions of partial differential equations, e.g., [Hör83].

1.8.3 Example. A Diffusion Equation and Image Processing

Combining the two-dimensionality of Example 1.8.2 with the time dependence of Example 1.8.1, let us consider the diffusion equation

$$\Delta u = \frac{\partial u}{\partial s} \tag{1.8.13}$$

on $\mathbb{R}^2 \times [0, \infty)$ with initial condition,

$$u(x, y, 0) = f(x, y), \qquad (1.8.14)$$

where Δ is defined by (1.8.8), f is a given function, and $(x, y, s) \in \mathbb{R}^2 \times [0, \infty)$. Since such problems are a well-established part of partial differential equations and classical physics, we choose to look at (1.8.13) and (1.8.14) in terms of a more recent interpretation in *image processing*.

Let

$$u_g(x, y, s) = g_{\sqrt{1/4s}}(x) g_{\sqrt{1/4s}}(y). \qquad (1.8.15)$$

We are using "s" instead of "t" in the dilations of the Gaussian on $\mathbb{R}$ to denote scale instead of time. Note that if we define the function k on $\mathbb{R}^2$ as

$$k(x, y) = g(x)g(y),$$

then

$$k_\lambda(x, y) = \lambda^2 k(\lambda x, \lambda y), \qquad \lambda > 0,$$

is the $L^1(\mathbb{R}^2)$-dilation of k by λ; and $\{k_\lambda\}$ is an approximate identity for $L^1(\mathbb{R}^2)$ as $\lambda \to \infty$. By (1.8.15), we see that u_g is the dilation

$$u_g(x, y, s) = k_{\sqrt{1/4s}}(x, y).$$

Further, for this u we have

$$\Delta u_g(x, y, s) = \frac{\partial u_g}{\partial s}, \qquad (1.8.16)$$

that is

$$\Delta k_{\sqrt{1/4s}}(x, y) = \frac{\partial}{\partial s} k_{\sqrt{1/4s}}(x, y).$$

It is easy to see that $u = u_g * f$ is a solution of the system (1.8.13) and (1.8.14) when f satisfies natural hypotheses. If f is an *image* to be processed, then $u(\cdot, \cdot, s)$ is a blurred image of f, more blurred for larger s. An idea in image processing is to reconstruct f from blurred versions of f, which may not require too much data or "expense", along with available detailed data. This is associated with the notion of multiresolution (in wavelet theory) due to Mallat [Mall89] and Meyer [Mey90], and the discrete version of this scheme due to Burt and Adelson [BA83].

The convolution $u_g * f$ is, in engineering terms, *filtering the image f by the Gaussian u_g*. Now, it also makes sense to filter the image f by Δu_g for the following reason [KJ55]. An elementary calculation shows that

$$\iint \Delta u_g(x, y, s) \, dx \, dy = 0.$$

For small s, we can think of Δu_g as being concentrated in a small region; and so if the image f is constant in a region R of comparable size (or larger), then $(\Delta u_g) * f(x, y) = 0$ for (x, y) near the center of R. Similarly, if f is not constant on R, then this convolution procedure can have the effect of detecting *edges* in the image.

Thus, since filtering by Δu_g is potentially important, we see the value of (1.8.13) in image analysis for the following reason. Because of (1.8.16), we can estimate Δu_g by considering a scaled difference of Gaussians, thereby reducing computational complexity, e.g., [HM89].

1.9. Gibbs Phenomenon

1.9.1 Remark. Description of Gibbs Phenomenon

Let f be a function on $\mathbb{R}$; and suppose f is continuous on

$$I = [t_0 - T, t_0 + T], \qquad T > 0,$$

except for a jump discontinuity at t_0, i.e., $f(t_0+)$ and $f(t_0-)$ exist and $f(t_0+) - f(t_0-) \neq 0$.

In the case $f \in L^1(\mathbb{R}) \cap BV(I)$, we showed in Jordan's Theorem (Theorem 1.7.6) that

$$\lim_{\Omega \to \infty} f * d_{2\pi\Omega}(t_0) = \frac{f(t_0+) + f(t_0-)}{2}. \tag{1.9.1}$$

We shall now investigate this limit more closely and shall detect a remarkable behavior of the "partial sums" $S_\Omega = f * d_{2\pi\Omega}$. This behavior is called *Gibbs phenomenon*.

To fix ideas let $f = H$, the Heaviside function. Even though $H \notin L^1(\mathbb{R})$, the method of Jordan's Theorem is valid; and the partial sums $S_\Omega \equiv H * d_{2\pi\Omega}$ exist and satisfy

$$\forall t \in \mathbb{R}, \qquad \lim_{\Omega \to \infty} S_\Omega(t) = \frac{H(t+) + H(t-)}{2}.$$

In fact,

$$S_\Omega(t) = H * d_{2\pi\Omega}(t) = \int_0^\infty \frac{\sin 2\pi\Omega(t-u)}{\pi(t-u)} \, du$$

$$= \int_{-\infty}^{2\pi\Omega t} \frac{\sin u}{\pi u} \, du = \frac{1}{2} + \int_0^{2\pi\Omega t} \frac{\sin u}{\pi u} \, du, \tag{1.9.2}$$

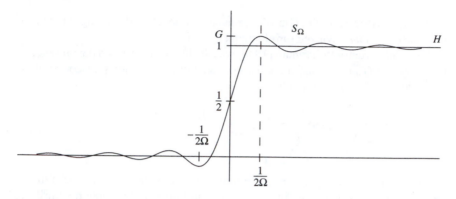

FIGURE 1.11.

noting that t can be negative. Clearly, for a fixed Ω, the last integral achieves its maximum (on $[0, \infty)$, say) at $t = 1/(2\Omega)$ since for larger t the integrand alternates between "decreasing" negative and positive values. Thus, for each Ω, $H * d_{2\pi\Omega}$ is maximized at $t = 1/(2\Omega)$, and, similarly, is minimized at $t = -1/(2\Omega)$. The values of $H * d_{2\pi\Omega}$ at these points are the two *constants*

$$G = H * d_{2\pi\Omega}\left(\frac{1}{2\Omega}\right) = \frac{1}{2} + \int_0^\pi \frac{\sin u}{\pi u}\, du > 1$$

and

$$H * d_{2\pi\Omega}\left(\frac{-1}{2\Omega}\right) = \frac{1}{2} - \int_0^\pi \frac{\sin u}{\pi u}\, du < 0.$$

These two inequalities are clear from our knowledge of the Dirichlet function d.

The fact that the convergence of $\{H * d_{2\pi\Omega}\}$ to H on $\mathbb{R}\backslash\{0\}$ involves the intrinsic "overshoot" G (and the corresponding behavior on the negative axis) is the *Gibbs phenomenon*, e.g., Figure 1.11. This pointwise convergence is uniform on closed bounded subintervals $K \subseteq \mathbb{R}\backslash\{0\}$; but the behavior of the "partial sums" $H * d_{2\pi\Omega}$ near the jump discontinuity always exhibits a fixed rise $G > H(t) = 1$ at $t = 1/(2\Omega)$ as Ω increases to infinity.

Note that

$$\forall \Omega > 0, \qquad H * d_{2\pi\Omega}(0) = \frac{1}{2}.$$

1.9.2 Remark. Historical Note

All of the early work on Gibbs phenomenon was in the context of Fourier series, e.g., [Car30], [HH79]. The term "Gibbs phenomenon" is due to Bôcher (1906), who also provided a proof of Gibbs' original assertion, cf., [HH79, pages 155–156] for a later less-than-civilized development.

Apparently, Henry Wilbraham (1848) was the first to understand the presence of the overshoot G in the Fourier analysis of functions having jump discontinuities. Wilbraham dealt with the function

$$\frac{4}{\pi} \sum_{k=0}^{\infty} (-1)^k \frac{\cos(2k+1)t}{2k+1} = \begin{cases} 1, & \text{if } |t| < \frac{1}{2}\pi, \\ -1, & \text{if } \frac{1}{2}\pi < |t| < \pi, \end{cases} \tag{1.9.3}$$

defined 2π-periodically on $\mathbb{R}$, e.g., Definition 3.1.1. Without knowing of Wilbraham's work, Gibbs made his fundamental contribution to the topic on April 27, 1899, as part of a lively interchange in *Nature* (Volumes 58–60, 1898–1899) initiated by Michelson, and also involving Love and Poincaré. They dealt with the function

$$\forall t \in (-\pi, \pi), \qquad 2 \sum_{k=1}^{\infty} (-1)^{k+1} \frac{\sin kt}{k} = t, \tag{1.9.4}$$

defined 2π-periodically on $\mathbb{R}$.

For perspective, it is interesting to note that in 1898 Michelson and Stratton designed their *harmonic analyzer*, complete with graphs of partial sums of the series in (1.9.3) and (1.9.4). This work probably inspired Michelson's letter to *Nature*. Michelson and Stratton had collaborated on X-ray research; and Stratton was the founder of the National Bureau of Standards. Their harmonic analyzer machine was used for the decomposition of sound and electrical waves into simpler components, cf., our remarks on *harmonic analysis* in the Preface. Harmonic analyzers *and* synthesizers were first designed by Lord Kelvin, cf., Exercise 1.47, for tidal analysis.

The study of tides received its scientific basis with Newton's law of gravitation for the sun, moon, and earth—even the ancient mariner knew that tidal phenomena were related to astronomical factors. Then Laplace was able to separate various cyclic influences of the sun and moon (on tides) by defining a model of the sun, moon, and earth having a number of tide-affecting satellites; and the Newtonian solution of this model associated an elementary tidal constituent with each satellite in such a way that the tide was viewed as a combination of these constituents. Lord Kelvin systematized this method, and the analyzing and synthesizing machines he designed were meant to determine the constituents and to reconstruct the tide (from the constituents), respectively, e.g., [God72], [Mac66]. These mechanical analyzers and synthesizers may be on view at your local science museum!

Instead of the standard partial sum, as in (1.9.2), it is possible to consider functions f having a jump discontinuity in the case that an approximate identity $\{k_{(\lambda)}\}$ replaces the Dirichlet kernel $\{d_{2\pi\Omega}\}$, e.g., Theorem 1.6.9. In this situation, if each $k_{(\lambda)}$ is nonnegative and even, then $H * k_{(\lambda)}(0) = \frac{1}{2}$ and

$$\forall t \in \mathbb{R}, \qquad 0 \le H * k_{(\lambda)}(t) \le 1,$$

i.e., *the Gibbs phenomenon is eliminated for nonnegative approximate identities.*

Using the Gaussian approximate identity, Weyl (1910) studied what he called *heat conduction partial sums,*

$$f * g_{\pi/(tk)^{1/2}}.$$

From our discussion in Example 1.8.1, these functions are solutions of the *heat equation*, i.e., the convection equation (1.8.1) for $c = 0$, cf., Exercise 1.47. Weyl was interested in a two-temperature boundary value function f, e.g., f is either $\alpha = 100°C$ or $\beta = 0°C$ [Wey50b], and the corresponding eigenfunction expansion, cf., Exercise 1.27. He made note of the Gibbs phenomenon for such expansions, and had to deal with rational approximation of irrational numbers (at specified rates) to complete his analysis for arbitrary α and β. Such approximations are part of Diophantine Approximation, a branch of number theory. Later, because of work by P. Bohl and the problem of mean motion in Lagrange's linear theory of perturbation for the planetary system, Weyl (1916) generalized the aforementioned approximation procedure; and gave his definition and characterization of *equidistribution mod 1*, e.g., Exercise 3.40, [HW65], [KK64], [Sal63]. This characterization, the *Weyl Equidistribution Theorem*, is not only an important result in number theory, but has a host of applications in a variety of fields including Wiener's Generalized Harmonic Analysis, e.g., [Bas84], [KK64, pages 23–28, by J.-P. Bertrandias], cf., Section 2.9. Our point in highlighting Weyl's work is to trace the intellectual excursion relating Gibbs phenomenon to heat conduction problems to number theory.

1.9.3 Example. Computation of G

We can estimate G by expanding $\sin u$ in a Taylor series, noting that $\pi \in (3.141, 3.142)$ and estimating the integral $\int_0^\pi (\sin u/\pi u)\,du$, e.g., Exercise 1.29. In fact,

$$G = \frac{1}{2} + \frac{1}{\pi} \int_0^\pi \frac{\sin u}{u}\,du$$

$$= \frac{1}{2} + \frac{1}{\pi} \int_0^\pi \frac{1}{u} \sum_{n=0}^\infty (-1)^n \frac{u^{2n+1}}{(2n+1)!}\,du$$

$$= \frac{1}{2} + \frac{1}{\pi} \sum_{n=0}^\infty (-1)^n \frac{\pi^{2n+1}}{(2n+1)(2n+1)!}$$

$$= \frac{1}{2} + \left[1 - \frac{\pi^2}{3 \cdot 3!} + \frac{\pi^4}{5 \cdot 5!} - \frac{\pi^6}{7 \cdot 7!} + \frac{\pi^8}{9 \cdot 9!} - \cdots \right].$$

1.9.4 Example. Perspective on Bandlimited Approximants

Let f be a function and let $\Omega > 0$. A basic approximation problem, associated with the notion of *aliasing* in engineering, is to approximate f in a realistic way by an $\Omega - BL$ function, this latter notion having been defined in Example 1.2.2. Of course, the *criterion* for approximation must be specified.

Suppose the criterion is pointwise convergence. Then our discussion of the Gibbs phenomenon shows that, because of the overshoot and undershoot, the natural $\Omega - BL$ approximant $f * d_{2\pi\Omega}$ of f is poor in a neighborhood of any jump discontinuity of f, even when the band $[-\Omega, \Omega]$ is large. As we saw in Remark 1.9.2 the overshoot and undershoot are obviated when dealing with the $\Omega - BL$ approximant $f * w_{2\pi\Omega}$, cf., Remark 1.10.8, Proposition 1.10.9, and Remark 1.10.10.

Another pointwise approach involves dealing with the continuous function $f * \left(\frac{1}{2T}\mathbf{1}_{[-T,T)}\right)$, and then defining the $\Omega - BL$ approximant

$$f_{T,\Omega} = f * \left(\frac{1}{2T}\mathbf{1}_{[-T,T)}\right) * d_{2\pi\Omega}.$$

Notice that $f * \left(\frac{1}{2T}\mathbf{1}_{[-T,T)}\right)(t)$ is the average value,

$$\frac{1}{2T}\int_{t-T}^{t+T} f(u)\, du,$$

of f on $[t - T, t + T]$. In this case, $f_{T,\Omega}$ can be a good pointwise (or even uniform) $\Omega - BL$ approximant of f if Ω is large and T is small. For example, if $f \in L^1(\mathbb{R}) \cap A(\mathbb{R})$, then

$$\forall t \in \mathbb{R}, \qquad |f(t) - f_{T,\Omega}(t)|$$

$$\leq \int_{|\gamma| \geq \Omega} |\hat{f}(\gamma)|\, d\gamma + \int_{-\Omega}^{\Omega} |\hat{f}(\gamma)| \left|1 - \frac{\sin 2\pi T\gamma}{2\pi T\gamma}\right| d\gamma.$$

An important case of our approximation problem deals with an approximation criterion motivated by physical considerations such as variance and energy. The mathematical setting is $L^2(\mathbb{R})$, the space of square integrable functions.

1.10. The $L^2(\mathbb{R})$ Theory

1.10.1 Definition. $L^2(\mathbb{R})$

$$L^2(\mathbb{R}) = \left\{ f : \mathbb{R} \to \mathbb{C} : \|f\|_{L^2(\mathbb{R})} = \left(\int |f(t)|^2\, dt\right)^{1/2} < \infty \right\}.$$

$L^2(\mathbb{R})$ is the space of *square-integrable functions f or signals f having finite energy* $\|f\|^2_{L^2(\mathbb{R})}$.

The major result about $L^2(\mathbb{R})$ is the following theorem [Pla10], [Pla15].

1.10.2 Theorem. Plancherel Theorem

There is a unique linear bijection $\mathcal{F} : L^2(\mathbb{R}) \longrightarrow L^2(\hat{\mathbb{R}})$ with the properties:

a) $\forall f \in L^2(\mathbb{R}), \quad \|f\|_{L^2(\mathbb{R})} = \|\mathcal{F}f\|_{L^2(\hat{\mathbb{R}})}$;

b) $\forall f \in L^1(\mathbb{R}) \cap L^2(\mathbb{R})$ and $\forall \gamma \in \hat{\mathbb{R}}, \quad \hat{f}(\gamma) = (\mathcal{F}f)(\gamma)$;

c) $\forall f \in L^2(\mathbb{R}), \exists \{f_n : n = 1, \ldots\} \subseteq L^1(\mathbb{R}) \cap L^2(\mathbb{R})$ *for which*

$$\lim_{n \to \infty} \|f_n - f\|_{L^2(\mathbb{R})} = 0 \quad and \quad \lim_{n \to \infty} \|\hat{f}_n - \mathcal{F}f\|_{L^2(\hat{\mathbb{R}})} = 0.$$

PROOF. Our outline of proof, which is one of the usual schemes to prove the Plancherel Theorem, has four steps: verification that $\|f\|_{L^2(\mathbb{R})} = \|\hat{f}\|_{L^2(\hat{\mathbb{R}})}$ for $f \in X \subseteq L^2(\mathbb{R})$ (part *i*), closure results on $\mathbb{R}$ and $\hat{\mathbb{R}}$ (parts *ii* and *iii*, respectively), and a routine functional analysis argument to obtain the result from parts *i, ii, iii* (part *iv*). There are trade-offs in difficulty between parts *i, ii,* and *iii* depending on which space X one chooses to use.

   i. Let $X = C_c(\mathbb{R})$, the space of continuous functions (on $\mathbb{R}$) having compact support, and consider the involution $\tilde{f}(t) \equiv \overline{f(-t)}$ of $f \in C_c(\mathbb{R})$. We shall prove that $\hat{f} \in L^2(\hat{\mathbb{R}})$. Clearly, $\hat{f} \in A(\hat{\mathbb{R}})$ since $C_c(\mathbb{R}) \subseteq L^1(\mathbb{R})$. Define $g = f * \tilde{f}$ so that g is continuous, $g \in L^1(\mathbb{R})$, e.g., Definition 1.5.1 or by a direct calculation, and

$$g(0) = \|f\|^2_{L^2(\mathbb{R})}. \tag{1.10.1}$$

Also,

$$\forall \gamma \in \hat{\mathbb{R}}, \qquad \hat{g}(\gamma) = |\hat{f}(\gamma)|^2 \tag{1.10.2}$$

by the Fubini–Tonelli Theorem and the translation invariance of Lebesgue measure (on the group $\mathbb{R}$). By Proposition 1.6.11 and (1.10.2), since $g \in L^1(\mathbb{R}) \cap L^\infty(\mathbb{R})$ is continuous, we have

$$\lim_{\lambda \to \infty} \int_{-\lambda/2\pi}^{\lambda/2\pi} \left(1 - \frac{2\pi|\gamma|}{\lambda}\right) |\hat{f}(\gamma)|^2 \, d\gamma = g(0). \tag{1.10.3}$$

The Beppo Levi Theorem and (1.10.3) allow us to assert that $\hat{f} \in L^2(\hat{\mathbb{R}})$ and that

$$\|\hat{f}\|^2_{L^2(\hat{\mathbb{R}})} = g(0) = \|f\|^2_{L^2(\mathbb{R})},$$

where the second equality follows from the definition of g.

ii. We shall now show that $L^1(\mathbb{R}) \cap L^2(\mathbb{R})$ is dense in $L^2(\mathbb{R})$, and prove that $C_c(\mathbb{R})$ is dense in $L^1(\mathbb{R}) \cap L^2(\mathbb{R})$ taken with the norm $\| \cdots \| = \| \cdots \|_{L^1(\mathbb{R})} + \| \cdots \|_{L^2(\mathbb{R})}$. These facts imply $\overline{C_c(\mathbb{R})} = L^2(\mathbb{R})$.

Let $f \in L^2(\mathbb{R})$ and define $f_T = f \mathbf{1}_{[-T,T]}$. For this proof, f_T does not designate dilation. $f_T \in L^2(\mathbb{R})$ since $|f_T| \le |f|$, and $f_T \in L^1(\mathbb{R})$ by Hölder's Inequality. Clearly,

$$\| f - f_T \|_{L^2(\mathbb{R})} = \left(\int_{|t|>T} |f(t)|^2 \, dt \right)^{1/2},$$

and so $L^1(\mathbb{R}) \cap L^2(\mathbb{R})$ is dense in $L^2(\mathbb{R})$.

Let $f \in L^1(\mathbb{R}) \cap L^2(\mathbb{R})$. Then $\lim_{T \to \infty} \| f - f_T \| = 0$ by the argument of the previous paragraph. Next, set $f_{S,T} = f_T * \Delta_S$ (Δ_S is the dilation of the triangle function Δ defined in Example 1.3.4), so that $f_{S,T} \in C_c(\mathbb{R})$. Letting $\epsilon > 0$, we shall find $S, T > 0$ such that $\| f - f_{S,T} \| < \epsilon$.

First, there is $T = T_\epsilon$ for which $\| f - f_T \| < \epsilon/2$. We keep this T fixed, and have the estimate

$$\| f_T - f_{S,T} \|_{L^1(\mathbb{R})} \le \iint \Delta_S(u) |f_T(t) - f_T(t-u)| \, du \, dt$$

$$= \int_{|u| \le 1/S} \Delta_S(u) \left(\int |f_T(t) - f_T(t-u)| \, dt \right) du.$$

Next, choose S_1 such that

$$\forall S \ge S_1 \text{ and } \forall |u| \le \frac{1}{S}, \qquad \| f_T - \tau_u f_T \|_{L^1(\mathbb{R})} < \frac{\epsilon}{4}.$$

Thus, for such S,

$$\| f_T - f_{S,T} \|_{L^1(\mathbb{R})} < \frac{\epsilon}{4}. \tag{1.10.4}$$

Finally, we use Minkowski's Inequality (Theorem A.16) to obtain the estimate

$$\| f_T - f_{S,T} \|_{L^2(\mathbb{R})} \le \int \left(\int |(f_T(t) - f_T(t-u)) \Delta_S(u)|^2 \, dt \right)^{1/2} du$$

$$= \int \Delta_S(u) \| f_T - \tau_u f_T \|_{L^2(\mathbb{R})} \, du.$$

We can choose $S \ge S_1$ for which

$$\forall |u| \le \frac{1}{S}, \qquad \| f_T - \tau_u f_T \|_{L^2(\mathbb{R})} < \frac{\epsilon}{4};$$

and, therefore,

$$\| f_T - f_{S,T} \|_{L^2(\mathbb{R})} < \frac{\epsilon}{4}. \tag{1.10.5}$$

Combining (1.10.4) and (1.10.5) with the above estimate $\| f - f_T \| < \epsilon/2$, we have the desired inequality, viz., $\| f - f_{S,T} \| < \epsilon$.

iii. We shall now prove that $C_c(\mathbb{R})\hat{} \subseteq A(\hat{\mathbb{R}}) \cap L^2(\hat{\mathbb{R}})$ is a dense subspace of $L^2(\hat{\mathbb{R}})$.

Let $G \in L^2(\hat{\mathbb{R}})$ and suppose

$$\forall f \in C_c(\mathbb{R}), \qquad \int \hat{f}(\gamma)\overline{G(\gamma)}\,d\gamma = 0. \qquad (1.10.6)$$

If $f \in C_c(\mathbb{R})$, then $\tau_u f \in C_c(\mathbb{R})$ and so (1.10.6) implies

$$\forall f \in C_c(\mathbb{R}) \text{ and } \forall u \in \mathbb{R}, \qquad \int \hat{f}(\gamma)\overline{G(\gamma)}e^{-2\pi i u \gamma}\,d\gamma = 0. \quad (1.10.7)$$

By Hölder's Inequality, $\hat{f}\bar{G} \in L^1(\hat{\mathbb{R}})$, and so (1.10.7) allows us to invoke the uniqueness theorem, Theorem 1.6.9c, to conclude that $\hat{f}\bar{G} = 0$ a.e. for each $f \in C_c(\mathbb{R})$.

Note that

$$\forall f \in C_c(\mathbb{R}) \text{ and } \forall \gamma \in \hat{\mathbb{R}}, \qquad e^{2\pi i t \gamma} f(t) \in C_c(\mathbb{R}).$$

Thus, $C_c(\mathbb{R})\hat{}$ is translation invariant, i.e.,

$$\forall f \in C_c(\mathbb{R}) \text{ and } \forall \gamma \in \hat{\mathbb{R}}, \qquad \tau_\gamma \hat{f} \in C_c(\mathbb{R})\hat{}.$$

From this we can conclude that *for each $\gamma_0 \in \hat{\mathbb{R}}$, there is $f = f_{\gamma_0} \in C_c(\mathbb{R})$ for which $|\hat{f}| > 0$ on an interval I centered about γ_0*. To verify this claim, suppose there is γ_0 such that for each $f \in C_c(\mathbb{R})$ and for each interval I centered at γ_0, $\hat{f}$ has a zero in I. Consequently, $\hat{f}(\gamma_0) = 0$ for each $f \in C_c(\mathbb{R})$. By the translation invariance of $C_c(\mathbb{R})\hat{}$, $\tau_\gamma \hat{f} \in C_c(\mathbb{R})\hat{}$ for each $\gamma \in \hat{\mathbb{R}}$, and so

$$\forall f \in C_c(\mathbb{R}) \text{ and } \forall \gamma \in \hat{\mathbb{R}}, \qquad (\tau_\gamma \hat{f})(\gamma_0) = 0.$$

i.e., $\hat{f} = 0$ on $\hat{\mathbb{R}}$ for each $f \in C_c(\mathbb{R})$. This contradicts the uniqueness theorem, Theorem 1.6.9c (for just one $f \in C_c(\mathbb{R})\backslash\{0\}$), and the claim is proved.

Therefore, if we assume (1.10.6) we can conclude that $G = 0$ a.e. Consequently, by the Hahn–Banach Theorem (Theorem B.12) and the fact that $L^2(\hat{\mathbb{R}})' = L^2(\hat{\mathbb{R}})$ (Theorem B.14), we have that $C_c(\mathbb{R})\hat{}$ is dense in $L^2(\hat{\mathbb{R}})$.

iv. We have shown that $\mathcal{F}$ is a continuous linear injection $C_c(\mathbb{R}) \to L^2(\hat{\mathbb{R}})$ (part *i*), when $C_c(\mathbb{R})$ is endowed with the L^2-norm, and so $\mathcal{F}$ has a unique linear injective extension to $L^2(\mathbb{R})$ (Theorem B.7) by (part *ii*). Also, $\mathcal{F}(C_c(\mathbb{R}))$ is closed and dense in $L^2(\hat{\mathbb{R}})$ by parts *i* and *iii*. Thus, $\mathcal{F}$ is also surjective. The remaining claims of the theorem are now immediate. ∎

Notationally, because of the Plancherel Theorem, we refer to $\mathcal{F}f$ as the *Fourier transform* of $f \in L^2(\mathbb{R})$, and we write the pairing between $f \in L^2(\mathbb{R})$ and $\mathcal{F}f$ in one of the following ways:

$$\mathcal{F}f = \hat{f} = F, \qquad f \longleftrightarrow F, \qquad f = \check{F}. \qquad (1.10.8)$$

1.10.3 Theorem. Parseval Formula

Consider $f \longleftrightarrow F$ *and* $g \longleftrightarrow G$, *where* $f, g \in L^2(\mathbb{R})$. *Then we have the formulas*

$$\|f\|_{L^2(\mathbb{R})} = \|F\|_{L^2(\hat{\mathbb{R}})}, \qquad (1.10.9)$$

$$\int f(t)\overline{g(t)}\,dt = \int F(\gamma)\overline{G(\gamma)}\,d\gamma, \qquad (1.10.10)$$

$$\int f(t)g(t)\,dt = \int F(\gamma)G(-\gamma)\,d\gamma, \qquad (1.10.11)$$

and

$$\forall \gamma \in \hat{\mathbb{R}}, \qquad \int f(t)g(t)e^{-2\pi it\gamma}\,dt = \int F(\lambda)G(\gamma - \lambda)\,d\lambda. \qquad (1.10.12)$$

PROOF.   Equation (1.10.9) is part of Theorem 1.10.2. Equation (1.10.10) is a consequence of (1.10.9) and the fact that $4f\bar{g} = |f + g|^2 - |f - g|^2 + i|f + ig|^2 - i|f - ig|^2$.

Equation (1.10.11) can be proved similarly or by the following formal calculation, which can be made valid, e.g., Exercise 1.44. This calculation actually gives (1.10.12), from which (1.10.11) follows for $\gamma = 0$. Note that $fg \in L^1(\mathbb{R})$ and $F * G \in A(\hat{\mathbb{R}})$. We compute

$$(F * G)\check{\,}(t) = \int F * G(\gamma)e^{2\pi it\gamma}\,d\gamma$$

$$= \iint F(\lambda)G(\gamma - \lambda)e^{2\pi it\gamma}\,d\gamma\,d\lambda$$

$$= g(t)\int F(\lambda)e^{2\pi it\lambda}\,d\lambda = f(t)g(t),$$

and, hence,

$$\int f(t)g(t)e^{-2\pi it\gamma}\,dt = \int F(\lambda)G(\gamma - \lambda)\,d\lambda. \qquad \blacksquare$$

We shall refer to Theorem 1.10.2 and Proposition 1.10.3 as the *Parseval–Plancherel Theorem*, and we shall refer to equations (1.10.10) or (1.10.11) as *Parseval's formula*. Parseval was a French engineer who gave a formal verification of the Fourier series version of (1.10.9) in 1799; the publication is dated 1805.

The following version of (1.10.10) is required in Example 2.4.8.

1.10.4 Proposition.

Consider $f \longleftrightarrow F$ *and* $g \longleftrightarrow G$, *where* $f \in L^1(\mathbb{R})$ *and* $g \in A(\mathbb{R})$, *i.e.,* $G \in L^1(\hat{\mathbb{R}})$ *and* $g(t) = \int G(\gamma)e^{2\pi it\gamma}\,d\gamma$. *Then*

$$\int f(t)\overline{g(t)}\,dt = \int F(\gamma)\overline{G(\gamma)}\,d\gamma$$

and

$$\int f(t)g(t)\,dt = \int F(\gamma)G(-\gamma)\,d\gamma.$$

PROOF.   Consider the first formula. The right side is

$$\iint f(t)e^{-2\pi it\gamma}\,\overline{G(\gamma)}\,dt\,d\gamma$$

$$= \int f(t)\left(\overline{\int G(\gamma)e^{2\pi it\gamma}\,d\gamma}\right)dt = \int f(t)\overline{g(t)}\,dt,$$

where the first equality is a consequence of the Fubini–Tonelli Theorem.   ∎

1.10.5 Example. Powers of the Dirichlet Kernel

Consider the pairing

$$\frac{\sin t}{t} \longleftrightarrow \pi\mathbf{1}_{[-1/2\pi,1/2\pi)},$$

noting that $(\sin t)/t \in L^2(\mathbb{R})\backslash(L^2(\mathbb{R}) \cap L^1(\mathbb{R}))$. We can compute

$$\int \frac{\sin^n t}{t^n}\,dt, \qquad n \geq 2,$$

by means of the Parseval–Plancherel Theorem.

For example,

$$\int \frac{\sin^3 t}{t^3} \, dt = \frac{3\pi}{4}$$

because of the pairing

$$\frac{\sin^2 t}{t^2} \longleftrightarrow \pi \max(1 - |\pi\gamma|, 0),$$

and by the Parseval–Plancherel Theorem.

Similarly,

$$\int \frac{\sin^4 t}{t^4} \, dt = \frac{2}{3}\pi,$$

e.g., Exercise 1.17.

1.10.6 Example. A Fourier Uniqueness Property

Do there exist $f \in (L^1(\mathbb{R}) \cap L^2(\mathbb{R})) \setminus \{0\}$ and $a \in \mathbb{C}$ for which

$$\forall t \in \mathbb{R}, \qquad af(t)\mathbf{1}_{[-T,T)}(t) = f * d_{2\pi\Omega}(t)? \qquad (1.10.13)$$

To answer this question we distinguish two cases, $a = 0$ and $a \neq 0$. Also, we say that a function k is *supported* by $[A, B] \subseteq \mathbb{R}$ if $k = 0$ on $[A, B]^c \equiv \mathbb{R} \setminus [A, B]$; in this case we write supp $k \subseteq [A, B]$, cf., Definition 2.2.1a.

a. If $a = 0$, then $\hat{f}\mathbf{1}_{[-\Omega,\Omega]} = 0$; and so any $f \in L^1(\mathbb{R}) \cap L^2(\mathbb{R})$, for which $\hat{f} = 0$ on $(-\Omega, \Omega)$, is a solution of (1.10.13).

b. If $a \neq 0$ and a solution f is constrained to be supported by $[-T, T]$, then (1.10.13) implies $a\hat{f} = \hat{f}\mathbf{1}_{[-\Omega,\Omega]}$. Thus, $a = 1$ and supp $\hat{f} \subseteq [-\Omega, \Omega]$. This contradicts the analyticity of f unless $f = 0$; in fact, if supp $f \subseteq [-T, T]$, then $\hat{f}$ is entire, and so supp $\hat{f} \subseteq [-\Omega, \Omega]$ can only occur if $\hat{f} = 0$ on $\hat{\mathbb{R}}$.

c. Generally, let $a \neq 0$, and let g denote $af\mathbf{1}_{[-T,T)}$ and $f * d_{2\pi\Omega}$. In particular, supp $g \subseteq [-T, T]$ and supp $\hat{g} \subseteq [-\Omega, \Omega]$, so that $g = 0$ as in part *b.* Consequently, $f = 0$ on $(-T, T)$ and $\hat{f} = 0$ on $(-\Omega, \Omega)$. It is a remarkable fact that there are functions $f \in (L^1(\mathbb{R}) \cap L^2(\mathbb{R})) \setminus \{0\}$ with this property, e.g., [Len72], [ABe77], [Bene84, Theorem 6]. This result is related to topics about the uncertainty principle and uniqueness, e.g., [Pric85, pages 149–170].

1.10.7 Example. An Idempotent Problem in $L^1(\mathbb{R})$ and $L^2(\mathbb{R})$

Consider the equation

$$f = f * f. \qquad (1.10.14)$$

a. If we ask whether (1.10.14) has a solution $f \in L^1(\mathbb{R})\backslash\{0\}$, the answer is "no" for the following reason, cf., Proposition 2.1.1. If there were such an f, then $\hat{f} = (\hat{f})^2$ so that $\hat{f}$ takes values 0 or 1. If $\hat{f} = 0$ on $\mathbb{R}$, then $f = 0$ by the uniqueness theorem. If $\hat{f} = 1$ on $\hat{\mathbb{R}}$, then $f \notin L^1(\mathbb{R})$ since $L^1(\mathbb{R})\hat{} \subseteq C_0(\hat{\mathbb{R}})$. If $\hat{f}$ takes both 0 and 1 values we contradict the continuity of $\hat{f}$.

b. If we ask whether (1.10.14) has a solution $f \in L^2(\mathbb{R})\backslash\{0\}$, the answer is "yes". In fact, let $\hat{f} = \mathbf{1}_E$ where $|E| < \infty$. (We are using the Parseval–Plancherel Theorem here to assert the existence of $f \in L^2(\mathbb{R})$ for which $\hat{f} = \mathbf{1}_E$.)

1.10.8 Remark. Root Mean Square (RMS) Problem

In Example 1.9.4 we defined an approximation problem, and alluded to a case that we now give.

For $\Omega > 0$, let $PW_\Omega = \{g \in L^2(\mathbb{R}) : \operatorname{supp} \hat{g} \subseteq [-\Omega, \Omega]\}$; this means that $g \in L^2(\mathbb{R})$ and $\hat{g} = 0$ on $[-\Omega, \Omega]^c \equiv \mathbb{R}\backslash[-\Omega, \Omega]$, cf., Definition 2.2.1a. PW_Ω is the *Paley–Wiener space*. Let $f \in L^2(\mathbb{R})$. The approximation problem in this setting is to find $g_f \in PW_\Omega$ such that

$$\forall g \in PW_\Omega, \qquad \|f - g_f\|_{L^2(\mathbb{R})} \leq \|f - g\|_{L^2(\mathbb{R})} = e_g^{1/2}. \qquad (1.10.15)$$

It is not clear that such a "minimizer" g_f exists. The quantity e_g is the *RMS error* corresponding to f and g.

1.10.9 Proposition.

For $f \in L^2(\mathbb{R})$ and $\Omega > 0$, let $g_f = f * d_{2\pi\Omega}$. Then $g_f \in PW_\Omega$, (1.10.15) is valid, and the RMS error corresponding to f and g_f is $\int_{|\gamma| \geq \Omega} |\hat{f}(\gamma)|^2\, d\gamma$.

PROOF. We take $g \in PW_\Omega$ and compute

$$\|f - g\|_{L^2(\mathbb{R})}^2 = \|\hat{f} - \hat{g}\|_{L^2(\hat{\mathbb{R}})}^2$$

$$= \int_{|\gamma| \geq \Omega} |\hat{f}(\gamma)|^2\, d\gamma \qquad (1.10.16)$$

$$+ \int_{-\Omega}^{\Omega} (\hat{f}(\gamma) - \hat{g}(\gamma))\overline{(\hat{f}(\gamma) - \hat{g}(\gamma))}\, d\gamma.$$

This last term is zero when $g = g_f$; and since the integrand, $(\hat{f} - \hat{g})\overline{(\hat{f} - \hat{g})}$, is nonnegative, we have verified (1.10.15). Also, because of (1.10.16), the RMS error corresponding to f and g is $\int_{|\gamma| \geq \Omega} |\hat{f}(\gamma)|^2\, d\gamma$. ∎

1.10.10 Remark. Perspective on Bandlimited Approximants

We saw in the case of a jump discontinuity that $f * w_{2\pi\Omega}$ was preferable to $f * d_{2\pi\Omega}$ as an approximant to f under pointwise convergence. On the other hand, Proposition 1.10.9 shows that $f * d_{2\pi\Omega}$ is the best approximant to f in the sense of minimizing the RMS error.

Another aspect of the approximation problem discussed in Example 1.9.4 is the following result.

1.10.11 Theorem. Timelimited and Bandlimited Approximation

Let $T, \Omega > 0$ and let

$$L_T^2(\mathbb{R}) = \{f \in L^2(\mathbb{R}) : \operatorname{supp} f \subseteq [-T, T]\},$$

i.e., $f \in L^2(\mathbb{R})$ and $f = 0$ on $[-T, T]^c \equiv \mathbb{R}\backslash[-T, T]$. Then

$$Y = \{F : [-\Omega, \Omega] \longrightarrow \mathbb{C} : \exists k \in L_T^2(\mathbb{R}) \text{ such that } \hat{k} = F \text{ a.e.} \quad \text{on } [-\Omega, \Omega]\}$$

is dense in $L^2[-\Omega, \Omega]$.

PROOF.   Suppose the result is not true. Then by the Hahn–Banach Theorem (Theorem B.12) there is $G \in L^2[-\Omega, \Omega]'\backslash\{0\} = L^2[-\Omega, \Omega]\backslash\{0\}$ such that for all $K \in Y$, where $k \longleftrightarrow K$, we have

$$0 = \int_{-\Omega}^{\Omega} G(\gamma)\overline{K(\gamma)}\,d\gamma$$

$$= \int (G\mathbf{1}_{[-\Omega,\Omega)})(\gamma)\overline{K(\gamma)}\,d\gamma = \int_{-T}^{T} (G\mathbf{1}_{[-\Omega,\Omega)})^{\vee}(t)\overline{k(t)}\,dt.$$

Since this is true for all $k \in L_T^2(\mathbb{R})$, we have $(G\mathbf{1}_{[-\Omega,\Omega)})^{\vee} = 0$ on $[-T, T]$, and, by definition, $G\mathbf{1}_{[-\Omega,\Omega)} = 0$ off $[-\Omega, \Omega]$, i.e., $\operatorname{supp} G\mathbf{1}_{[-\Omega,\Omega)}$ is compact. This last property implies $(G\mathbf{1}_{[-\Omega,\Omega)})^{\vee}$ is entire and so $(G\mathbf{1}_{[-\Omega,\Omega)})^{\vee} = 0$ on $\mathbb{R}$. By the L^1-uniqueness theorem and the fact that $G\mathbf{1}_{[-\Omega,\Omega)} \in L^1(\hat{\mathbb{R}})$ we have $G\mathbf{1}_{[-\Omega,\Omega)} = 0$ on $\hat{\mathbb{R}}$ and so $G = 0$ on $[-\Omega, \Omega]$, a contradiction. Thus, we have the desired density.   ∎

One proof of Theorem 1.10.2, due to Norbert Wiener, e.g., [Wie33], involves defining $\hat{f}$, for given $f \in L^2(\mathbb{R})$, as an eigenfunction expansion where the eigenfunctions are the so-called Hermite functions. For this reason we give the following example.

1.10.12 Example. Eigenvalues of the Fourier Transform

a.   Suppose we are given the pairing $f \longleftrightarrow F$ and that we can compute $\hat{\hat{f}}$, $\hat{\hat{\hat{f}}}$, etc. Formally, we have

$$\hat{f}\hat{}(\gamma) = f(-\gamma)$$

and

$$\hat{f}\hat{}\hat{}(\gamma) = f(-t)\hat{}\hat{}(\gamma) = \hat{f}(-\gamma);$$

and, hence,

$$\hat{f}\hat{}\hat{}\hat{} = f.$$

b. Next, consider the operator,

$$\mathcal{F} : L^2(\mathbb{R}) \longrightarrow L^2(\hat{\mathbb{R}}),$$

$$f \longmapsto \hat{f},$$

defined on a sufficiently well-behaved subset of $L^2(\mathbb{R})$, e.g., the space $\mathcal{S}(\mathbb{R})$ defined in Definition 2.4.3. Consider the eigenvalue problem,

$$\mathcal{F}f = \lambda f.$$

We have $\hat{f} = \lambda f$, $\hat{f}\hat{} = \lambda \hat{f} = \lambda^2 f$, $\hat{f}\hat{}\hat{} = \lambda^2 \hat{f} = \lambda^3 f$; and so, from part *a*, we obtain $f = \hat{f}\hat{}\hat{}\hat{} = \lambda^3 \hat{f} = \lambda^4 f$. Consequently, the eigenvalues of $\mathcal{F}$ are $\lambda = 1, i, -1, -i$; and the Hermite functions arise in this setting as the eigenfunctions of $\mathcal{F}$, e.g., Exercise 1.26, Exercise 1.27, and Remark 2.4.11.

c. If g is the Gaussian, we saw in Example 1.3.3 that $\hat{g}_{\sqrt{\pi}} = g_{\sqrt{\pi}}$. By part *a*, we see that there are many functions k for which $\hat{k} = k$. In fact, for any $f \in L^1(\mathbb{R}) \cap A(\mathbb{R})$, let

$$k = f + \hat{f} + \hat{f}\hat{} + \hat{f}\hat{}\hat{},$$

cf., Exercise 1.28.

1.10.13 Example. Computation of Integrals

Using the Parseval–Plancherel Theorem we can compute integrals of the form

$$\int \frac{dt}{(t^2 + a^2)(t^2 + b^2)}, \qquad a, b > 0.$$

[Hint: $p_{1/c} \longleftrightarrow e^{-2\pi c|\gamma|}$.]

1.10.14 Example. Computation of L^2-Fourier Transforms

a. Let $f \in L^2(\mathbb{R})$ and let $f_n = f \mathbf{1}_{[-n,n]}$. Clearly, $f_n \in L^1(\mathbb{R}) \cap L^2(\mathbb{R})$ (by Hölder's Inequality) and $\lim_{n \to \infty} \| f - f_n \|_{L^2(\mathbb{R})} = 0$. Thus, by the Parseval–

Plancherel Theorem,

$$\lim_{n \to \infty} \left\| \hat{f}(\gamma) - \int_{-n}^{n} f(t)e^{-2\pi i t \gamma} \, dt \right\|_{L^2(\mathbb{R})} = 0. \qquad (1.10.17)$$

b. We saw that the Dirichlet function has the property that

$$d_{2\pi} \in L^2(\mathbb{R}) \backslash (L^1(\mathbb{R}) \cap L^2(\mathbb{R})).$$

By (1.10.17),

$$\lim_{n \to \infty} \left\| \hat{d}_{2\pi}(\gamma) - \int_{-n}^{n} \frac{\sin 2\pi t}{\pi t} \cos 2\pi t \gamma \, dt \right\|_{L^2(\mathbb{R})} = 0; \qquad (1.10.18)$$

and it is easy to see, by Theorem 1.7.6, that

$$\forall \gamma \in \hat{\mathbb{R}}, \quad \lim_{n \to \infty} \int_{-n}^{n} \frac{\sin 2\pi t}{\pi t} \cos 2\pi t \gamma \, dt$$

$$= \begin{cases} 1, & \text{if} \quad |t| < 1, \\ \dfrac{1}{2}, & \text{if} \quad |t| = 1, \\ 0, & \text{if} \quad |t| > 1, \end{cases} \qquad (1.10.19)$$

e.g., Exercise 1.23*b*. Combining (1.10.18), (1.10.19), and F. Riesz's result used at the end of Theorem 1.7.8, we see that the L^2-Fourier transform of $d_{2\pi}$ is $\hat{d}_{2\pi} = 1_{[-1,1]}$ a.e., which, of course, is the way it had to turn out!

Exercises

Exercises 1.1–1.30 are appropriate for Course I. Recall that H is the Heaviside function.

1.1. Suppose that $pv \int g(t) \, dt$ exists. Prove that if g is even or g is nonnegative, then $\int g(t) \, dt$ exists and

$$\int g(t) \, dt = pv \int g(t) \, dt.$$

1.2. Consider the formal pairing $f \longleftrightarrow F$. Verify (formally) the following.
   a) If f is imaginary, then $\overline{F(\gamma)} = -F(-\gamma)$ and

$$f(t) = 2i \operatorname{Im} \int_0^{\infty} F(\gamma)e^{2\pi i t \gamma} \, d\gamma.$$

   b) f is imaginary and odd if and only if F is real and odd.

1.3. Let $f \in L^1(\mathbb{R})$ be real-valued, $f \longleftrightarrow F$.

    a) Verify whether or not $|F|^2$ is even, odd, or neither.

    b) Verify whether or not $|F|^3$ is even, odd, or neither.

1.4. Prove that

$$\frac{1}{\pi} \int_{-\pi}^{\pi} \frac{\sin u}{u}\, du > 1.$$

1.5. Prove that $\hat{f} \notin L^1(\hat{\mathbb{R}})$ for $f(t) = H(t)e^{-2\pi rt}$, where $r > 0$.

1.6. Verify whether or not the following functions f defined on $\mathbb{R}$ are elements of $L^1(\mathbb{R})$. For those that are not, verify whether or not they are elements of $L^1_{loc}(\mathbb{R})$.

    a) $f(t) = H(t)$. You may assign $H(0)$ any value you like; it will not affect the validity of your answer.

    b) $f(t) = 1/(1+t^2)$.

    c) $f(t) = \begin{cases} 1/t^2, & \text{if } t \neq 0, \\ 0, & \text{if } t = 0. \end{cases}$

    d) $f(t) = \begin{cases} 1/t, & \text{if } t \neq 0, \\ 0, & \text{if } t = 0. \end{cases}$

    e) $f(t) = \cos 2\pi t\gamma_0 + i \sin 2\pi t\gamma_0 = e^{2\pi it\gamma_0}$.

    f) $f(t) = \operatorname{sgn} t$, where $\operatorname{sgn} t \equiv H(t) - H(-t)$.

    g) $f(t) = d(t)$, the Dirichlet function.

1.7. Let $f : \mathbb{R} \longrightarrow \mathbb{C}$ be a function. The *even part* of f is the function $f_e(t) = \frac{1}{2}(f(t) + f(-t))$, and the *odd part* of f is the function $f_o(t) = \frac{1}{2}(f(t) - f(-t))$.

    a) Compute $f_e + f_o$.

    b) Compute the even and odd parts of the functions in Exercise 1.6.

    c) For functions f and g, verify that

$$(fg)_e = f_e g_e + f_o g_o.$$

What is the corresponding decomposition for $(fg)_o$?

1.8. Compute the Fourier transforms of the following functions.

    a) $f = \tau_u 1_{[-T,T)}$.

    b) $f(t) = -\cos t/(\pi(t - \pi/2))$. [Hint. Evaluate $\tau_{\pi/2}\, d$.]

    c) $f = \tau_u (1_{[-T,T)} H)$.

    d) $f = 1_{[-T,T)}(\tau_u H)$.

    e) $f = \tau_u g_\lambda$, where g is the Gaussian.

1.9. Solve the differential equation $F'(\gamma) = (-2\pi\gamma/r) F(\gamma)$, $r > 0$, in Example 1.3.3.

1.10. Prove that $\int f(t)\, dt = 0$ in Proposition 1.4.3.

    [Hint. $\underline{\lim}_{t\to\infty} |g(t)| > 0$ implies $g \notin L^1(\mathbb{R})$.]

1.11. Let $f \in L^1(\mathbb{R})$; and let $g(t) = f(at+b)$ for fixed $a, b \in \mathbb{R}, a \neq 0$. Compute $\hat{g}$ in terms of $\hat{f}$.

1.12. Compute the Fourier transforms of the following functions.
   a) $f(t) = (t^n/n!)e^{-2\pi rt}H(t),\quad r > 0$ and $n \in \mathbb{N}$.
   b) $f(t) = e^{-2\pi rt}H(t)\sin 2\pi t\gamma_0,\quad r > 0$.

1.13. Let $m, n \geq 0$ and let $f \in L^1(\mathbb{R})$. Suppose that $t^n f(t) \in L^1(\mathbb{R})$ and that the mth derivative $(t^n f(t))^{(m)}$ exists everywhere and is an element of $L^1(\mathbb{R})$. Verify the Fourier transform pairing

$$(t^n f(t))^{(m)} \longleftrightarrow (-1)^n (2\pi i)^{m-n} \gamma^m (\hat{f})^{(n)}(\gamma).$$

Use this result to show that

$$\forall \gamma \in \hat{\mathbb{R}}\backslash\{0\}, \qquad |\hat{f}(\gamma)| \leq \frac{1}{(2\pi|\gamma|)^m}\|f^{(m)}\|_{L^1(\mathbb{R})}.$$

1.14. Using the methods of this chapter, compute

$$\int \frac{dt}{t^4 + 7t^2 + 6}.$$

1.15. We have defined the Fourier transform in Definition 1.1.2 and Remark 1.1.3 using the kernels $e^{-2\pi it\gamma}$ and $e^{2\pi it\gamma}$, where the latter is required to ensure the validity of the inversion formula. These are the kernels of choice in major theoretical treatises, such as Stein and Weiss' masterpiece on Euclidean harmonic analysis [SW71], as well as in fundamental tools and algorithms such as the Discrete Fourier Transform (DFT) and the Fast Fourier Transform (FFT), e.g., [Walk91] and MATLAB. We find the above pairing to be computationally convenient. Other pairings in the literature include $e^{-it\gamma}$ and $(1/2\pi)e^{it\gamma}$, and $(1/\sqrt{2\pi})e^{-it\gamma}$ and $(1/\sqrt{2\pi})e^{it\gamma}$. For this exercise, do not use any of these latter kernels or their variations in any of the other exercises!

1.16. Let $f(t) = e^t H(t),\ g(t) = e^{-t}H(t)$, and $k(t) = e^t H(-t)$.
   a) Compute $f * \cdots * f$, where there are n factors.
   b) Compute $g * \cdots * g$, where there are n factors.
   c) Compute $k * \cdots * k$, where there are n factors.
   d) Compute $f * g$ and $g * k$. Comment on $f * k$.
   e) Compute $H * f$, $H * g$, and $H * k$.

1.17. a) Compute $d_{2\pi\Omega} * \cdots * d_{2\pi\Omega}$, where there are n factors and where d is the Dirichlet function.
   b) Compute and graph $\left(1_{[-\Omega,\Omega)}\right)^{*n} \equiv 1_{[-\Omega,\Omega)} * \cdots * 1_{[-\Omega,\Omega)}$, where there are n factors, for the cases $n = 1, \ldots, 6$.

   The n-fold convolution of part b is an example of a *spline* supported by $[-n\Omega, n\Omega]$, e.g., [dHR93], [Scho73], cf., Exercise 2.48. The general formula,

$$\left(1_{[-\Omega,\Omega)}\right)^{*n}(t) = \sum_{k=0}^{n} \binom{n}{k}\frac{(-1)^{n-k}}{(n-1)!}(t-(n-2k)\Omega)^{n-1}1_{[(n-2k)\Omega,\infty)}(t),$$

   is not difficult to verify.

1.18. Find the values of $n \in \mathbb{N}$ for which $d_{2\pi\Omega}^{(n)} \in L^2(\mathbb{R})$. [Hint. Use Theorem 1.4.1e and the Parseval–Plancherel Theorem.]

1.19. Let $g(t) = (1/\sqrt{\pi})e^{-t^2}$, the Gaussian.

a) Compute $g_a * g_b$ for $a, b > 0$.

b) Determine the points of inflection and regions of convexity and concavity of $\{g_\lambda\}$.

1.20. Let $f_{a,\lambda} = \tau_a d_\lambda + \tau_{-a} d_\lambda$, where d is the Dirichlet function.

a) Compute $f_{a,\lambda} * f_{a,\lambda}$.

b) Compute $\lim_{a \to a_0} f_{a,\lambda} * f_{a,\lambda}$ for $a_0 = 0, \infty$.

1.21. a) Compute the Fourier transforms of $tg_\lambda(t)$ and $tw_\lambda(t)$, where g is the Gaussian and w is the Fejér function. [Hint. The fact that $tw_\lambda(t) \notin L^1(\mathbb{R})$ is not a problem. Begin by guessing at the answer using Theorem 1.4.1e, and then justify your guess.]

b) Graph the functions of part a and their Fourier transforms.

1.22. Let $g(t) = (1/\sqrt{\pi})e^{-t^2}$, the Gaussian. Compute the Fourier transforms of the functions

$$\frac{d^{10}}{dt^{10}}(g_a * g_b)$$

and

$$\left(\frac{d^5}{dt^5}g_a\right) * \left(\frac{d^5}{dt^5}g_b\right),$$

where $a, b > 0$.

1.23. Use the inversion theorems to prove the following:

a) $\dfrac{2r}{\pi}\displaystyle\int_0^\infty \dfrac{\cos 2\pi t\gamma}{r^2 + \gamma^2}\,d\gamma = e^{-2\pi r|t|}, r > 0.$

b) $\dfrac{2}{\pi}\displaystyle\int_0^\infty \dfrac{\sin 2\pi\gamma \cos 2\pi t\gamma}{\gamma}\,d\gamma = \begin{cases} 1, & \text{and} \quad |t| < 1; \\ 1/2, & \text{and} \quad |t| = 1; \\ 0, & \text{and} \quad |t| > 1. \end{cases}$

1.24. Let $f_{(\Omega)}(t) = \mathbf{1}_{[-1,1)}(t)\cos 2\pi t\Omega$. Compute

$$\lim_{\Omega \to 0} \int |\hat{f}_{(\Omega)}(\gamma) - \hat{\mathbf{1}}_{[-1,1)}(\gamma)|^2\,d\gamma.$$

1.25. Compute the Fourier transform of

$$f(t) = g_a(t)\sin 2\pi bt, \qquad a > 0,$$

in terms of a hyperbolic trigonometric function. g is the Gaussian.

1.26. The first six *Hermite polynomials* are

$$H_0(t) = 1, \qquad H_1(t) = t, \qquad H_2(t) = t^2 - 1,$$
$$H_3(t) = t^3 - 3t, \qquad H_4(t) = t^4 - 6t^2 + 3,$$
$$H_5(t) = t^5 - 10t^3 - 15t.$$

In general,

$$\forall n \geq 0, \qquad H_n(t) = (-1)^n e^{t^2/2} \frac{d^n}{dt^n} e^{-t^2/2},$$

e.g., [Wie33], [CH53], [Jac41].

a) Verify, in fact, that H_n is a polynomial of nth degree with the coefficient of t^n equal to 1. [Hint. $H_{n+1}(t) = t H_n(t) - H_n'(t)$.]

b) Prove the orthogonality relations, viz., if $0 \leq m < n$, then

$$\int H_m(t) H_n(t) e^{-t^2/2} \, dt = 0.$$

1.27. If the convection term in (1.8.1) is replaced by $V(x)u$, where V is a potential energy, then (with an adjustment of constants) we obtain the one-dimensional *Schrödinger equation*,

$$\frac{\partial^2 u}{\partial x^2} - V(x)u - c\frac{\partial u}{\partial t} = 0,$$

e.g., [Wey50a, pages 54–60]. Assuming solutions of the form $u(x, t) = X(x)T(t)$, the *separation of variables method* leads to

$$\frac{1}{X(x)} \frac{d^2 X(x)}{dx^2} - V(x) = \frac{c}{T(t)} \frac{dT(t)}{dt};$$

and if $-\lambda$ is a constant common value of both sides of this equation, we are led to the problem of determining *eigenvalues* λ and *eigenfunctions* X_λ for which the *oscillator equation*,

$$\frac{d^2 X_\lambda(x)}{dx^2} + (\lambda - V(x))X_\lambda(x) = 0,$$

is valid, cf., [Str88, pages 243–253] to see the relationship between the solution of differential equations and eigenvalue problems. Solve this particular problem for the potential energy $V(x) = x^2/4$. [Hint. Let $X_n(x) = e^{-x^2/4} H_n(x)$ (Exercise 1.26), and obtain $\lambda_n = (n + \frac{1}{2})$, $n \geq 0$.]

1.28. Let $f(t) = e^{-\pi(t/r)^2} + re^{-\pi(rt)^2}$. For what numbers $r \in \mathbb{R}\setminus\{0\}$ do we have $f = \hat{f}$? Note the case $r = -1$.

1.29. Compute the Gibbs "overshoot" G (from Remark 1.9.1 and Example 1.9.3) accurately to six decimal places.

1.30. Compute $(H(u)\sin u) * (H(u)\sin u)(t)$ for $t = \pm\pi/2$.

1.31. Let $f, g \in L^1(\mathbb{R})$.

    a) Verify that

$$\int |f(t-u)||g(u)|\, du < \infty \qquad t\text{-a.e. in } \mathbb{R}. \qquad (E1.1)$$

    b) For the set X of $t \in \mathbb{R}$ that satisfies (E1.1) in part a, define

$$f * g(t) = \int f(t-u)g(u)\, du.$$

    Prove that $f * g \in L^1(\mathbb{R})$, and that

$$\|f * g\|_{L^1(\mathbb{R})} \leq \|f\|_{L^1(\mathbb{R})}\|g\|_{L^1(\mathbb{R})}.$$

1.32. The set $\mathbb{R}$ of real numbers satisfies the commutative and associative laws under both addition and multiplication, and it satisfies the distributive law under addition and multiplication. Prove that $L^1(\mathbb{R})$ satisfies the same properties when multiplication is replaced by convolution, i.e., show that

$$\begin{aligned} f + g &= g + f, & f * g &= g * f, \\ (f + g) + h &= f + (g + h), & f * (g * h) &= (f * g) * h, \\ f * (g + h) &= f * g + f * h \end{aligned}$$

    for $f, g, h \in L^1(\mathbb{R})$.

1.33. Let $f, g \in L^1(\mathbb{R})$ and define the L^1-*cross-correlation*

$$f \circledast g(t) = \int f(t+u)\overline{g(u)}\, du$$

    of f and g. Which properties of Exercise 1.32 are valid? Note that $f \circledast g = f * \tilde{g}$, where $\tilde{g}$ is the *involution* defined as $\tilde{g}(t) = \overline{g(-t)}$. Clearly, $\tilde{\tilde{f}} = f$ and $f * \tilde{g} = \widetilde{(g * \tilde{f})}$.

1.34. Using the methods of this chapter, verify that

$$4\int \frac{\sin \pi \gamma}{\gamma} e^{-2\pi|\gamma|+\pi i\gamma}\, d\gamma = \pi.$$

1.35. a) Let $f \in L^1(\mathbb{R}) \cap L^\infty(\mathbb{R})$. Prove that $f \in L^2(\mathbb{R})$.

    b) Construct $f \in L^2(\mathbb{R}) \cap L^\infty(\mathbb{R})$, resp., $f \in L^1(\mathbb{R}) \cap L^2(\mathbb{R}) \cap L^\infty_{\text{loc}}(\mathbb{R})$, such that $f \notin L^1(\mathbb{R})$, resp., $f \notin L^\infty(\mathbb{R})$.

    c) Clearly, if $f \equiv 1$, then $f \in L^\infty(\mathbb{R})$ and $f \notin L^1(\mathbb{R}) \cup L^2(\mathbb{R})$. Construct $f \in L^1(\mathbb{R})$, resp., $f \in L^2(\mathbb{R})$, such that $f \notin L^2(\mathbb{R}) \cup L^\infty(\mathbb{R})$, resp., $f \notin L^1(\mathbb{R}) \cup L^\infty(\mathbb{R})$.

    d) Are the constructions of part c possible if $L^1(\mathbb{R})$, resp., $L^2(\mathbb{R})$, is replaced by $L^1(\mathbb{R}) \cap L^\infty_{\text{loc}}(\mathbb{R})$, resp., $L^2(\mathbb{R}) \cap L^\infty_{\text{loc}}(\mathbb{R})$?

1.36. Verify *Viète's formula*

$$\prod_{k=1}^{\infty} \cos \frac{\gamma}{2^k} = \frac{\sin \gamma}{\gamma}. \qquad (E1.2)$$

In particular,

$$d_{2\pi T}(\gamma) = 2T \prod_{k=0}^{\infty} \cos\left(\frac{\pi T \gamma}{2^k}\right).$$

Viète proved the case $\gamma = \pi/2$:

$$\frac{2}{\pi} = \frac{\sqrt{2}}{2} \frac{\sqrt{2 + \sqrt{2}}}{2} \frac{\sqrt{2 + \sqrt{2 + \sqrt{2}}}}{2} \cdots.$$

François Viète was a lawyer from Poitou, who became one of the foremost mathematicians of the sixteenth century because of his contributions to algebra, e.g., [Kli72]. He was also the major "codebreaker" for Henry IV of France [Kahn67, pages 116–117] amidst the Holy League, the Huguenots, the Hapsburgs, the demise of the House of Valois, and the death of the Dark Eminence, Catherine de Medici.

1.37. Prove the following results, where $f \in L^1(\mathbb{R})$ and $\{k_{(\lambda)}\}$ is an approximate identity.

a) There is $\{\lambda_n\} \subseteq (0, \infty)$ such that

$$\lim_{n \to \infty} f * k_{(\lambda_n)} = f \quad \text{a.e.}$$

b) For each $\gamma \in \hat{\mathbb{R}}$, $\lim_{\lambda \to \infty} \hat{k}_{(\lambda)}(\gamma) = 1$.

c) Assume $\hat{f} \in L^1(\hat{\mathbb{R}})$, $\hat{k}_{(\lambda)} \in L^1(\hat{\mathbb{R}})$, and

$$\forall t \in \mathbb{R}, \qquad k_{(\lambda)}(t) = \int \hat{k}_{(\lambda)}(\gamma) e^{2\pi i t \gamma} \, d\gamma. \qquad (E1.3)$$

Then

$$\lim_{\lambda \to \infty} \left\| \int \hat{f}(\gamma) e^{2\pi i t \gamma} \, d\gamma - f * k_{(\lambda)}(t) \right\|_{L^\infty(\mathbb{R})} = 0.$$

[Hint. Add and subtract $\int \hat{f}(\gamma) \hat{k}_{(\lambda)}(\gamma) e^{2\pi i t \gamma} \, d\gamma$; use part b and LDC for one part and the Fubini–Tonelli Theorem for the other part.]

The Gauss kernel $\{g_\lambda\}$ is an example of an approximate identity where we have already verified that (E1.3) is satisfied. There are also direct calculations for the Poisson and Fejér kernels.

The point of this exercise is to illustrate the idea behind the proof of Theorem 1.7.8. In fact, we obtain that proof by combining parts *a* and *c*.

1.38. Let $f \in L^1(\mathbb{R})$ and assume that $|\int f(t)dt| = 1$. Prove that

$$\{f_\lambda * \tilde{f}_\lambda\}$$

is an approximate identity, where f_λ, $\lambda > 0$, is the L^1-dilation of f.

1.39. Let $f \in L^1(\mathbb{R})$, and let $u(x, t)$ be defined by (1.8.7). Prove that u is a solution of the system (1.8.1)–(1.8.3). For the case of (1.8.2), prove only that

$$\lim_{t \to 0} \left\| u(x, t) - f(x) \right\|_{L^1(\mathbb{R})} = 0.$$

1.40. a) Prove that $A(\hat{\mathbb{R}})$ is dense in $C_0(\hat{\mathbb{R}})$.

   b) Prove that $A(\hat{\mathbb{R}})$ is a set of first category in $C_0(\hat{\mathbb{R}})$.

1.41. Let $f \in L^1(\mathbb{R})$, $g \in L^2(\mathbb{R})$. Prove that $f * g \in L^2(\mathbb{R})$ and that

$$\|f * g\|_{L^2(\mathbb{R})} \le \|f\|_{L^1(\mathbb{R})} \|g\|_{L^2(\mathbb{R})}. \qquad (E1.4)$$

[Hint. Use Minkowski's inequality.] Verify that $(f * g)\hat{} = \hat{f}\hat{g}$.

   (E1.4) should be compared with Exercise 1.31*b*. The inequality (E1.4) is a special case of *W. H. Young's inequality*,

$$\|f * g\|_{L^q(\mathbb{R})} \le \|f\|_{L^p(\mathbb{R})} \|g\|_{L^r(\mathbb{R})},$$

where $1/q = 1/p + 1/r - 1$, e.g., [SW71, pages 178–183].

1.42. Let $f \in C_c(\mathbb{R})$. Verify whether or not $f * \tilde{f}$ is a function of bounded variation. If $f * \tilde{f}$ is a function of bounded variation, then we can use the Jordan inversion theorem in place of Proposition 1.6.11 in Theorem 1.10.2. [Hint. Using the Cantor function $f = f_C$, e.g., [Ben76, page 22] and Remark 1.4.2, we can show that $f * \tilde{f}$ need not be absolutely continuous. It is more difficult to prove that $f * \tilde{f}$ need not have bounded variation, cf., the remarks on continuous nowhere differentiable functions in Section 3.2.4.]

1.43. The *de la Vallée–Poussin function* v is defined as $v = 2w_2 - w_1$, where w is the Fejér function. Note that $\int v(t)dt = 1$.

   a) Verify that the dilations v_λ are equal to $v_{(\lambda)}$, where $v_{(\lambda)} \equiv 2w_{2\lambda} - w_\lambda$. Clearly, the *de la Vallée–Poussin kernel* $\{v_\lambda\}$ is an approximate identity.

   b) Graph $\hat{v}_\lambda$.

   c) Clearly, $1 \le \|v_\lambda\|_{L^1(\mathbb{R})} = \|v\|_{L^1(\mathbb{R})}$. Estimate $\|v\|_{L^1(\mathbb{R})}$.

1.44. Prove (1.10.11) and (1.10.12).

1.45. Verify that

$$\lim_{\Omega \to \infty} \int \left(\frac{\sin t}{t} \right)^j d_{2\pi\Omega}(t)\, dt = 1$$

for $j \in \mathbb{N}$. [Hint. Consider the integral over $(0, \infty)$, verify that

$$\frac{1}{t}\left[\left(\frac{\sin t}{t}\right)^j - 1\right] \in L^1(0, 1),$$

and use the Riemann–Lebesgue Lemma.]

1.46. Compute

$$I = \int \left(\frac{\sin t}{t}\right)^j \frac{\sin rt}{t} dt$$

for $j \in \mathbb{N}$ and real $r \geq j$. This generalizes the $n = 2$ case of Example 1.10.5, cf., Exercises 1.17, 1.34, and 1.45. [Hint. By Parseval's formula,

$$I = \pi^{j+1} \int_{-r/2\pi}^{r/2\pi} \mathbf{1}_{[-1/2\pi, 1/2\pi]} * \cdots * \mathbf{1}_{[-1/2\pi, 1/2\pi]}(\gamma) \, d\gamma.$$

The support of the integrand is contained in $[-j/2\pi, j/2\pi]$, e.g., Exercise 2.11. Thus,

$$I = \pi^{j+1} \hat{\mathbf{1}}_{[-1/2\pi, 1/2\pi]}(0)^j.]$$

1.47. If the potential energy V of the Schrödinger equation in Exercise 1.27 is a constant b, we obtain the fundamental equation of *linear cable theory*. Equations such as this were used by Lord Kelvin (1856) to analyze electric current flow in submarine cables, cf., [Kör88, pages 332–337] where the heat equation, i.e., $b = 0$, is analyzed vis á vis the success of the transatlantic cable. Solve the *cable equation*

$$\frac{\partial^2 u}{\partial x^2} - bu - c\frac{\partial u}{\partial t} = f$$

for a given forcing function $f(x, t)$. Nowadays, the cable equation plays a critical role in electric models of neuron excitation, e.g., [JNT75].

1.48. Prove Lemma 1.7.3. [Hint. The proof is based on the following argument in the case that f is not only increasing but also absolutely continuous. Without loss of generality, let g be real-valued and let $G(t) = \int_a^t g(u) \, du$. Then FTC and integration by parts give

$$\int_a^b f(t)g(t) \, dt = G(b)f(b) - \int_a^b G(t)f'(t) \, dt.$$

Since $f' \geq 0$ a.e. and G is continuous on $[a, b]$, we have

$$m \int_a^b f'(t) \, dt \leq \int_a^b G(t)f'(t) \, dt \leq M \int_a^b f'(t) \, dt,$$

where $m = \inf\{G(t)\}$ and $M = \sup\{G(t)\}$. Thus, by the intermediate value theorem there is $\xi \in [a, b]$ such that

$$\int_a^b f(t)g(t)\,dt = G(b)f(b) - G(\xi)\int_a^b f'(t)\,dt.$$

Another application of FTC gives the result.]

1.49. Let $g \in L^1(\mathbb{R})$; and, for $s > 0$ and $(t, \gamma) \in \mathbb{R} \times \hat{\mathbb{R}}$, set

$$g_{s,t,\gamma}(u) = g_s(u - t)e^{2\pi i u \gamma}.$$

Compute $\hat{g}_{s,t,\gamma}$, cf., Theorem 1.2.1d, e, f and Exercise 1.11.

The algebraic (and geometric) operations of dilation, modulation, and translation are the fundamental "invariants" under the action of the Fourier transform on $\mathbb{R}$ (Theorem 1.2.1). As such it is natural to attempt to "synthesize" or reconstruct functions f in terms of scale-time-frequency harmonics $\{g_{s,t,\gamma} : s > 0 \text{ and } (t, \gamma) \in \mathbb{R} \times \hat{\mathbb{R}}\}$ for a given "analyzing" function g. For a fixed frequency $\gamma \in \hat{\mathbb{R}}$, this program of signal reconstruction is a fundamental part of wavelet theory [Dau92], [Mey90]; and for a fixed scale $s > 0$, this program becomes the Weyl–Heisenberg or Gabor theory [Gabo46], [vN55], [BW94]. The signal processing program for the complete set $\{g_{s,t,\gamma}\}$ has important applications, e.g., [MZ93], and theoretical developments in terms of the metaplectic group, e.g., [HL95]. There have been many contributors to the whole program, and we refer to [HL95] for an overview. The metaplectic group contains the Heisenberg group as a subgroup as well as an isomorphic copy of the affine group. These two groups correspond to the Gabor and wavelet theories, respectively.

1.50. Let $0 < \alpha < \beta$ and let K be the trapezoid

$$K(\gamma) = \begin{cases} 1, & \text{if} \quad |\gamma| \leq \alpha, \\ 0, & \text{if} \quad |\gamma| \geq \beta, \\ \text{linear}, & \text{if} \quad \alpha \leq |\gamma| \leq \beta. \end{cases}$$

a) Compute $k \equiv K^\vee$, cf., Exercise 1.43.

b) Show that $\|K\|_{A(\hat{\mathbb{R}})} \leq (\beta + \alpha)/(\beta - \alpha)$, and hence $\|K\|_{A(\hat{\mathbb{R}})} \leq 3$ when $\beta = 2\alpha$, cf., (3.5.4) in Example 3.5.3.

c) Besides the estimate in part b, it is obvious that $\|K\|_{L^\infty(\hat{\mathbb{R}})} = 1$. Compute $\|K\|_{L^2(\hat{\mathbb{R}})}$.

1.51. Let $f \in PW_\Omega$. Prove that

$$\forall t \in \mathbb{R}, \qquad f(t) = \int \hat{f}(\gamma)e^{2\pi i t \gamma}\,d\gamma,$$

cf., Theorems 1.1.6 and 1.1.7.

2

Measures and Distribution Theory

2.1. Approximate Identities and δ

In $\mathbb{R}$, we know that $7 \times 1 = 7$, $\pi \times 1 = \pi$, $0 \times 1 = 0, \ldots$. 1 is the *multiplicative unit* for $\mathbb{R}$. In $L^1(\mathbb{R})$, convolution is the multiplication (Section 1.5), and the norm $\| \cdots \|_{L^1(\mathbb{R})}$ is the absolute value (measure of distance).

2.1.1 Proposition.

$L^1(\mathbb{R})$ *does not have a unit under convolution.*

PROOF. Suppose $u \in L^1(\mathbb{R})$ were a unit. Choose $f \in L^1(\mathbb{R})$ for which $\hat{f}$ never vanishes, e.g., $f(t) = e^{-2\pi|t|}$. Then $\| \hat{f} - \hat{f}\hat{u} \|_{L^\infty(\hat{\mathbb{R}})} \le \| f - f * u \|_{L^1(\mathbb{R})} = 0$, and so $\hat{u}$ is identically 1 on $\hat{\mathbb{R}}$. This contradicts the Riemann–Lebesgue Lemma, and so $u \notin L^1(\mathbb{R})$. ∎

On the other hand, we did show in Theorem 1.6.9 that there are families $\{k_{(\lambda)}\} \subseteq L^1(\mathbb{R})$ of functions, appropriately called approximate identities, with the property that

$$\forall f \in L^1(\mathbb{R}), \qquad \lim_{\lambda \to \infty} \| f - f * k_{(\lambda)} \|_{L^1(\mathbb{R})} = 0. \qquad (2.1.1)$$

Similarly, recall from Proposition 1.6.11 that if $f \in L^\infty(\mathbb{R})$ is continuous on $\mathbb{R}$ and $\{k_{(\lambda)}\}$ is an approximate identity, then

$$\forall t \in \mathbb{R}, \qquad \lim_{\lambda \to \infty} f * k_{(\lambda)}(t) = f(t). \qquad (2.1.2)$$

2.1.2 Remark. Motivation for δ

We can think of "$f * k_{(\lambda)} \longrightarrow f$" in (2.1.1) or (2.1.2) as a set $\{f * k_{(\lambda)}\}$ of functions approximating the function f; or we can think of the family $\{k_{(\lambda)}\}$ as approximating something, call it δ, which plays the role of an identity under convolution for elements in $L^1(\mathbb{R})$, i.e., "$f * \delta = f$" even though $\delta \notin L^1(\mathbb{R})$.

In order to quantify this latter interpretation, assume each $k_{(\lambda)}$ is an even function. Then fix the point $t = 0$ in (2.1.2), and think of the integral $f * k_{(\lambda)}(0)$ as a function,

$$
\begin{aligned}
k_{(\lambda)} : C_b(\mathbb{R}) &\longrightarrow \mathbb{C} \\
f &\longmapsto f * k_{(\lambda)}(0) = \int k_{(\lambda)}(t) f(t) \, dt,
\end{aligned}
\qquad (2.1.3)
$$

whose domain $C_b(\mathbb{R})$ is the set of bounded, continuous functions on $\mathbb{R}$.

2.1.3 Definition. δ

a. δ is the function

$$
\begin{aligned}
\delta : C_b(\mathbb{R}) &\longrightarrow \mathbb{C} \\
f &\longmapsto f(0),
\end{aligned}
\qquad (2.1.4)
$$

i.e., for each $f \in C_b(\mathbb{R})$, $\delta(f) \equiv f(0)$. A function (such as (2.1.4)), whose domain is a set of functions and whose range is a set of numbers, is called a *functional*, cf., Definition 2.2.3c.

δ is often called the *Dirac δ-function* in spite of the fact that it was neither discovered by Dirac nor is it an ordinary function on $\mathbb{R}$. It is also called the *unit impulse*. We shall refer to δ as the *Dirac measure*. The concept of a *measure* generalizes that of a locally integrable function, i.e., an element of $L^1_{loc}(\mathbb{R})$, and is a special type of distribution. We shall see that δ is a measure (Definition 2.3.6c).

b.   We sometimes write $\delta = \delta_0$ since the definition of δ in (2.1.4) specifies values of functions f (in the domain of δ) at 0. Similarly, we define the *Dirac measure δ_r at* $r \in \mathbb{R}$ as the function, with domain $C_b(\mathbb{R})$ (*not* $\mathbb{R}$),

$$
\begin{aligned}
\delta_r : C_b(\mathbb{R}) &\longrightarrow \mathbb{C} \\
f &\longmapsto f(r),
\end{aligned}
$$

i.e., for each $f \in C_b(\mathbb{R})$, $\delta_r(f) \equiv f(r)$. We also write $\tau_r \delta$ in place of δ_r for reasons that will become apparent in Definition 2.5.3.

c.   Recall our introduction of δ in Remark 1.1.4 and the "formula" (δ) there. The point of that formula was that if δ were an ordinary function, then it would be 0 everywhere except at the origin, where it would be so large that "$\int \delta(u) \, du = 1$". This is of course nonsense because of the definition of the integral. It is not nonsense from the point of view and needs of engineers and physicists (such as Dirac!); and the definition (2.1.4), motivated by the theorems expressed in (2.1.1) and (2.1.2), is a mathematically sound way to legitimize the ingenuity of these scientists in their formulas such as (δ).

d.   To be consistent, (2.1.3) also tells us that any element $g \in L^1_{loc}(\mathbb{R})$ can be thought of as a function whose domain $C_c(\mathbb{R})$ is the vector space of all compactly supported, continuous functions on $\mathbb{R}$. In fact, in this case, we can formulate the notation $g(f)$ to mean

$$\forall f \in C_c(\mathbb{R}), \qquad g(f) = \int g(t) f(t)\, dt, \qquad\qquad (2.1.5)$$

cf., Section 2.2. The domain $C_c(\mathbb{R})$ is used instead of $C_b(\mathbb{R})$ so that the integral on the right side of (2.1.5) is well defined for $g \in L^1_{\text{loc}}(\mathbb{R})$.

   e.   This approach of defining objects such as δ, which beg to exist but do not exist as ordinary functions, as functions (functionals) whose domain is a space of functions, is rooted in ideas associated with Parseval's formula and weak solutions in physics. These generalized objects are called *distributions* or *generalized functions*; and in this chapter we shall develop their elementary properties, especially those used in applications.

   The following result is expected in light of Remark 2.1.2; its proof is part of Exercise 2.38.

2.1.4 Proposition.

Let $\{k_{(\lambda)}\} \subseteq L^1(\mathbb{R})$ be an approximate identity. Then

$$\forall f \in C_b(\mathbb{R}), \qquad \lim_{\lambda \to \infty} \int k_{(\lambda)}(t) f(t)\, dt = \delta(f).$$

2.1.5 Remark. $\delta(t)$ and Composition

In spite of the fact that the domain of δ is a space of functions, we shall sometimes write $\delta(t)$ instead of δ; and in this case we replace the notation $\delta(f)$ (after (2.1.4)) by the seemingly more cumbersome notation

$$\delta(t)(f(t)).$$

The reasons for doing this are that δ is intuitively perceived as an ordinary function by the (technically false) description of it in Definition 2.1.3c, and that δ is also approximated by ordinary functions as in Proposition 2.1.4.

   The theoretical rationale in the previous paragraph would be an effete exercise if it were not for the importance of *composition*, for example, in mathematics, e.g., [AZ90], in neural nets, e.g., [Hay94], and for compression problems in signal processing. We shall illustrate a role of composition for data compression in Example 2.3.10, but for now are interested in giving an intelligent meaning to the compelling notation "$\delta \circ g$", noting that if $g(t) = t$, then we are really dealing with δ itself, cf., [AMS73], [Jon82].

2.1.6 Example. $\delta(at + b)$

We give a reasonable meaning to "$\delta(at + b)$", where $a \neq 0$; and in the process we show that

$$\delta(at + b) = \frac{1}{|a|}\delta\left(t + \frac{b}{a}\right) = \frac{1}{|a|}\tau_{-b/a}\delta(t), \qquad (2.1.6)$$

cf., Exercises 2.10 and 2.21. Let $\{k_{(\lambda)}\}$ be an approximate identity. Then, by (2.1.5) and Proposition 2.1.4, it is reasonable to *define* $\delta(at + b)$ by the limit

$$\forall f \in C_b(\mathbb{R}), \qquad \lim_{\lambda \to \infty} \int k_{(\lambda)}(at + b) f(t)\, dt = \delta(at + b)(f(t)).$$

The integral on the left side is

$$\frac{1}{|a|}\int_{-\infty}^{\infty} k_{(\lambda)}(u) f\left(\frac{u - b}{a}\right) du,$$

and this converges to $(1/|a|)\, f(-b/a)$. (2.1.6) follows by definition of $\delta_{-b/a}$.

2.2. Definition of Distributions

The purpose of the theory of distributions is to provide a unified setting and *calculus* for many of the objects arising in analysis. These objects include the customary functions, viz., the elements of the space $L^1_{\text{loc}}(\mathbb{R})$ of locally integrable functions. They also include impulses (Dirac measures), dipoles, and other notions from the sciences, whose role and mathematical identity could not be assimilated by the seventeenth century calculus, or the spectacular eighteenth and nineteenth century developments of this calculus, and the nineteenth and early twentieth century theory of real analysis.

A key feature of the theory of distributions is that all of these objects (distributions) can be differentiated in a natural way inspired by the integration by parts formula, e.g., Section 2.3 and Theorem A.22. Further, some of the important results from real analysis allowing for switching of operations, such as summation and differentiation, are true without hypotheses in the case of distributions, e.g., Exercise 2.46, cf., Example 2.2.2 *b*.

2.2.1 Definition. $C_c^{\infty}(\mathbb{R})$

a. $C^{\infty}(\mathbb{R})$ denotes the space of infinitely differentiable complex-valued functions on $\mathbb{R}$.

If $f : \mathbb{R} \longrightarrow \mathbb{C}$ is any function on $\mathbb{R}$, then the *support* of f, denoted by supp f, is the smallest closed set outside of which f vanishes, e.g., Example 1.2.2 *b*.
Let

$$C_c^\infty(\mathbb{R}) = \{f : f \in C^\infty(\mathbb{R}) \text{ and } \operatorname{supp} f \text{ is compact}\}.$$

(*Compact sets* of $\mathbb{R}$ are precisely the closed and bounded sets of $\mathbb{R}$, cf., Definition B.1. A set $B \subseteq \mathbb{R}$ is *bounded* if there is $T > 0$ such that $B \subseteq [-T, T]$. $C \subseteq \mathbb{R}$ is *closed* if its complement $\mathbb{R}\backslash C$ is open; and $U \subseteq \mathbb{R}$ is *open* if U is the countable union of disjoint (open) intervals (a, b). The structure of open sets in $\mathbb{R}^2$ is more complicated, e.g., [Ben76, pages 15–16]. Clearly, closed intervals $[-T, T]$ are compact.)

b. Just as $C_b(\mathbb{R})$ allowed us to define the object δ, the even smaller space $C_c^\infty(\mathbb{R})$ will allow us to define more unusual distributions T than δ. T exists as an element of a large space of distributions (which we shall soon define), but can generally only be realized or evaluated by operating on a "test function" $f \in C_c^\infty(\mathbb{R})$.

2.2.2 Example. $C_c^\infty(\mathbb{R})$

$C_c^\infty(\mathbb{R})$ is not the trivial set $\{0\}$. This statement is not so absurd since there are no analytic functions in $C_c^\infty(\mathbb{R})$. Why? (Exercise 2.5). In fact, $C_c^\infty(\mathbb{R})$ is an infinite-dimensional space, so let us write down at least one element.

a. Let

$$\phi(t) = \begin{cases} e^{1/t}, & \text{if } t < 0, \\ 0, & \text{if } t \geq 0, \end{cases}$$

and define $f(t) = c\phi(|t|^2 - 1)$. Clearly, $f(t) = 0$ if $|t| \geq 1$ so that supp $f \subseteq [-1, 1]$. The constant c is chosen so that $\int f(t)\, dt = 1$, and it is a straightforward calculation to show that f is infinitely differentiable, e.g., Exercise 2.5.

Of course, the set $\{f_\lambda\}$ of dilations is an approximate identity since $\int f(t)\, dt = 1$.

b. To generate other examples of elements in $C_c^\infty(\mathbb{R})$, take any $g \in L^1(\mathbb{R})$ having compact support. Then $f * g \in C_c^\infty(\mathbb{R})$. To see this we must verify that supp $f * g$ is compact and that $f * g$ is infinitely differentiable. The first fact is routine to prove, and the second involves checking conditions to switch the operations of differentiation and integration, e.g., [Apo57] for the conditions and [Fey86, pages 72 and 93] for motivation.

c. We shall have occasion to need elements of $C_c^\infty(\mathbb{R})$ with special properties such as vanishing moments. In this regard let $N \in \mathbb{N}$. We shall construct real-valued, even functions $k \in C_c^\infty(\mathbb{R})$ such that

$$\forall 0 \leq n \leq N, \qquad \int t^n k(t)\, dt = 0 \qquad (2.2.1)$$

and

$$\forall \gamma \in \hat{\mathbb{R}} \backslash \{0\}, \qquad \int_0^\infty \hat{k}(\lambda \gamma)^2 \frac{d\lambda}{\lambda} = 1. \tag{2.2.2}$$

Let $h = f^{(2j)}$ for some $j > \frac{N}{2}$, where $f \in C_c^\infty(\mathbb{R})$ is real and even. Clearly, $h \in C_c^\infty(\mathbb{R})$, h is real-valued, and supp h is compact. By definition of the derivative, one computes that h is even. Noting that $n \leq N < 2j$, we compute

$$\int t^n f^{(2j)}(t)\, dt = -n \int t^{n-1} f^{(2j-1)}(t)\, dt$$

$$= (-1)^2 n(n-1) \int t^{n-2} f^{(2j-2)}(t)\, dt$$

$$= (-1)^n n! \int f^{(2j-n)}(t)\, dt = 0$$

by integration by parts, using the facts that $2j - n > 0$ and that $f \in C_c^\infty(\mathbb{R})$. This is (2.2.1) for $k = h$.

Since $h \in C_c^\infty(\mathbb{R})$, then $\hat{h}(\lambda \gamma)$ is rapidly decreasing as $\lambda \longrightarrow \infty$ for any fixed $\gamma \in \hat{\mathbb{R}} \backslash \{0\}$, e.g., Exercise 1.13. Thus, $\int_1^\infty (\hat{h}(\lambda \gamma)^2/\lambda) d\lambda$ is an absolutely convergent integral. By (2.2.1), $\int h(t)\, dt = 0$ and so $\hat{h}(0) = 0$. Clearly, $\hat{h} \in C^\infty(\mathbb{R})$, and so, for a fixed $\gamma \in \hat{\mathbb{R}} \backslash \{0\}$, $\hat{h}(\lambda \gamma)^2/\lambda$ is bounded in a neighborhood of 0 since $\hat{h}(0) = 0$ and by the definition of derivative. Thus, $\int_0^1 (\hat{h}(\lambda \gamma)^2/\lambda)\, d\lambda$ is an absolutely convergent integral. We compute

$$\forall \gamma \in \hat{\mathbb{R}} \backslash \{0\}, \qquad \int_0^\infty \hat{h}(\lambda \gamma)^2 \frac{d\lambda}{\lambda} = \int_0^\infty \hat{h}(\lambda)^2 \frac{d\lambda}{\lambda} = c^2, \tag{2.2.3}$$

using the fact that $\hat{h}$ is even. Note that $c \neq 0$ since the integrands of (2.2.3) are positive. (2.2.2) is therefore obtained by setting $k = c^{-1}h$.

2.2.3 Definition. Distributions

a. $C_c^\infty(\mathbb{R})$ is a vector space. A linear function,

$$
\begin{aligned}
T : C_c^\infty(\mathbb{R}) \;&\longrightarrow\; \mathbb{C} \\
f \;&\longmapsto\; T(f),
\end{aligned}
$$

is a *distribution* or *generalized function* if $\lim_{n\to\infty} T(f_n) = 0$ for every sequence $\{f_n\} \subseteq C_c^\infty(\mathbb{R})$ satisfying the properties:

i. $\exists K \subseteq \mathbb{R}$, compact, such that $\forall n$, supp $f_n \subseteq K$;

ii. $\forall k \geq 0$, $\lim_{n \to \infty} \|f_n^{(k)}\|_{L^{\infty}(\mathbb{R})} = 0$.

b. A distribution T is *positive*, written $T \geq 0$, if $T(f) \geq 0$ for all nonnegative functions $f \in C_c^{\infty}(\mathbb{R})$.

c. The space of all distributions on $\mathbb{R}$ is denoted by $D'(\mathbb{R})$. We incorporate the prime "'" in this notation to continue established notation and to emphasize the fact that $D'(\mathbb{R})$ is the dual space of $C_c^{\infty}(\mathbb{R})$, i.e., the space of all continuous linear functionals on $C_c^{\infty}(\mathbb{R})$.

We use the word "functional" just as we use the word "function"; the only reason we make any distinction is because of our new situation dealing with a space of functions as domain and $\mathbb{C}$ as range, e.g., (2.1.4) and (2.1.5). We denote the operation of T on f by $T(f)$. The *linearity* of T means that

$$T(c_1 f_1 + c_2 f_2) = c_1 T(f_1) + c_2 T(f_2)$$

for all $c_1, c_2 \in \mathbb{C}$ and $f_1, f_2 \in C_c^{\infty}(\mathbb{R})$.

d. $D'(\mathbb{R})$ is a vector space. In fact, if $T_1, T_2 \in D'(\mathbb{R})$ and $c_1, c_2 \in \mathbb{C}$, then $c_1 T_1 + c_2 T_2$ is a well-defined distribution, defined by the rule

$$\forall f \in C_c^{\infty}(\mathbb{R}), \qquad (c_1 T_1 + c_2 T_2)(f) = c_1 T_1(f) + c_2 T_2(f).$$

e. Occasionally we shall have to keep track of the underlying variable $t \in \mathbb{R}$ in dealing with the notation $T(f)$. In such cases, as in Remark 2.1.5 and Example 2.1.6, we shall denote $T(f)$ by

$$T(t)(f(t)).$$

2.2.4 Remark. Historical and Bibliographical Note

Laurent Schwartz received the Fields medal in 1950 for developing the theory of distributions. His classic book is *Théorie des distributions* [Sch66], [Sch61]. The first edition was published in two volumes in 1950 and 1951. These volumes are a compendium of diverse past accomplishments, a unification of technologies, an original formulation of ideas both new and old, and a research manual leading to new mathematics and applications.

Two other monumental contributions are Gelfand and Shilov's *Generalized Functions*, in five volumes, and Hörmander's three volumes, *The Analysis of Linear Partial Differential Equations* [Hör83].

The origins of distribution theory are based in the operational calculus from engineering, e.g., Section 2.6, and the concepts of "turbulent" and "weak" solutions of partial differential equations from physics. Schwartz' Introduction and Gelfand and Shilov's bibliographic notes give a nice overview.

A great number of books has been written on the theory of distributions, running the gamut from pure topological vector space presentations [Hor66] to applications in optics and supersonic wing theory [deJ64]. We hesitate listing excellent books that we know, since our omissions would surely include comparably excellent ones.

2.2.5 Definition. Equality of Distributions

a. Let $T_1, T_2 \in D'(\mathbb{R})$. T_1 *equals* T_2, i.e., T_1 is the same distribution as T_2, if

$$\forall f \in C_c^\infty(\mathbb{R}), \qquad T_1(f) = T_2(f).$$

Notationally, in this case, we write $T_1 = T_2$. In particular, $T = 0$ if $T(f) = 0$ for all $f \in C_c^\infty(\mathbb{R})$.

b. This notion of equality can be explained mathematically in functional analytic terms, e.g., Definition B.6. Intuitively, however, the idea is clear: $T(f) = 0$ for all f in the domain $C_c^\infty(\mathbb{R})$ of T implies T is the 0-distribution just as $g(t) = 0$ for all $t \in \mathbb{R}$ implies g is the 0-function.

Another compelling reason to accept this notion of equality is that if $g \in L^1_{\text{loc}}(\mathbb{R})$ and $g(f) = 0$ for all $f \in C_c^\infty(\mathbb{R})$, then g is the 0-function, e.g., Exercise 2.8.

2.2.6 Example. Locally Integrable Distributions

a. Let $g \in L^1_{\text{loc}}(\mathbb{R})$ and define the functional

$$
\begin{aligned}
T_g : C_c^\infty(\mathbb{R}) &\longrightarrow \quad \mathbb{C} \\
f &\longmapsto \quad T_g(f),
\end{aligned}
$$

where T_g is defined as

$$\forall f \in C_c^\infty(\mathbb{R}), \qquad T_g(f) = \int g(t) f(t)\, dt, \qquad (2.2.4)$$

cf., (2.1.5) where we wrote $g(f)$ instead of $T_g(f)$. It is easy to check that $T_g \in D'(\mathbb{R})$.

b. Not only does T_g define an element of $D'(\mathbb{R})$, but the linear mapping

$$
\begin{aligned}
L : L^1_{\text{loc}}(\mathbb{R}) &\longrightarrow \quad D'(\mathbb{R}) \\
g &\longmapsto \quad L(g) = T_g
\end{aligned}
\qquad (2.2.5)
$$

allows us to identify $L^1_{\text{loc}}(\mathbb{R})$ with a subset of $D'(\mathbb{R})$.

To see this, we must show that the mapping L is injective, i.e., a one-to-one function. This means we must prove that if $T_{g_1} = T_{g_2}$ in $D'(\mathbb{R})$, then $g_1 = g_2$ a.e., or, equivalently (by the linearity), if $g(= g_1 - g_2)$ is not the 0-function in $L^1_{\text{loc}}(\mathbb{R})$, then there is $f \in C_c^\infty(\mathbb{R})$ for which $\int g(t) f(t)\, dt \neq 0$. As pointed out in Definition 2.2.5*b*, this fact is a consequence of Exercise 2.8.

Since the mapping is injective, there is no ambiguity in identifying $g \in L^1_{loc}(\mathbb{R})$ with $L(g) = T_g \in D'(\mathbb{R})$.

   c.   Examples of such distributions T_g arise from $g = H$ and $g(t) = 1/|t|^{1/2}$. On the other hand, $g(t) = 1/t \notin L^1_{loc}(\mathbb{R})$, cf., Example 2.3.8c.

   d.   The domain of the Dirac measure defined in (2.1.4) can be restricted to $C^\infty_c(\mathbb{R})$, in which case we have

$$
\begin{aligned}
\delta : C^\infty_c(\mathbb{R}) &\longrightarrow \mathbb{C} \\
f &\longmapsto \delta(f) = f(0).
\end{aligned}
\tag{2.2.6}
$$

It is easy to check that δ defined by (2.2.6) is a distribution. We hasten to point out that the Dirac measure is not an element of $L^1_{loc}(\mathbb{R})$, i.e., δ is not in the range of the function L defined in (2.2.5), e.g., Exercise 2.39.

2.2.7 Definition. Support

   a.   We have defined the 0-distribution. We now "localize" this definition in the following way. $T \in D'(\mathbb{R})$ is *zero on the open set* $U \subseteq \mathbb{R}$, written $T = 0$ on U, if

$$
\forall f \in C^\infty_c(\mathbb{R}), \quad \text{such that } \operatorname{supp} f \subseteq U, \qquad T(f) = 0.
$$

The *support* of T, denoted by $\operatorname{supp} T$, is the smallest closed set $C = \mathbb{R} \backslash U$ outside of which T is 0.

   b.   The definition in part *a* is usually easier to check than you might imagine. In particular, we have

$$
\forall r \in \mathbb{R}, \quad \operatorname{supp} \delta_r = \{r\}
\tag{2.2.7}
$$

and

$$
\operatorname{supp} T_g = \operatorname{supp} g,
\tag{2.2.8}
$$

where $g \in L^1_{loc}(\mathbb{R})$ and where the right side of (2.2.8) was defined in Definition 2.2.1, e.g., Exercise 2.25, cf., Exercise 2.33.

2.3. Differentiation of Distributions

2.3.1 Definition. Distributional Differentiation

   a.   The duality between the small space $C^\infty_c(\mathbb{R})$ and the large space $D'(\mathbb{R})$, allowing us to define so many objects T in $D'(\mathbb{R})$, can be coupled with the integration by parts formula to provide a definition of "T'", the distributional derivative of T, in part *b* below.

Let $C^1(\mathbb{R})$ be the space of continuously differentiable functions on $\mathbb{R}$. If $f \in C_c^\infty(\mathbb{R})$ and g is sufficiently smooth, e.g., if $g \in C^1(\mathbb{R})$ or even if g is only an element of $AC_{\mathrm{loc}}(\mathbb{R})$, then

$$\int g'(t) f(t)\, dt = -\int g(t) f'(t)\, dt. \tag{2.3.1}$$

The integration by parts formula in (2.3.1) is the distributional "duality formula"

$$\forall f \in C_c^\infty(\mathbb{R}), \qquad g'(f) = -g(f'). \tag{2.3.2}$$

Since the right side of (2.3.2) is well defined when g is replaced by any $T \in D'(\mathbb{R})$, we are motivated to make the following definition.

b. The *distributional derivative* T' of $T \in D'(\mathbb{R})$ is defined by the formula

$$\forall f \in C_c^\infty(\mathbb{R}), \qquad T'(f) = -T(f'). \tag{2.3.3}$$

c. To establish the viability of (2.3.3) as an effective definition of the notion of derivative we must prove the following:

$$T' \in D'(\mathbb{R}), \tag{2.3.4}$$

and

$$\forall g \in AC_{\mathrm{loc}}(\mathbb{R}), \qquad T_g' = T_{Dg}, \tag{2.3.5}$$

where Dg denotes the ordinary pointwise derivative. The verification of (2.3.4) and (2.3.5) is routine, e.g., Exercise 2.7. In concert with (2.3.3) we write $T_g' = g'$.

d. For each $T \in D'(\mathbb{R})$ we define $T^{(n)}$, the *nth distributional derivative* of T, as the distribution defined by the formula,

$$\forall f \in C_c^\infty(\mathbb{R}), \qquad T^{(n)}(f) = (-1)^n T(f^{(n)}). \tag{2.3.6}$$

2.3.2 Remark. Definition by Duality

a. The use of "duality formulas" such as (2.3.2) to define an analytic operation (such as differentiation) on an arbitrary distribution T is a critical aspect of the theory. The idea is to define an operation on T in terms of the same (or similar) operation on $C_c^\infty(\mathbb{R})$, where it makes perfectly good sense; the vehicle for effecting this definition is a "duality formula". As we shall soon see, Parseval's formula is the "duality formula" that allows us to define the Fourier transform of distributions.

b. In general, we cannot extend (2.3.5) to arbitrary elements $g \in L_{\mathrm{loc}}^1(\mathbb{R})$ even though T_g is a well-defined generalized function. In fact, if g is infinitely differentiable on $\mathbb{R}\backslash\{t_0\}$ in the ordinary pointwise sense and Dg is the ordinary pointwise derivative of g defined everywhere except at t_0, then, in general, $T_g{}'$ and T_{Dg} are distributions but $T_g{}' \neq T_{Dg}$, e.g., Example 2.3.3.

2.3.3 Example. $H' = \delta$

Let H be the Heaviside function.

   a.   The ordinary pointwise derivative DH exists and takes the value 0 on $\mathbb{R}\backslash\{0\}$. Thus, $DH \in L^1_{\text{loc}}(\mathbb{R})$ and DH is the 0 distribution.

   b.   The distributional derivative H' is evaluated as follows. Choose $f \in C_c^\infty(\mathbb{R})$ and compute

$$H'(f) = -H(f') = -\int_0^\infty f'(t)\,dt = f(0) = \delta(f). \qquad (2.3.7)$$

Since $H'(f) = \delta(f)$ for all $f \in C_c^\infty(\mathbb{R})$, we can conclude that H' and δ are the same distribution, i.e.,

$$H' = \delta,$$

where H' is the distributional derivative of H.

2.3.4 Remark. Notation for Differentiation

Let $g : \mathbb{R} \longrightarrow \mathbb{C}$ be a function, and suppose the ordinary pointwise derivative of g exists at t. As indicated above, we denote this value by $Dg(t)$; and if $Dg(t)$ exists a.e., we denote the resulting function by Dg.

   On the other hand, if g defines a distribution T_g, then g' denotes the distributional derivative $T_g{}'$.

   In general, the distinction will be clear; and when no confusion arises we shall be less than compulsive about the "D" notation.

2.3.5 Proposition.

Let $g \in C^1(\mathbb{R}\backslash\{0\})$, i.e., g is continuously differentiable on $\mathbb{R}\backslash\{0\}$. Assume that g has a jump discontinuity at the origin with jump $\sigma_0 = g(0+) - g(0-)$. Then we have the distributional equation

$$g' = Dg + \sigma_0\delta,$$

i.e.,

$$\forall f \in C_c^\infty(\mathbb{R}), \qquad g'(f) = \int (Dg)(t)f(t)\,dt + \sigma_0 f(0),$$

where g corresponds to $T_g \in D'(\mathbb{R})$, Dg corresponds to T_{Dg}, and $g' = T_g{}'$.

PROOF.   For $f \in C_c^\infty(\mathbb{R})$ we compute

$$g'(f) = -g(f') = -\int g(t) f'(t)\, dt$$

$$= -\int_{-\infty}^{0} g(t) f'(t)\, dt - \int_{0}^{\infty} g(t) f'(t)\, dt$$

$$= -\left[g(t) f(t) \, \Big|_{-\infty}^{0} - \int_{-\infty}^{0} (Dg)(t) f(t)\, dt \right]$$

$$\quad - \left[g(t) f(t) \, \Big|_{0}^{\infty} - \int_{0}^{\infty} (Dg)(t) f(t)\, dt \right]$$

$$= \sigma_0 f(0) + \int (Dg)(t) f(t)\, dt. \quad \blacksquare$$

We cannot conclude from Proposition 2.3.5 that $Dg \in L_{\text{loc}}^1(\mathbb{R})$. For example, let

$$g(t) = \begin{cases} t \exp(i e^{-1/t^2}), & \text{if } t > 0, \\ 0, & \text{if } t < 0. \end{cases}$$

2.3.6 Definition. Radon Measures

a.   The space $M(\mathbb{R}) \subseteq D'(\mathbb{R})$ of *Radon measures* is defined as

$$M(\mathbb{R}) = \{ F' \in D'(\mathbb{R}) : F \in BV_{\text{loc}}(\mathbb{R}) \}, \tag{2.3.8}$$

i.e., $T \in M(\mathbb{R})$ if there is $F \in BV_{\text{loc}}(\mathbb{R})$ for which $T = F'$, the first distributional derivative of F. The space $M_b(\mathbb{R}) \subseteq M(\mathbb{R})$ of *bounded Radon measures* is defined as

$$M_b(\mathbb{R}) = \{ F' \in D'(\mathbb{R}) : F \in BV(\mathbb{R}) \}. \tag{2.3.9}$$

(2.3.8) and (2.3.9) are well defined since $BV_{\text{loc}}(\mathbb{R}) \subseteq L_{\text{loc}}^1(\mathbb{R})$, e.g., Exercise 2.23.

The elements of $M(\mathbb{R})$, resp., $M_b(\mathbb{R})$, are often referred to as *measures*, resp., *bounded measures*. It is also often the case that measures are denoted by Greek letters such as μ and ν. Further, if $\mu \in M(\mathbb{R})$ then, notationally, we write

$$\forall f \in C_c^\infty(\mathbb{R}), \qquad \mu(f) = \int f(t)\, d\mu(t). \tag{2.3.10}$$

For a given measure μ the domain of functions f for which $\mu(f)$ can be defined is much larger than $C_c^\infty(\mathbb{R})$.

b. A measure, resp., bounded measure, μ is *positive* if $\mu(f) \geq 0$ for all nonnegative functions $f \in C_c^\infty(\mathbb{R})$. The space of positive measures, resp., bounded positive measures, is denoted by $M_+(\mathbb{R})$, resp., $M_{b+}(\mathbb{R})$. It turns out that $\mu \in M(\mathbb{R})$ is positive if and only if each of the corresponding functions F (for which $F' = \mu$) is increasing on $\mathbb{R}$, e.g., Section 2.7.

c. Since the Heaviside function H is an element of $BV(\mathbb{R})$, we see that $\delta = H' \in M_b(\mathbb{R})$. *Lebesgue measure* $\mu \in M(\mathbb{R}) \backslash M_b(\mathbb{R})$ is $\mu = F'$, where $F \in BV_{\text{loc}}(\mathbb{R})$ is defined as $F(t) = t$, i.e., Lebesgue measure can be identified with the distribution which is the constant 1, cf., Definition A.4.

Further, note that $L^1(\mathbb{R}) \subseteq M_b(\mathbb{R})$ and $L_{\text{loc}}^1(\mathbb{R}) \subseteq M(\mathbb{R})$, e.g., Exercise 2.39. The *norm* of $\mu \in M_b(\mathbb{R})$ is defined as

$$\|\mu\|_1 = \sup\{|\mu(f)| : f \in C_0(\mathbb{R}) \text{ and } \|f\|_{L^\infty(\mathbb{R})} \leq 1\}.$$

If $g \in L^1(\mathbb{R})$, then $\|g\|_{L^1(\mathbb{R})} = \|g\|_1$.

d. The definition of "measure" in part *a* is equivalent to that from real analysis. The underlying idea establishing this equivalence is the Riesz Representation Theorem, e.g., Section 2.7, cf., [Ben76, Appendix III], [Bou65, Chapter III.1], [Sch66, Chapitre II.4, Théorème II].

2.3.7 Remark. Multipoles and $\delta^{(n)}$

The first distributional derivative of δ, viz., δ', is not an element of $M(\mathbb{R})$ since $\delta \notin BV_{\text{loc}}(\mathbb{R})$. δ' is the *dipole* at the origin, and the distributions $\delta^{(n)}, n \geq 1$, are the *multipoles* that arise in several important applications. For example, in fluid mechanics, a dipole is the limiting case in fluid flow of a source and a sink of equal strength approaching each other under the constraint that the product of the distance between them and their strength is constant.

To quantify the fact that multipoles arise in applications, let us consider the case of electromagnetism and the potential due to point charges. The laws or equations of electromagnetism can be developed and formulated beginning with the notions of *length, mass, time*, and the electrical quantity of *charge*. Coulomb's law asserts that the force F_{jk} between two charges q_j, q_k at points $\mathbf{v}_j, \mathbf{v}_k \in \mathbb{R}^3$ is

$$F_{jk} = c \frac{q_j q_k}{|\mathbf{v}_j - \mathbf{v}_k|^2} \mathbf{u}_{j,k},$$

where c is the permittivity constant and $\mathbf{u}_{j,k} \in \mathbb{R}^3$ is the unit vector directed from $\mathbf{v}_j$ to $\mathbf{v}_k$. If there are charges $q_1, \ldots, q_n$ at $\mathbf{v}_1, \ldots, \mathbf{v}_n \in \mathbb{R}^3$, then the electric field $F : \mathbb{R}^3 \longrightarrow \mathbb{R}^3$ due to these charges can be formulated in terms of Coulomb's law and linear superposition. For conservative fields there is a potential function $V : \mathbb{R}^3 \longrightarrow \mathbb{R}$ whose gradient is F. In the case the charges are close to the origin and $r = |\mathbf{x}|$ is much larger than any $|\mathbf{v}_j|$, then $V(\mathbf{x})$ is of the form

$$V(\mathbf{x}) = c \left\{ \frac{1}{r} \sum_{j=1}^{n} q_j + \frac{1}{r^2} \left[\frac{x_1}{r} \sum_{j=1}^{n} q_j v_{j1} + \frac{x_2}{r} \sum_{j=1}^{n} q_j v_{j2} + \frac{x_3}{r} \sum_{j=1}^{n} q_j v_{j3} \right] \right.$$
$$\left. + \frac{1}{r^3} [\cdots] + \cdots \right\},$$
(2.3.11)

where $\mathbf{x} = (x_1, x_2, x_3)$, $\mathbf{v}_j = (v_{j1}, v_{j2}, v_{j3}) \in \mathbb{R}^3$, e.g., [Con58, Chapter 8]. The first sum, $\sum q_j$, is the total charge or *monopole moment* of the charge distribution. The next three sums of (2.3.11) are the components of the *dipole moment* of the charge distribution, and we have omitted writing the quadrupole moment, octupole moment, and so on.

Now consider the particular example of two charges on the x-axis, viz., q at $\epsilon/2$ and $-q$ at $-\epsilon/2$. The monopole moment vanishes, and the dipole moment is $q\epsilon$. If $q = q(\epsilon) = -1/\epsilon$, then it is natural to formulate the notion of the *dipole of moment* 1 *at the origin* to be

$$- \lim_{\epsilon \to 0} \frac{1}{\epsilon} \left(\delta_{\epsilon/2} - \delta_{-\epsilon/2} \right),$$

which, in turn, is δ'. In fact,

$$- \lim_{\epsilon \to 0} \frac{1}{\epsilon} \left(\delta_{\epsilon/2} - \delta_{-\epsilon/2} \right)(f) = - \lim_{\epsilon \to 0} \frac{f(\frac{\epsilon}{2}) - f(-\frac{\epsilon}{2})}{\epsilon}$$
$$= -f'(0) = -\delta(f') = \delta'(f).$$

2.3.8 Example. The Principal Value Distribution

a. We define the functional T by the formula

$$\forall f \in C_c^\infty(\mathbb{R}), \qquad T(f) = \lim_{\epsilon \to 0} \int_{|t| \geq \epsilon} \frac{f(t)}{t} \, dt.$$

We shall verify that T is well defined (part b), observe that $T \in D'(\mathbb{R})$ (Exercise 2.43), and show that $T \notin M(\mathbb{R})$. T is the *principal value distribution* and is denoted by

$$pv \left(\frac{1}{t} \right).$$

b. To show that T is well defined, note that

$$\int_{|t| \geq \epsilon} \frac{f(t)}{t} \, dt = \int_{R \geq |t| \geq \epsilon} \frac{f(t) - f(0)}{t - 0} \, dt + f(0) \int_{R \geq |t| \geq \epsilon} \frac{dt}{t}$$
$$= \int_{R \geq |t| \geq \epsilon} \frac{f(t) - f(0)}{t - 0} \, dt$$

for $f \in C_c^\infty(\mathbb{R})$ and some $R > 0$, depending on f. If

$$h_\epsilon(t) = \frac{f(t) - f(0)}{t} \mathbf{1}_{[-R,-\epsilon] \cup [\epsilon, R]}(t),$$

then $|h_\epsilon(t)| \leq \|f'\|_{L^\infty(\mathbb{R})}$ by the mean value theorem, the constant function $g \equiv \|f'\|_{L^\infty(\mathbb{R})}$ belongs to $L^1[-R, R]$, and, for each $t \neq 0$,

$$\lim_{\epsilon \to 0} h_\epsilon(t) = \frac{f(t) - f(0)}{t} \mathbf{1}_{[-R,R]}(t).$$

LDC applies, and, thus, $T(f)$ is well defined.

    c.   Let $g(t) = \log|t|$, $t \in \mathbb{R} \setminus \{0\}$. Clearly, $g \in L^1_{\text{loc}}(\mathbb{R})$. In fact, if $a \geq 1$, then we have

$$\int_0^a |\log|t|| \, dt = -\int_0^1 \log t \, dt + \int_1^a \log t \, dt \qquad (2.3.12)$$

and $D(t \log t - t) = \log t$, so that the right side of (2.3.12) is finite. Obvious adaptations of this calculation yield the local integrability of g. Thus, $g \in D'(\mathbb{R})$ and we compute g'. For any $f \in C_c^\infty(\mathbb{R})$ we have

$$g'(f) = -g(f') = -\int \log|t| \, f'(t) \, dt$$

$$(2.3.13)$$

$$= -\int_{-\infty}^0 \log|t| \, f'(t) \, dt - \int_0^\infty \log|t| \, f'(t) \, dt.$$

The second term on the right side of (2.3.13) is

$$-\int_0^\infty (\log t) f'(t) \, dt = -\lim_{\epsilon \to 0} \int_\epsilon^\infty (\log t) f'(t) \, dt$$

$$= -\lim_{\epsilon \to 0} \left[(\log t) f(t) \Big|_\epsilon^\infty - \int_\epsilon^\infty \frac{1}{t} f(t) \, dt \right] \qquad (2.3.14)$$

$$= \lim_{\epsilon \to 0} \left[f(\epsilon) \log \epsilon + \int_\epsilon^\infty \frac{1}{t} f(t) \, dt \right].$$

Note that $\log \epsilon \, (f(\epsilon) - f(0)) \longrightarrow 0$ as $\epsilon \longrightarrow 0$ since

$$\log \epsilon \, (f(\epsilon) - f(0)) = \epsilon (\log \epsilon) \frac{f(\epsilon) - f(0)}{\epsilon} \longrightarrow 0 \cdot f'(0) = 0, \qquad \epsilon \longrightarrow 0.$$

Thus, replacing $f(\epsilon) \log \epsilon$ by

$$\log \epsilon \, (f(\epsilon) - f(0)) + f(0) \log \epsilon$$

in (2.3.14), we obtain

$$-\int_0^\infty (\log t) \, f'(t) \, dt = \lim_{\epsilon \to 0} \left[f(0) \log \epsilon + \int_\epsilon^\infty \frac{f(t)}{t} \, dt \right]_+,$$

where we have used the fact that $\log t \in L^1_{\mathrm{loc}}(0, \infty)$ (a primitive of $\log t$ is $t \log t - t$, and therefore $\int_0^a \log t \, dt = a \log a - a$) and properties of sums of limits. $[\cdots]_+$ designates the fact that we are integrating over $(0, \infty)$ on the left side.

Similarly, for the first term on the right side of (2.3.13), we compute

$$-\int_{-\infty}^0 \log|t| \, f'(t) \, dt = \lim_{\epsilon \to 0} \left[-f(0) \log \epsilon + \int_{-\infty}^{-\epsilon} \frac{f(t)}{t} \, dt \right]_-,$$

where $[\cdots]_-$ designates that we are integrating over $(-\infty, 0)$ on the left side.

Since $\lim_{\epsilon \to 0}[\cdots]_+$ and $\lim_{\epsilon \to 0}[\cdots]_-$ exist, we see that

$$\lim_{\epsilon \to 0}[\cdots]_+ + \lim_{\epsilon \to 0}[\cdots]_- = \lim_{\epsilon \to 0}([\cdots]_+ + [\cdots]_-)$$

$$= \lim_{\epsilon \to 0} \int_{|t| > \epsilon} \frac{f(t)}{t} \, dt, \tag{2.3.15}$$

cf., part *b*. On the other hand, equation (2.3.13) shows that the left side of (2.3.15) is $g'(f)$, where $g(t) \equiv \log|t|$, $t \in \mathbb{R} \backslash \{0\}$. Therefore, since the right side of (2.3.15) defines the principal value distribution, we obtain

$$(\log|t|)' = pv\left(\frac{1}{t}\right). \tag{2.3.16}$$

Clearly, $\log|t| \notin BV_{\mathrm{loc}}(\mathbb{R})$ and so $pv(1/t) \notin M(\mathbb{R})$, e.g., Exercise 2.30.

2.3.9 Example. First Moment of a Measure

a.   We shall verify that

$$\int t \, d\mu(t) = 1 - \int_0^1 F(t) \, dt, \tag{2.3.17}$$

where $\mu = F' \in M_{b+}(\mathbb{R})$ has the property that $F = 0$ on $(-\infty, 0]$, $F = 1$ on $[1, \infty)$, and F is increasing on $[0, 1]$. In fact, if $f \in C_c^\infty(\mathbb{R})$, then

$$(tF'(t))(f(t)) = F'(t)(tf(t)) = -F(t)(tf'(t) + f(t))$$

$$= -\int_0^1 tf'(t)F(t)\,dt - \int_1^\infty tf'(t)\,dt - \int_0^1 f(t)F(t)\,dt$$

$$- \int_1^\infty f(t)\,dt = -\int_0^1 tf'(t)F(t)\,dt + f(1) - \int_0^1 f(t)F(t)\,dt$$

$$= (\delta_1 - \mathbf{1}_{[0,1]}F)(f) - \int_0^1 tf'(t)F(t)\,dt.$$

$$(2.3.18)$$

Consequently, if $f = 1$ on $[0, 1]$ we obtain (2.3.17) since the left side of (2.3.18) is $\int t\,d\mu(t)$ in this case.

b. Let F in part a be defined on $[0, 1]$ as the Cantor functions f_C for the $\frac{1}{3}$-Cantor set C, cf., Remark 1.4.2 a. In this case, $DF = 0$ a.e. and $F' = \mu_C \in M_{b+}(\mathbb{R})$. μ_C is the *Cantor measure*, and it has the property that $\int d\mu_C(t) = 1$ and supp $\mu_C = C$, e.g., [Ben75, pages 28–29], [KS63, Chapitre I], [Zyg59, Volume I, pages 194–196]. $F \in BV(\mathbb{R}) \cap C(\mathbb{R})$, but $F \notin AC_{\mathrm{loc}}(\mathbb{R})$.

Using (2.3.17) we have the fact that

$$\int t\,d\mu_C(t) = \frac{1}{2},$$

e.g., Exercise 2.44.

2.3.10 Example. Data Compression and Lateral Networks

Suppose f is a signal that cannot be understood by direct examination or that must be reconstructed after less than ideal transmission. f could be, for example, a speech signal, radar data, an image, or MRI or EEG data. Let $F(t, \omega)$ be data collected from the signal. F could be, for example, a spectrogram, a scalogram, or a radar ambiguity function, e.g., [BF94], [Mey91], [Rih85]. The goal is to understand or reconstruct f from F or a modified version of F. A possible modification of F, because of excessive volume, is the compressed data

$$H \circ F(t, \omega)$$

consisting of 0's and 1's. Here, H is the Heaviside function. In some systems, the variables (t, ω), which could be time-frequency or time-scale, for example, are interrelated because of physical constraints. For example, in the auditory system and for ω being a scale variable, the modified data can be of the form

$$\partial_\omega(H \circ F)(t, \omega) = [\delta \circ F(t, \omega)]\partial_\omega F(t, \omega),$$

e.g., [BT93]; in particular, compositions of the form $\delta \circ F$ arise in a natural way.

2.4. The Fourier Transform of Distributions

In (2.3.1) and (2.3.2) we used the integration by parts formula to motivate the definition of the distributional derivative. In this section we shall use the Parseval formula to motivate the definition of the distributional Fourier transform. In this role, such classical formulas become *creative formulas* in the sense of the following remark.

2.4.1 Remark. Creative Formulas

What do we mean by a creative formula? The Pythagorean Theorem is an example of such a formula. It leads to the proper definition of distance between points in Euclidean space. It is fundamental in defining lengths of curves and areas of surfaces, for example. It extends to defining the notion of distance by means of the L^2-norm in the space of square integrable functions. It provides a fundamental guideline in the geometry of Hilbert space. It is a backdrop from which various non-Euclidean geometries are assessed. It inspires new developments in the sciences, with concepts such as *energy* or fields of study such as *quantum mechanics*, e.g., [vN55].

2.4.2 Definition. Fourier Transform of Distributions

a. Recall the Parseval formula (1.10.10),

$$\forall f, g \in L^2(\mathbb{R}), \qquad \int \hat{g}(\gamma)\overline{\hat{f}(\gamma)}\,d\gamma = \int g(t)\overline{f(t)}\,dt. \qquad (2.4.1)$$

In the notation of (2.1.5), and thinking of f as a test function, we rewrite (2.4.1) as

$$\hat{g}(\overline{\hat{f}}) = g(\overline{f}). \qquad (2.4.2)$$

In the notation of (2.2.4), we rewrite (2.4.2) as

$$T_{\hat{g}}(\overline{\hat{f}}) = T_g(\overline{f}). \qquad (2.4.3)$$

Since $L^2(\mathbb{R})$ is a Hilbert space, e.g., [GG81], it is customary to introduce the notation of inner products and write

$$\langle g, f \rangle = \int g(t)\overline{f(t)}\,dt.$$

Thus, (2.4.1) is

$$\langle \hat{g}, \hat{f} \rangle = \langle g, f \rangle,$$

and (2.4.3) is

$$\langle T_{\hat{g}}, \hat{f} \rangle = \langle T_g, f \rangle.$$

b. Motivated by part *a*, the *Fourier transform* $\hat{T}$ of a distribution T is *formally defined* by the equation

$$\hat{T}(\overline{\hat{f}}) = T(\overline{f}) \tag{2.4.4}$$

or, equivalently,

$$\langle \hat{T}, \hat{f} \rangle = \langle T, f \rangle \tag{2.4.5}$$

for all functions f in an appropriate space of test functions, cf., Definition 2.2.1*b*.

c. We have been purposely vague about specifying the distributions and test functions in part *b*, since the meaning of (2.4.4) is difficult to formulate for *all* distributions. In fact, if $T \in D'(\mathbb{R})$ and $f \in C_c^{\infty}(\mathbb{R})$, then the right side of (2.4.4) is well defined, whereas $\hat{f} \notin C_c^{\infty}(\hat{\mathbb{R}})$, e.g., Exercise 2.5. Thus, $\hat{T}$ on the left side is not necessarily defined on $C_c^{\infty}(\hat{\mathbb{R}})$, and would not necessarily be a distribution. Even if we worked with $\hat{f} \in C_c^{\infty}(\hat{\mathbb{R}})$ and $\hat{T} \in D'(\hat{\mathbb{R}})$, then T would not be well defined on $C_c^{\infty}(\mathbb{R})$ for the same reason.

This quandary is resolved by introducing the *Schwartz space* of test functions and the space of *tempered distributions* in the following material.

2.4.3 Definition. The Schwartz Space

a. An infinitely differentiable function $f : \mathbb{R} \rightarrow \mathbb{C}$ is an element of the *Schwartz space* $\mathcal{S}(\mathbb{R})$ if

$$\forall n = 0, 1, \ldots, \qquad \|f\|_{(n)} = \sup_{0 \le j \le n} \sup_{t \in \mathbb{R}} (1 + |t|^2)^n |f^{(j)}(t)| < \infty. \tag{2.4.6}$$

b. Note that the Gaussian $g(t) = (1/\sqrt{\pi}) e^{-t^2} \in \mathcal{S}(\mathbb{R})$, and that

$$C_c^{\infty}(\mathbb{R}) \subseteq \mathcal{S}(\mathbb{R}) \subseteq L^1(\mathbb{R}) \cap L^2(\mathbb{R}) \cap A(\mathbb{R}).$$

c. Using (2.4.6) we define the function $\rho : \mathcal{S}(\mathbb{R}) \times \mathcal{S}(\mathbb{R}) \rightarrow \mathbb{R}^+$ as

$$\forall f, g \in \mathcal{S}(\mathbb{R}), \qquad \rho(f, g) = \sum_{n=0}^{\infty} \frac{1}{2^n} \frac{\|f - g\|_{(n)}}{1 + \|f - g\|_{(n)}}. \tag{2.4.7}$$

The following theorem is a basic result about the Schwartz space, and its verification is left as Exercise 2.45.

2.4.4 Theorem. $S(\mathbb{R})^{\wedge} = S(\hat{\mathbb{R}})$

a.   *The mapping* $S(\mathbb{R}) \longrightarrow S(\hat{\mathbb{R}})$, $f \longmapsto \hat{f}$, *is a bijection.*

b.   *The mapping* ρ *defined by (2.4.7) is a metric on* $S(\mathbb{R})$, *and, as such, the metric space* $S(\mathbb{R})$ *is complete.*

c.   *The Fourier transform mapping of part a, where* $S(\mathbb{R})$ *is given the metrizable topology of part b is bicontinuous. Thus, the Fourier transform,*

$$S(\mathbb{R}) \longrightarrow S(\hat{\mathbb{R}})$$
$$f \longmapsto \hat{f},$$

is a metric space isomorphism, cf., Theorem 1.10.2 and Exercise 2.47.

2.4.5 Definition. Tempered Distributions

a.   A linear functional,

$$T : S(\mathbb{R}) \longrightarrow \mathbb{C}$$
$$f \longmapsto T(f),$$

is a *tempered distribution* if $\lim_{n\to\infty} T(f_n) = 0$ for every sequence $\{f_n\} \subseteq S(\mathbb{R})$ satisfying the properties

$$\forall k, m \geq 0, \qquad \lim_{n\to\infty} \|t^m f_n^{(k)}(t)\|_{L^\infty(\mathbb{R})} = 0. \qquad (2.4.8)$$

The space of all tempered distributions on $\mathbb{R}$ is denoted by $S'(\mathbb{R})$.

b.   It is not difficult to prove that "$f_n \to 0$" in the sense of (2.4.8) if and only if $\lim_{n\to\infty} \rho(f_n, 0) = 0$, where ρ is defined in (2.4.7), e.g., Exercise 2.45.

c.   In light of our terminology, it is natural to ask if $S'(\mathbb{R}) \subseteq D'(\mathbb{R})$. The answer is "yes". The proof of this fact depends on functional analysis duality results (Remark B.16) and the following easily verifiable facts: the mapping

$$C_c^\infty(\mathbb{R}) \longrightarrow S(\mathbb{R})$$
$$f \longmapsto f \qquad\qquad (2.4.9)$$

is continuous, and

$$\overline{C_c^\infty(\mathbb{R})} = S(\mathbb{R}), \qquad\qquad (2.4.10)$$

cf., Example 2.4.6 *h*. The mapping (2.4.9) means that if "$f_n \to 0$" in the sense of Definition 2.2.3*a*, then "$f_n \to 0$" in the sense of (2.4.8). The density (2.4.10) means that if $f \in S(\mathbb{R})$, then there is $\{f_n\} \subseteq C_c^\infty(\mathbb{R})$ for which "$f - f_n \to 0$" in the sense of (2.4.8). To prove (2.4.10), let $f_n = f g_n$ where $g_n \in C_c^\infty(\mathbb{R})$ is even,

$g_n = 1$ on $[-n, n]$, supp $g_n \subseteq [-(n+1), n+1]$, and g_n has the same shape on $[n, n+1]$ as g_m has on $[m, m+1]$.

The following section contains some advanced material and relevant references, but no details.

2.4.6 Example. Distributions: Potpourri and Titillation

a. Let $T \in D'(\mathbb{R})$. Then $T \in S'(\mathbb{R})$ if and only if there is $g \in C_b(\mathbb{R})$, and $k, m \geq 0$, such that

$$T = f^{(k)} \text{ (distributionally)},$$

where $f(t) = (1 + t^2)^{m/2} g(t)$ [Sch66, Chapitre VII, §4, Théorème VI]. For example, if $f \in C_b(\mathbb{R})$, then $f^{(k)} \in S'(\mathbb{R})$. In particular, taking $f(t) = tH(t)$, we see that $\delta^{(k)} \in S'(\mathbb{R})$ for each $k \geq 0$.

b. If $T \in D'(\mathbb{R})$ and supp $T \subseteq \mathbb{R}$ is compact, then $T \in S'(\mathbb{R})$ [Sch66]. The space of distributions having compact support is denoted by $\mathcal{E}'(\mathbb{R})$, and so

$$\mathcal{E}'(\mathbb{R}) \subseteq S'(\mathbb{R}).$$

c. If $g(t) = e^{|t|}$, then $g \notin S'(\mathbb{R})$.
d. For each $p \in [1, \infty]$,

$$L^p(\mathbb{R}) \subseteq S'(\mathbb{R}).$$

e. $S'(\mathbb{R})$ is the smallest subspace of $D'(\mathbb{R})$ that contains $L^1(\mathbb{R})$ and is invariant under differentiation and multiplication by polynomials, cf., Definition 2.5.7.

f. Spectral synthesis. If the norm of $f \in A(\hat{\mathbb{R}})$ is defined as $\|\check{f}\|_{L^1(\mathbb{R})}$, then $A(\hat{\mathbb{R}})$ is a Banach space (in fact, it is a Banach algebra under pointwise multiplication of functions). The dual space $A'(\hat{\mathbb{R}})$ of $A(\hat{\mathbb{R}})$, i.e., the space of continuous linear functionals $T : A(\hat{\mathbb{R}}) \to \mathbb{C}$, is the space of *pseudomeasures*. We have the inclusion

$$M_b(\hat{\mathbb{R}}) \subseteq A'(\hat{\mathbb{R}}) \subseteq S'(\hat{\mathbb{R}}).$$

An important area of harmonic analysis, rooted in physical considerations and with many unsolved problems, is *spectral synthesis*, e.g., [Ben75], [Beu89], [KS63], [Kah70], [Kat76]. Its creation, canonicity, and depth were revealed to mankind by Beurling. In deceptive form, an essential *problem of spectral synthesis* is to determine the pseudomeasures $T \in A'(\hat{\mathbb{R}})$ for which there is a sequence $\{\mu_n\} \subseteq M_b(\hat{\mathbb{R}})$ satisfying the properties that supp $\mu_n \subseteq$ supp T for each n, and

$$\forall f \in A(\hat{\mathbb{R}}), \qquad \lim_{n \to \infty} \mu_n(f) = T(f).$$

g. *Riemann Hypothesis.* The most celebrated problem in analytic number theory is to settle the validity or not of the *Riemann Hypothesis.*

The *Riemann zeta function* $\zeta(s)$ is defined as

$$\zeta(s) = \sum_{n=1}^{\infty} \frac{1}{n^s}, \qquad \mathrm{Re}\, s > 1,$$

and it has an analytic continuation, whereby it is analytic on $\mathbb{C}\backslash\{1\}$ and has a simple pole at $s = 1$. The *Riemann Hypothesis* is the statement that the complex zeros of $\zeta(s)$ all have real part equal to $\frac{1}{2}$, e.g., [Edwa74], [Tit51].

The *Weil distribution* $W \in D'(\mathbb{R})$ is defined as

$$\forall f \in C_c^{\infty}(\mathbb{R}), \qquad W(f) = \Sigma_\rho \Phi_f(\rho),$$

where $\zeta(\rho) = 0$ for $0 \leq \mathrm{Re}\, \rho \leq 1$, and where

$$\Phi_f(s) = \int f(t)e^{(s-(1/2))t}\, dt;$$

this integral is the bilateral Laplace transform of $f(t)e^{-(1/2)t}$. Tempered distributions arise in the following result. *The Riemann Hypothesis is valid if and only if* $W \in S'(\mathbb{R})$ [Ben80], [Joy86, page 6].

The Riemann Hypothesis can be considered a strong form of the Prime Number Theorem; and Wiener's Tauberian Theorem, a fundamental result in harmonic analysis, is an indispensable tool in this type of analysis, e.g., [Ben75, Section 2.3].

h. If $D'(\mathbb{R})$ is taken with the canonical dual space topology from $C_c^{\infty}(\mathbb{R})$, then the natural injection $C_c^{\infty}(\mathbb{R}) \longrightarrow D'(\mathbb{R})$, $f \longmapsto T_f$, is continuous. Further, if the topological vector space X has the properties that $C_c^{\infty}(\mathbb{R}) \subseteq X \subseteq D'(R)$, the inclusions are continuous, and $C_c^{\infty}(\mathbb{R})$ is dense in X, then the dual X' is a subspace of $D'(\mathbb{R})$.

2.4.7 Definition. The Fourier Transform of Tempered Distributions

a. The *Fourier transform* of $T \in S'(\mathbb{R})$ is $\hat{T}$, defined by

$$\forall f \in S(\mathbb{R}), \qquad \hat{T}(\overline{\hat{f}}) = T(\overline{f}). \qquad (2.4.11)$$

b. Equation (2.4.11) is a quantified version of (2.4.4), and there is the equivalent quantified analogue of (2.4.5). Because of Theorem 2.4.4, we know that $\hat{f} \in S(\mathbb{R})$, and, thus, it is straightforward to show that $\hat{T} \in S'(\mathbb{R})$. In fact, the mapping

$$
\begin{array}{ccc}
S'(\mathbb{R}) & \longrightarrow & S'(\hat{\mathbb{R}}) \\
T & \longmapsto & \hat{T}
\end{array}
\qquad (2.4.12)
$$

is a linear bijection. Further, there is a natural convergence criterion (topology) on S' so that (2.4.12) is bicontinuous, e.g., [Hor66], [Sch66], and, hence, (2.4.12) is a topological vector space isomorphism.

We now want to compute Fourier transforms with this new definition; and, in particular, in light of Example 2.4.6 *d*, we want to make sure that the new definition reduces to the classical definition of Chapter 1.

2.4.8 Example. The Fourier Transform of $L^1(\mathbb{R})$ and $\delta^{(n)}$

a. Let $g \in L^1(\mathbb{R})$. We shall verify that $(T_g)\hat{} = T_{\hat{g}}$ so that the distributional Fourier transform is a generalization of the usual, i.e., $L^1(\mathbb{R})$, definition. For each $f \in S(\mathbb{R})$ we compute

$$\langle (T_g)\hat{}, \hat{f} \rangle = \langle T_g, f \rangle = \int g(t)\overline{f(t)}\, dt$$

$$= \int \hat{g}(\gamma)\overline{\hat{f}(\gamma)}\, d\gamma = \langle T_{\hat{g}}, \hat{f} \rangle,$$

(2.4.13)

where we have used the Parseval formula for $g \in L^1(\mathbb{R})$ and $\hat{f} \in L^1(\hat{\mathbb{R}})$, e.g., Proposition 1.10.4. By our discussion of equality for distributions we can conclude that $(T_g)\hat{} = T_{\hat{g}}$ since (2.4.13) is true for each $\hat{f} \in S(\hat{\mathbb{R}})$.

b. Let $T = \delta^{(n)}$. From Example 2.4.6 *a* or *b*, we know that $T \in S'(\mathbb{R})$. To evaluate $\hat{T}$ we compute

$$\langle \hat{\delta}^{(n)}, \hat{f} \rangle = \langle \delta^{(n)}, f \rangle = (-1)^n \langle \delta, f^{(n)} \rangle$$

$$= (-1)^n \left\langle \delta(t), \int (2\pi i \gamma)^n \hat{f}(\gamma) e^{2\pi i t\gamma}\, d\gamma \right\rangle$$

$$= (-1)^n \langle (\bar{i})^n (2\pi\gamma)^n, \hat{f}(\gamma) \rangle = \langle (2\pi i \gamma)^n, \hat{f}(\gamma) \rangle$$

for each $\hat{f} \in S(\hat{\mathbb{R}})$. Consequently, we have

$$\left(\delta^{(n)} \right)\hat{}(\gamma) = (2\pi i \gamma)^n$$

(2.4.14)

for each $n \in \mathbb{N} \cup \{0\}$. In particular, $\hat{\delta} = 1$, which is compatible with our discussion of approximate identities.

An important point about (2.4.14) is that $(\delta^{(n)})\hat{}$ has polynomial growth at $\pm\infty$ as opposed to the behavior of $\hat{f}$ at $\pm\infty$ for $f \in L^1(\mathbb{R})$ (or $L^2(\mathbb{R})$).

2.4.9 Example. The Fourier Transform of the Heaviside Function

The Heaviside function $H \in L^1_{\text{loc}}(\mathbb{R}) \cap S'(\mathbb{R})$, $\delta \in M_b(\mathbb{R}) \subseteq S'(\mathbb{R})$, and $pv(1/t) \in S'(\mathbb{R})$, whereas $pv(1/t) \notin M(\mathbb{R})$. The Fourier transforms of any one of these distributions have a nice formulation in terms of the others or their Fourier transforms. We shall verify

$$H(t) \longleftrightarrow \frac{1}{2\pi i} pv\left(\frac{1}{\gamma}\right) + \frac{1}{2}\delta(\gamma),$$

$$(2.4.15)$$

$$\frac{1}{2}\delta(t) - \frac{1}{2\pi i} pv\left(\frac{1}{t}\right) \longleftrightarrow H(\gamma).$$

For $f \in S(\mathbb{R})$ and $\hat{f} = F$, we compute

$$\left\langle \left(pv\left(\frac{1}{t}\right)\right)^{\wedge}(\gamma), F(\gamma)\right\rangle = \left\langle pv\left(\frac{1}{t}\right), f(t)\right\rangle = \lim_{\epsilon \to 0} \int_{|t| \geq \epsilon} \frac{\overline{f(t)}}{t}\, dt$$

$$= \lim_{\epsilon \to 0} \iint_{|t| \geq \epsilon} \overline{F(\gamma)} \frac{e^{-2\pi i t \gamma}}{t}\, d\gamma\, dt \qquad (2.4.16)$$

$$= \lim_{\epsilon \to 0} \int \overline{F(\gamma)} \int_{|t| \geq \epsilon} \frac{e^{-2\pi i t \gamma}}{t}\, dt\, d\gamma.$$

Now note that if $\gamma \neq 0$, then

$$\lim_{\epsilon \to 0,\, T \to \infty} \int_{1/\epsilon}^{T} \frac{\cos 2\pi t \gamma}{t}\, dt = 0,$$

where $1/\epsilon < T$. In fact, by the second mean value theorem for integrals (Lemma 1.7.3), we have

$$\int_{1/\epsilon}^{T} \frac{\cos 2\pi t \gamma}{t}\, dt = \epsilon \int_{1/\epsilon}^{\xi} \cos 2\pi t \gamma\, dt + \frac{1}{T} \int_{\xi}^{T} \cos 2\pi t \gamma\, dt$$

$$= \frac{\epsilon \sin 2\pi t \gamma}{2\pi \gamma}\bigg|_{t=1/\epsilon}^{\xi} + \frac{\sin 2\pi t \gamma}{2\pi T \gamma}\bigg|_{t=\xi}^{T};$$

and for fixed $\gamma \neq 0$ this last term tends to 0 as $\epsilon \to 0$ and $T \to \infty$. A similar calculation for the domain $[-S, -1/\epsilon]$ (where $S > 1/\epsilon$) combined with an application of LDC allow us to write the right side of (2.4.16) as

$$\int \overline{F(\gamma)} \left[\lim_{\epsilon \to 0} \int_{\epsilon \leq |t| \leq 1/\epsilon} \frac{e^{-2\pi i t \gamma}}{t}\, dt\right] d\gamma. \qquad (2.4.17)$$

(The use of LDC requires certain hypotheses, e.g., Exercise 2.19.)

Next, we compute

$$\lim_{\epsilon \to 0} \int_{\epsilon \le |t| \le 1/\epsilon} \frac{e^{-2\pi i t \gamma}}{t} \, dt = -\lim_{\epsilon \to 0} i \int_{\epsilon \le |t| \le 1/\epsilon} \frac{\sin 2\pi t \gamma}{t} \, dt$$

(2.4.18)

$$= \begin{cases} -\pi i, & \text{if } \gamma > 0, \\ \pi i, & \text{if } \gamma < 0. \end{cases}$$

We can write the right side of (2.4.18) as

$$\pi i - 2\pi i H(\gamma) = \pi i \hat{\delta}(\gamma) - 2\pi i H(\gamma)$$

for $\gamma \ne 0$, and thus we have proved that

$$\forall f \in S(\mathbb{R}), \qquad \left\langle \left(pv \left(\frac{1}{t} \right) \right)^{\widehat{}} (\gamma), F(\gamma) \right\rangle = \langle \pi i \hat{\delta}(\gamma) - 2\pi i H(\gamma), F(\gamma) \rangle,$$

where $\hat{f} = F$. Therefore, $(pv(1/t))^{\widehat{}}(\gamma) = \pi i \hat{\delta}(\gamma) - 2\pi i H(\gamma)$ and (2.4.15) is obtained.

2.4.10 Remark. Perspective on $\hat{H}$

Our calculation of (2.4.15) is relatively honest but too long-winded because of the distributional setup. The formula was certainly known and used long before distribution theory, and several formal, short, and essentially correct calculations give the result, e.g., (2.4.18) contains the essential details.

2.4.11 Remark. Eigenfunctions

a. We defined the Hermite polynomials H_n, $n \ge 0$, in Exercise 1.26, and computed the eigenvalues of the Fourier transform mapping $\mathcal{F} : L^2(\mathbb{R}) \to L^2(\hat{\mathbb{R}})$ in Example 1.10.12. The *Hermite functions* $h_n(t) \equiv e^{-\pi t^2} H_n(2\sqrt{\pi} \, t)$ are the eigenfunctions of $\mathcal{F}$, and, in fact, $\hat{h}_n = (-i)^n h_n$. Further, $\{h_n / \|h_n\|_{L^2(\mathbb{R})} : n \ge 0\}$ is an orthonormal basis of $L^2(\mathbb{R})$. This material was developed by Norbert Wiener [Wie33, pages 51–71] to give his proof of the Parseval–Plancherel Theorem, cf., [Wie33, page 70] for an interesting historical note and [Wie81, article 29d], which is related to earlier work of Hermann Weyl and which establishes the Fourier transform of fractional order in terms of Hermite polynomials.

b. Since the Hermite functions are contained in $S(\mathbb{R})$, they are also eigenfunctions of the mapping $\mathcal{F} : S(\mathbb{R}) \longrightarrow S(\hat{\mathbb{R}})$. In this context it is natural to determine the eigenfunctions of the Fourier transform mapping $\mathcal{F} : S'(\mathbb{R}) \longrightarrow S'(\hat{\mathbb{R}})$. In this regard, we do note now that $\Sigma \delta_n \in M(\mathbb{R}) \cap S'(\mathbb{R})$ and

$$\left(\sum \delta_n \right)^{\widehat{}} = \sum \delta_n.$$

(2.4.19)

Equation (2.4.19) is a form of the *Poisson Summation Formula*, e.g., Theorem 3.10.8.

2.5. Convolution of Distributions

In Section 1.5 we defined the convolution $g * h$ for $g, h \in L^1(\mathbb{R})$. In this section, we shall see to what extent we can define the *convolution* $S * T$ for $S, T \in D'(\mathbb{R})$. In Proposition 1.5.2 we established *the exchange formula* $(g * h)\hat{} = \hat{g}\hat{h}$ for $g, h \in L^1(\mathbb{R})$. In this section we shall see to what extent this formula is valid for distributions; and, in view of the right side of the exchange formula, we shall see to what extent the *multiplication of distributions* is well defined.

Convolution and the exchange formula are major components of the operational calculus (Section 2.6). Further, the formulation of convolution and multiplication of distributions is a formidable mathematical task that we shall really only address at the motivational and computational levels.

2.5.1 Definition. Convolution

a.   If $g, h \in L^1(\mathbb{R})$, then for each $f \in C_c^\infty(\mathbb{R})$ we have

$$(g * h)(f) = \int g * h(t) f(t) \, dt = \iint g(t - u) h(u) f(t) \, du \, dt$$

$$= \int h(u) \left(\int g(t - u) f(t) \, dt \right) du \tag{2.5.1}$$

$$= \int h(u) \int g(v) f(u + v) \, dv \, du.$$

b.   Because of the calculation in part *a*, we define the *convolution* $S * T$ of certain distributions S and T by

$$\forall f \in C_c^\infty(\mathbb{R}), \qquad (S * T)(f) = T(u)\big(S(v)(f(u + v))\big), \tag{2.5.2}$$

where, although T is not necessarily an ordinary function on $\mathbb{R}$, we write $T(u)$ to indicate its dependence on the u variable.

c.   We have been deliberately noncommital in part *b* about which distributions S and T we can convolve; and whether, in case $S * T$ exists, it is an element of $D'(\mathbb{R})$, cf., Theorem 2.5.4. In fact,

$$\forall f \in C_c^\infty(\mathbb{R}) \setminus \{0\}, \qquad f(u + v) \notin C_c^\infty(\mathbb{R} \times \mathbb{R}); \tag{2.5.3}$$

and $S(v)(f(u + v))$ is not necessarily an element of $C_c^\infty(\mathbb{R})$ (as a function of u), so that the right side of (2.5.2) may not make sense. To visualize (2.5.3) it is instructive to let supp $f = [a, b]$, $a > 0$, and then note that the support X of $f(u + v)$ as a function of two variables is the diagonal strip of $\mathbb{R} \times \mathbb{R}$ whose intersection with both the u- and v-axes is $[a, b]$, i.e.,

$$X = \{(u, v) \in \mathbb{R} \times \mathbb{R} : a \leq u + v \leq b\}. \tag{2.5.4}$$

2.5.2 Example. Properties of Convolution

a. The calculation in part *a* of Definition 2.5.1 shows that (2.5.2) generalizes the definition of convolution in $L^1(\mathbb{R})$. It can also be shown that $\mu * \nu \in M_b(\mathbb{R})$ for $\mu, \nu \in M_b(\mathbb{R})$. This can be proved measure theoretically, e.g., [Rud66], or by our definition of $M_b(\mathbb{R})$ in Definition 2.3.6 combined with the *Riesz Representation Theorem* in Section 2.7, e.g., Exercise 2.49.

b. If $S, T \in D'(\mathbb{R})$ and $S * T$, defined by (2.5.2), is a well-defined element of $D'(\mathbb{R})$, then $T * S$ is also a well-defined element of $D'(\mathbb{R})$, and $S * T = T * S$. Thus, convolution is a *commutative* operation.

c. Convolution is *not* necessarily *associative*. In fact,

$$(1 * \delta') * H = 0 * H = 0$$

and

$$1 * (\delta' * H) = 1 * \delta = 1.$$

d. Let $S = T = 1$. Then $S, T \in \mathcal{S}'(\mathbb{R}) \cap M(\mathbb{R})$ and

$$1(v)(f(u + v)) = \int f(t)\, dt.$$

Therefore, $S * T(f) = \infty$ for each nonnegative $f \in C_c^\infty(\mathbb{R}) \backslash \{0\}$, and, in particular, $S * T$ does not exist.

e. If $T \in D'(\mathbb{R})$ and $g \in C_c^\infty(\mathbb{R})$, then $T * g \in C^\infty(\mathbb{R})$.

2.5.3 Definition. Translation

a. If $g \in L_{\mathrm{loc}}^1(\mathbb{R})$, then for each fixed $t \in \mathbb{R}$ and each $f \in C_c^\infty(\mathbb{R})$, we have

$$(T_{\tau_t g})(f) = \int g(v - t) f(v)\, dv = \int g(u) f(u + t)\, du$$

$$= T_g(u)(f(u + t)) = T_g(u)\Big(\delta_t(v)(f(u + v))\Big)$$

$$= (\delta_t * T_g)(f).$$

b. Because of the calculation in part *a*, we define the *translation* $\tau_t T$ of $T \in D'(\mathbb{R})$ by $t \in \mathbb{R}$ to be

$$\tau_t T = \delta_t * T. \tag{2.5.5}$$

c. It should be noted that the right side of (2.5.5) is a well-defined element of $D'(\mathbb{R})$, although there is something to prove, e.g., Exercise 2.51.

We have seen that $\delta_t * T \in D'(\mathbb{R})$ for arbitrary distributions T, whereas $S * T$ is not necessarily a distribution for $S, T \in \mathcal{S}'(\mathbb{R})$ (Example 2.5.2 *d*). The following result records some of the elementary theory for the existence of $S * T$, e.g., [Hor66, pages 365–401, especially pages 382–388], [Sch66, Chapitre VI].

2.5.4 Theorem. Existence of Convolution

a. *Let $S, T \in D'(\mathbb{R})$ satisfy the property that for each compact set $C \subseteq \mathbb{R}$,*

$$((\operatorname{supp} S) \times (\operatorname{supp} T)) \cap C^{\Delta}$$

is compact in $\mathbb{R} \times \mathbb{R}$, where $C^{\Delta} = \{(u, v) \in \mathbb{R} \times \mathbb{R} : u + v \in C\}$, cf., (2.5.3) and (2.5.4). Then $S * T \in D'(\mathbb{R})$.
b. *Let $S \in \mathcal{E}'(\mathbb{R})$ and $T \in D'(\mathbb{R})$. Then $S * T \in D'(\mathbb{R})$.*

2.5.5 Proposition.

Let $T \in D'(\mathbb{R})$ and let $n \geq 0$. Then

$$T * \delta^{(n)} = T^{(n)};$$

in particular, $T * \delta = T$.

PROOF.   For each $f \in C_c^{\infty}(\mathbb{R})$, we compute

$$T * \delta^{(n)}(f) = T(u)\left(\delta^{(n)}(v)(f(u + v))\right)$$
$$= (-1)^n T(u)\left(\delta(v)(f^{(n)}(u + v))\right)$$
$$= (-1)^n T(u)(f^{(n)}(u)) = T^{(n)}(f). \quad \blacksquare$$

2.5.6 Definition. Exchange Formula

a.   If $g, h \in L^1(\mathbb{R})$, then, as mentioned at the beginning of this section, we have the *exchange formula*,

$$(g * h)\hat{} = \hat{g}\hat{h}. \tag{2.5.6}$$

Besides the direct proof in Section 1.5, we could also prove it "distributionally" as follows. For each $f \in C_c^{\infty}(\mathbb{R})$,

$$(T_g * T_h)\hat{}\,(\overline{\hat{f}}) = T_g * T_h(\overline{\hat{f}}) = T_g(u)\left(T_h(v)\left(\overline{\hat{f}(u + v)}\right)\right)$$

$$= T_g(u)\left(T_h(v)\left(\int \overline{\hat{f}(\gamma)} e^{-2\pi i(u+v)\gamma} \, d\gamma\right)\right) \tag{2.5.7}$$

$$= \left[T_g(u)\left(e^{-2\pi i u\gamma} T_h(v)(e^{-2\pi i v\gamma})\right)\right]\left(\overline{\hat{f}(\gamma)}\right) = [(T_g)\hat{}(T_h)\hat{}](\overline{\hat{f}}).$$

The calculation (2.5.7) is valid for each $f \in C_c^{\infty}(\mathbb{R})$, and so we can conclude that $(T_g * T_h)\hat{} = (T_g)\hat{}(T_h)\hat{}$, which is precisely (2.5.6).

Although the notation in (2.5.7) is labyrinthine, it does inspire part *b* and possible proof of the formula therein, when T_g is replaced by S and T_h is replaced by T.

b. The *exchange formula* for certain distributions S and T is

$$(S * T)\widehat{} = \hat{S}\hat{T}. \tag{2.5.8}$$

c. In the best of all worlds, if $S, T \in \mathcal{S}'(\mathbb{R})$, then we could conclude that $S * T$ exists and is an element of $\mathcal{S}'(\mathbb{R})$, that the multiplication $\hat{S}\hat{T}$ is well defined and $\hat{S}\hat{T} \in \mathcal{S}'(\mathbb{R})$, and finally that (2.5.8) is valid, cf., [Hor66, page 424], [Sch66, Chapitre VII.8] for the first distributional version of the exchange formula. Alas, we already saw that $S * T$ need not exist for $S, T \in \mathcal{S}'(\mathbb{R})$. For the multiplication of distributions we refer to Definition 2.5.7, and for a reasonable *theorem* yielding the validity of the exchange formula we refer to Theorem 2.5.9.

2.5.7 Definition. Multiplication of Distributions

a. Let $g \in C^\infty(\mathbb{R})$, resp., $\mathcal{S}(\mathbb{R})$ or $C^\infty(\mathbb{R})$, and let $T \in D'(\mathbb{R})$, resp., $\mathcal{S}'(\mathbb{R})$ or $\mathcal{E}'(\mathbb{R})$. The *product* $gT \in D'(\mathbb{R})$, resp., $\mathcal{S}'(\mathbb{R})$ or $\mathcal{E}'(\mathbb{R})$, is defined by

$$\forall f \in C_c^\infty(\mathbb{R}), \quad \text{resp.,} \ \mathcal{S}(\mathbb{R}) \text{ or } C^\infty(\mathbb{R}), \qquad (gT)(f) = T(gf). \quad (2.5.9)$$

It is easy to see that gT, defined by the right side of (2.5.9), is an element of $D'(\mathbb{R})$, resp., $\mathcal{S}'(\mathbb{R})$ or $\mathcal{E}'(\mathbb{R})$.

Similarly, $gT \in \mathcal{E}'(\mathbb{R})$ for $g \in C_c^\infty(\mathbb{R})$ and $T \in D'(\mathbb{R})$.

b. Let $S, T \in D'(\mathbb{R})$, and let $\{f_n\} \subseteq C_c^\infty(\mathbb{R})$ be an approximate identity. If the limit

$$\lim_{n \to \infty} [(S * f_n)(T * f_n)](f)$$

exists for each $f \in C_c^\infty(\mathbb{R})$ and is independent of $\{f_n\}$, then S, T are *multiplicable* with *product* $ST \in D'(\mathbb{R})$ defined by

$$\forall f \in C_c^\infty(\mathbb{R}), \qquad (ST)(f) = \lim_{n \to \infty} [(S * f_n)(T * f_n)](f) \quad (2.5.10)$$

for any fixed $\{f_n\}$.

If $S = g \in C^\infty(\mathbb{R})$, then (2.5.10) reduces to (2.5.9).

c. Defining the multiplication of distributions is not an annoying technical problem, but rather a large theoretical program motivated by quantum electrodynamics, nonlinear shock waves, and the harmonic analysis associated with Sobolev spaces and the Littlewood–Paley theory, e.g, [Col85], [Mey81], [Obe92]. In quantum field theory, the products of distributions that arose led to Feynman integrals and renormalization theory, cf., [Bre65], [deJ64] for classical background and [Glei92] for a thrilling layman's approach to this material.

The definition of multiplication in (2.5.10) is due to Mikusiński (1960) and is equivalent to one of Hirata and Ogata (1958), cf., [SI64].

2.5.8 Definition. $\mathcal{S}'$-Convolution

a. Let $S, T \in \mathcal{S}'(\mathbb{R})$. The $\mathcal{S}'$-*convolution* $S * T$ exists if

$$\forall f, g \in \mathcal{S}(\mathbb{R}), \qquad (S * f)(\widetilde{T} * g) \in L^1(\mathbb{R}),$$

where $\widetilde{T}(f) = T(t)(f(-t))$. In this case, $S * T$ is the unique element of $\mathcal{S}'(\mathbb{R})$ for which

$$\forall f, g \in \mathcal{S}(\mathbb{R}), \qquad ((S * T) * f)(g) = \int (S * f)(t)(\widetilde{T} * g)(t)\, dt.$$

b. This definition is due to Hirata and Ogata (1958). That the analogous definition for $D'(\mathbb{R})$ reduces to (2.5.2) was proved by results due to L. Schwartz (1954), Shiraishi (1959), and Horváth (1974), cf., [DV78].

The first part of the following theorem was given by Hirata and Ogata (1958), with a lovely proof in [SI64]; and there is a generalization of the result due to Oberguggenberger (1986), e.g., [Obe92]. The second part of the theorem can be proved directly [Ben75] or as a corollary of the first part.

2.5.9 Theorem. Exchange Formula

a. *Assume the $\mathcal{S}'$-convolution of $S, T \in \mathcal{S}'(\mathbb{R})$ exists. Then $\hat{S}, \hat{T}$ are multiplicable and*

$$(S * T)\hat{\ } = \hat{S}\hat{T}.$$

b. *Let $g \in L^1(\mathbb{R})$ and $h \in L^\infty(\mathbb{R})$. Then $g * h \in L^\infty(\mathbb{R})$ and*

$$(g * h)\hat{\ } = \hat{g}\hat{h}.$$

Further, $\hat{g} \in A(\hat{\mathbb{R}})$, $\hat{h} \in A'(\hat{\mathbb{R}})$ (the space of pseudomeasures defined in Example 2.4.6f), and $\hat{g}, \hat{h}$ are not only multiplicable in the sense of Definition 2.5.7 but in the sense of (2.5.9) where the domain space of functions is $A(\hat{\mathbb{R}})$.

2.5.10 Example. Convolution and Multiplication

a. Recall that $L^1_{\mathrm{loc}}(\mathbb{R})$ is not closed under multiplication pointwise a.e. For example, let $g(t) = 1/|t|^{1/2}$, $t \neq 0$, so that $g \in L^1_{\mathrm{loc}}(\mathbb{R})$ and $g^2 \notin L^1_{\mathrm{loc}}(\mathbb{R})$.

b. Does the product δ^2 exist as a distribution? It would be a success of the theory if $\delta^2 \in D'(\mathbb{R})$, since formulas such as

$$\delta^2 - \frac{1}{(\pi t)^2} = -\left(\frac{\pi}{t}\right)^2$$

arise in the surreal quantum world.

First, the product δ^2 doesn't exist in the sense of (2.5.10), as dealing with the approximate identity $\{\frac{n}{2}\mathbf{1}_{[-1/n,1/n]}\}$ shows.

Even with this setback, in light of the "naturalness" of the exchange formula, which afterall could be one of our "creative formulas", we are tempted to define δ^2 as $(\check{\delta} * \check{\delta})\hat{}$. However,

$$\check{\delta} * \check{\delta}(\bar{f}) = \check{\delta}(u)\left(\check{\delta}(v)\left(\overline{f(u+v)}\right)\right) = 1\left(1\left(\overline{f(u+v)}\right)\right)$$

$$= 1\left(\int \overline{f(u+v)}\,dv\right) = \iint \overline{f(u+v)}\,dv\,du = \int\left(\int \overline{f(v)}\,dv\right)du,$$

which is divergent for test functions f for which $\int f(v)\,dv \neq 0$.

Fortunately, δ^2 *does* fit into Colombeau's theory [Col85]!

c. By the exchange formula it is easy to see that if $T(t) = (1/\pi)pv(1/t)$, then

$$T * T = \delta. \tag{2.5.11}$$

Note that supp $T = \mathbb{R}$, and that, even though convolution is intuitively a smoothing operation, supp $T * T = \{0\}$! It is instructive to graph $1/t$ and its translates to see how the cancellation in (2.5.11) can occur.

d. In Section 1.5 we saw that $L^1(\mathbb{R})$ is an algebra with convolution as the multiplicative operation; and in Proposition 2.1.1 we noted that $L^1(\mathbb{R})$ does not have a unit under convolution. On the other hand, we saw that $T * \delta = T$ for all $T \in D'(\mathbb{R})$, and, hence, $\mathcal{E}'(\mathbb{R})$ is a convolution algebra with unit δ.

e. δ, H are multiplicable in the sense of (2.5.10), and

$$\delta H = \frac{1}{2}\delta. \tag{2.5.12}$$

To verify (2.5.12), let $\{f_n\} \subseteq L^1(\mathbb{R})$ be an approximate identity, and note that

$$(\delta * f_n)(H * f_n) = \left(\frac{(H * f_n)^2}{2}\right)'.$$

Clearly, $\lim_{n\to\infty}[(H * f_n)^2/2](f) = (H/2)(f)$ for all $f \in C_c^\infty(\mathbb{R})$, and so the result is obtained.

The exchange formula and formula (2.4.15) for $\hat{H}$ also yield (2.5.12), when we take note of the fact that

$$\frac{1}{2\pi i}pv\left(\frac{1}{\gamma}\right)\left(1(\lambda)(f(\lambda+\gamma))\right) = \frac{1}{2\pi i}pv\left(\frac{1}{\gamma}\right)\left(\int f(\lambda)\,d\lambda\right) = 0.$$

f. It turns out that μ, $\mathbf{1}_{[a,b]}$ are multiplicable for $\mu \in M(\mathbb{R})$, but that the product $T\mathbf{1}_{[a,b]}$ is more elusive and is related to spectral synthesis in the case T

is a pseudomeasure, e.g., [Ben75]. A related and equally challenging issue is to define the notion of the point value of a distribution, e.g., [Łoj57].

We shall conclude Section 2.5 with a discussion of the Hilbert transform, which is a special but far-reaching convolution.

2.5.11 Definition. Hilbert Transform

a. The *Hilbert transform* $\mathcal{H}T$ of a distribution T is the convolution

$$\mathcal{H}T(t) = \left(\frac{1}{\pi} pv \left(\frac{1}{u} \right) * T(u) \right)(t),$$

where we have used the notation of point functions to deal with "$pv(1/u)$". Thus, *formally*, the Hilbert transform $\mathcal{H}g$ of a function g is

$$\mathcal{H}g(t) = \lim_{\epsilon \to 0} \frac{1}{\pi} \int_{|t-u| \geq \epsilon} \frac{g(u)}{t-u} \, du.$$

We are purposely vague about which distributions (or functions) yield a well-defined Hilbert transform; and, in fact, some of the fundamental theory of Hilbert transforms is associated with the domain X and range Y of the Hilbert transform operator $\mathcal{H} : X \longrightarrow Y$.

b. The Hilbert transform opens the door to a large and profound area of harmonic analysis associated with the theory, relevance, and importance of *singular integral operators*, e.g., [Ste70]. A magnificent exposition of the basic theory of Hilbert transforms is due to Neri [Ner71]. The following result is fundamental.

2.5.12 Theorem. $\mathcal{H} : L^2(\mathbb{R}) \longrightarrow L^2(\mathbb{R})$

a. $\mathcal{H} : L^2(\mathbb{R}) \longrightarrow L^2(\mathbb{R})$ *is a well-defined linear bijection with the property that*

$$\forall f \in L^2(\mathbb{R}), \qquad \|\mathcal{H}f\|_{L^2(\mathbb{R})} = \|f\|_{L^2(\mathbb{R})}.$$

b. *Let $\sigma(\mathcal{H})(\gamma) \equiv -i \operatorname{sgn} \gamma$. Then*

$$\mathcal{H} = \mathcal{F}^{-1} \sigma(\mathcal{H}) \mathcal{F},$$

where $\mathcal{F} : L^2(\mathbb{R}) \longrightarrow L^2(\hat{\mathbb{R}})$ is the Fourier transform mapping, e.g., Theorem 1.10.2.

c. $\mathcal{H} \circ \mathcal{H} = -I$, where I is the identity mapping on $L^2(\mathbb{R})$.

Parts *b* and *c* of Theorem 2.5.12 are easy to prove. Using (2.4.15) we see that

$$\frac{1}{\pi} pv\left(\frac{1}{t}\right) \longleftrightarrow \sigma(\mathcal{H})(\gamma),$$

where $\sigma(\mathcal{H})$ is called the *symbol* of $\mathcal{H}$. By the exchange formula, we have

$$(\mathcal{H}f)\hat{} = \sigma(\mathcal{H})\hat{f}, \tag{2.5.13}$$

and part *b* follows. In the context of (2.5.13), $\sigma(\mathcal{H})$ is called a *multiplier*, e.g., [Ste70], [Lar71]. Part *c* is a consequence of part *b* and the calculation

$$(\mathcal{H}(\mathcal{H}f))\hat{} = \sigma(\mathcal{H})(\mathcal{H}f)\hat{} = \sigma(\mathcal{H})^2\,\hat{f} = -\hat{f}.$$

2.5.13 Example. A Distributional Domain for $\mathcal{H}$

a. It can be shown by direct calculation that $\mathcal{H}f \in L^\infty(\mathbb{R})$ for all $f \in \mathcal{S}(\mathbb{R})$, cf., Exercise 2.53 for an ingenious calculation due to Logan yielding a stronger result [Log83].

Then, by Definition 2.5.8, $\mathcal{H}T \in \mathcal{S}'(\mathbb{R})$ exists for those $T \in \mathcal{S}'(\mathbb{R})$ for which $T * g \in L^1(\mathbb{R})$ whenever $g \in \mathcal{S}(\mathbb{R})$, cf., Exercise 2.54.

b. Besides our method in part *a*, the Hilbert transform of distributions can be defined by other methods, which lead to general real variable formulations, e.g., [Cart91], [Jon82], as well as complex variable formulations, e.g., [Bre65] and research by Lauwerier, Martineau, Orton, and Tillmann.

Also note that if $T \in \mathcal{S}'(\mathbb{R})$ and $\hat{T}$ is 0 on $(-a, a)$, then Theorem 2.5.12 *b* allows us to define $\mathcal{H}T$ as $\mathcal{F}^{-1}\sigma(H)\mathcal{F}T$.

2.5.14 Remark. Perspective on $\mathcal{H}$

Suppose $f \in L^2(\mathbb{R})$. Then it is easy to check that

$$\forall t \in \mathbb{R}, \qquad \mathcal{H}(\tau_t f) = \tau_t(\mathcal{H}f) \tag{2.5.14}$$

and

$$\forall \lambda > 0, \qquad \mathcal{H}f_\lambda = (\mathcal{H}f)_\lambda, \tag{2.5.15}$$

i.e., the Hilbert transform $\mathcal{H} : L^2(\mathbb{R}) \longrightarrow L^2(\mathbb{R})$, which is continuous by Theorem 2.5.12, commutes with translations and positive dilations. The algebraic facts (2.5.14) and (2.5.15) should be juxtaposed with the fact that the only continuous

operators $L : L^2(\mathbb{R}) \longrightarrow L^2(\mathbb{R})$ that commute with translations and both positive
and negative dilations are constant multiples of the identity, e.g., [Ste70, pages
55–56].

The fact, that $\mathcal{H}$ commutes with translations and has a multiplier associated with
it, is a feature in common with a large class of operators, some of which play an
important role in the next section.

2.6. Operational Calculus

As Norbert Wiener points out in his paper on the operational calculus [Wie81,
26c], "the operational calculus owes its inception to Leibniz, who was struck by
the resemblance between the formula for the n-fold differentiation of a product
and the nth power of a sum", e.g., Exercise 2.28. To fix ideas, we shall think
of the *operational calculus* as a symbolic method of solving an equation. For
example, if we are given the differential equation $Lf = g$ for a given forcing
function g and differential operator L, the goal is to design an operator L^{-1} so
that $L^{-1}g \equiv f$ is a solution of the equation. Major contributions in this area are
due to Lagrange, Boole, and Pincherle; but the most spectacular, nonrigorous, and
successful "formal theory" is Oliver Heaviside's operational calculus [Hea1894],
cf., Heaviside's biographies [Nah88], [Sea87], the former professional, the latter
personal, and both exquisite.

Before developing his operational calculus, Heaviside had made a profound
contribution to the Atlantic submarine cable problem, cf., Exercise 1.47. The operators in his operational calculus applied to voltages and currents, gave a meaning
to fractional differentiation, led to asymptotic series that could be successfully
applied in computations, and infuriated the "wooden-headed" (Heaviside's appel-
lation) mathematicians of his day. Lest we mathematicians become a splinter (sic)
group, Laurent Schwartz came along with his theory of distributions to legitimize
Heaviside's ideas—which, of course, may have infuriated Heaviside! A concept
in Wiener's paper on the operational calculus [Wie81, 26c, Section 8] played a
role in Schwartz's motivation to define distributions; although, at the time of his
research, Schwartz was unaware of Wiener's contribution and attributed it to related work done by others after Wiener, e.g., [Wie81, Volume II, pages 426–427],
[Sch66, page 4].

Other twentieth century work on the operational calculus includes contributions
by Volterra (1920), Carson (1925), van der Pol and Bremmer (1955), Doetsch
(1958), and Mikusiński (1959), as well as books such as [Erd62], [Sch61, pages
123–140 and 230–235], [Yos84], and [Zem87].

In the following result the distributional proof is correct up to the last step. The
subtleties are discussed in Remark 2.6.2.

2.6.1 Theorem. Differential Equation and Operational Solution

Let $\{a_n : n = 0, \ldots, N\} \subseteq \mathbb{C}$, $a_N \neq 0$, and let $S \in \mathcal{S}'(\mathbb{R})$. A solution $X \in \mathcal{S}'(\mathbb{R})$ of the differential equation

$$LX = \sum_{n=0}^{N} a_n X^{(n)} = S \tag{2.6.1}$$

is

$$X = \left(\frac{\hat{S}}{P}\right)^{\vee}, \quad \text{where} \quad P(\gamma) = \sum_{n=0}^{N} (2\pi i)^n a_n \gamma^n. \tag{2.6.2}$$

PROOF. By Proposition 2.5.5, (2.6.1) can be written as

$$X * \left(\sum_{n=0}^{N} a_n \delta^{(n)}\right) = S.$$

Thus, by the exchange formula, $\hat{X} P = \hat{S}$. Dividing by P and taking the inverse transform yield (2.6.2). ∎

2.6.2 Remark. Division of Distributions

a. If the polynomial P of Theorem 2.6.1 does not have real zeros, then the reasoning in the last step of the proof (of Theorem 2.6.1) is correct, and little further detail is needed. If P has real zeros, then a more elaborate argument is required that depends on the relatively elementary fact that *if $T \in D'(\mathbb{R})$ and P is a polynomial on $\mathbb{R}$, then there is a distribution $X \in D'(\mathbb{R})$ for which $PX = T$*, e.g., [Sch66, Chapitre V.4, pages 123–126].

b. The *division problem* for $\mathbb{R}^d$, $d \geq 2$, i.e., the solution of the distributional equation $PX = T$ on $\mathbb{R}^d$ for given distributions T and polynomials P, is difficult. It was solved by Hörmander and Lojasiewicz independently in the late 1950s. There is a wonderful exposition of the topic including its relationship with partial differential equations in [Sch63].

c. Let L be a partial differential operator on $\mathbb{R}^d$. A distribution E for which $LE = \delta$ is a *fundamental solution* of L. A great success of distribution theory is the theorem that *if L has constant coefficients, then it has a fundamental solution.* This result was proved by Ehrenpreis and Malgrange independently in 1953, e.g., [Hör83]. There is now a elementary proof of this theorem in [Ros91].

Suppose L has constant coefficients and $S \in \mathcal{E}'(\mathbb{R}^d)$ is a "forcing function". Then $X = E * S$ is a solution of the equation $LX = S$ since, formally,

$$L(E * S) = \delta * S = S, \tag{2.6.3}$$

cf., the proof of Theorem 2.6.3 *b*. Every other solution is of the form $E * S + T$, where T is a solution of the corresponding homogeneous equation, i.e., $LT = 0$.

We shall not get involved in aspects of Heaviside's calculus, such as his *expansion theorem*, which caused "gorges to rise" in yesteryear. However, we shall prove the following result, which quantifies the point of view of the first paragraph of this section. In light of the proof of Theorem 2.6.1, let us first clarify the notation for a differential operator L. If LX has the designation

$$LX = \sum_{n=0}^{N} a_n X^{(n)},$$

then

$$L\delta = \sum_{n=0}^{N} a_n \delta^{(n)} \in D'(\mathbb{R}),$$

and so

$$LX = (L\delta) * X.$$

2.6.3 Theorem. Differential Equation and Fundamental Solution

Let $\{a_n : n = 0, \dots, N\} \subseteq \mathbb{C}$, $a_N = 1$, and let $h \in C^\infty(\mathbb{R})$ be the solution of the initial value problem

$$(L\delta) * h = Lh = 0 \quad \text{on } \mathbb{R},$$
$$h(0) = \dots = h^{(N-2)}(0) = 0, \qquad h^{(N-1)}(0) = 1. \tag{2.6.4}$$

(*The method to find the solution of (2.6.4) is elementary and well known in ordinary differential equations, e.g., [Swe96].*)

    a. $(L\delta) * (Hh) = \delta$, where H is the Heaviside function. Thus, $E \equiv Hh$ is a fundamental solution of L.

    b.   Let $S \in \mathcal{E}'(\mathbb{R})$. Then $X = E * S$ is a solution of the differential equation $LX = S$.

PROOF.   *a.*   We shall evaluate $(Hh)^{(n)}, n = 1, \dots, N$. Note that $(Hh)^{(1)} = Hh^{(1)} + \delta h$, e.g., Exercise 2.28, and $\delta h = h(0)$. The second fact results from the calculation $(\delta h)(f) = \delta(hf) = h(0)f(0) = (h(0)\delta)(f)$. By hypothesis, then,

$$(Hh)^{(1)} = Hh^{(1)}.$$

Continuing to use the product rule and (2.6.4), in conjunction with the calculation from the previous step, we obtain

$$(Hh)^{(2)} \quad = \quad Hh^{(2)} + \delta h^{(1)} \qquad = \quad Hh^{(2)},$$

$$\vdots$$

$$(Hh)^{(N-1)} \quad = \quad Hh^{(N-1)} + \delta h^{(N-2)} \quad = \quad Hh^{(N-1)},$$

$$(Hh)^{(N)} \quad = \quad Hh^{(N)} + \delta h^{(N-1)} \quad = \quad Hh^{(N)} + \delta.$$

Thus,

$$(L\delta) * (Hh) = \sum_{n=0}^{N} a_n (Hh)^n = H(Lh) + \delta = \delta.$$

b. We calculate

$$L(E * S) = (L\delta) * (E * S) = (L\delta * E) * S = \delta * S = S.$$

Note that we have assumed associativity of convolution in the second equality, cf., Example 2.5.2 c where we showed such associativity is not generally valid. We are justified in this case since the distributions $L\delta$, $E = Hh$, and S are each supported in an interval of the $[c, \infty)$, e.g., [Sch66, page 172]. ∎

2.6.4 Example. $L\delta = \sum_{n=0}^{N} a_n \delta^{(n)}$

It is easy to see that $\operatorname{supp} L\delta = \{0\}$. Conversely, *if $T \in D'(\mathbb{R})$ and $\operatorname{supp} T = \{0\}$, then there is $\{a_0, \ldots, a_N\} \subseteq \mathbb{C}$ such that $T = L\delta$.*

This important fact is a corollary of a theorem, e.g., [Hör83, Volume I, pages 46–47], that depends on a "philosophy" (a big word, but "point of view" is not quite accurate in this case) underlying significant parts of spectral synthesis [Ben75], [KS63] and potential theory [Hed80]. To describe this "philosophy", first note that *if $g \in L^1(\mathbb{R})$ and $f \in C_0(\mathbb{R})$ vanishes on $\operatorname{supp} g$, then $\int g(t)f(t)\,dt = 0$.* Similarly, *if $\mu \in M_b(\mathbb{R})$ and $f \in C_0(\mathbb{R})$ vanishes on $\operatorname{supp} \mu$, then $\mu(f) = \int f(t)d\mu(t) = 0$.* However, *if $T \in A'(\mathbb{R})$ and $f \in A(\mathbb{R})$ vanishes on $\operatorname{supp} T$, then it is not necesssarily true that $T(f) = 0$.* This last fact is a deoderized version of Malliavin's profound example of nonsynthesis. Suppose that $f \in X$ and $T \in X'$, where $X \supseteq C_c^\infty(\mathbb{R})$ and $X' \subseteq D'(\mathbb{R})$ satisfy the natural condition of Example 2.4.6 h. The "philosophy", alluded to above, is that if $f = 0$ on $\operatorname{supp} T$ and if the set $\operatorname{supp} T$ is regular enough vis a vis the smoothness of f near the boundary of $\operatorname{supp} T$, then $T(f) = 0$.

2.6.5 Definition. Linear Translation-Invariant Systems

a. Let X be a linear subspace of $D'(\mathbb{R})$ with the properties that $\delta \in X$ and that if $T \in X$, then

$$\forall t \in \mathbb{R}, \qquad \tau_t T \in X.$$

A *linear translation-invariant (LTI) system* is a linear operator $L : X \to X$ for which

$$\forall T \in X \text{ and } \forall t \in \mathbb{R}, \qquad L(\tau_t T) = \tau_t L(T). \qquad (2.6.5)$$

Property (2.6.5) is *translation invariance*, and for many applications this reflects *time invariance*. The property

$$\forall S, T \in X \text{ and } \forall a, b \in \mathbb{C}, \qquad L(aS + bT) = aL(S) + bL(T)$$

is the *linearity* of L.

The *impulse response* of the system L is

$$L\delta = h \in X.$$

The *filter* corresponding to L is $\hat{h}$; $\hat{h}$ is also referred to as the *frequency response* or *transfer function* corresponding to L.

b.   Norbert Wiener argues persuasively for studying translation-invariant operators on physical grounds [Wie33, Introduction], i.e., although time shifts may distort some astronomical observations, laboratory experiments should generally be time invariant. Wavelets and nonstationary methods can be used in dealing with *time-varying* events.

The notion of an LTI system L is an important and basic engineering concept, e.g., [OS75], [OW83], [Pap77], that also has a long history in mathematics, e.g., [Ben75, pages 216–217]. Suppose X is closed under convolution, e.g., $X = M_b(\mathbb{R})$. Then, if L satisfies certain natural conditions, e.g., part c below, the impulse response h and filter $\hat{h}$ play a central role in quantifying L, viz.,

$$\forall f \in X, \qquad Lf = h * f. \qquad (2.6.6)$$

c.   We have already defined the notion of a causal signal. We now say that an LTI system L is *causal* if, whenever $T \in \mathbb{R}$ and $f \in X$ vanishes on $(-\infty, T)$, then Lf vanishes on $(-\infty, T)$. This means that there cannot be an output signal from the system L before there is an input signal—a reasonable point of view.

Definition 2.6.5 can obviously be extended to operators $L : X \longrightarrow Y$. The following result is due to [AN79], and a precursor depending on the continuity of L is due to [Sch66, pages 197–198]. *Let $L : C_c^\infty(\mathbb{R}) \to D'(\mathbb{R})$ be a causal LTI system. Then there is a unique distribution $h \in D'(\mathbb{R})$ such that $Lf = h * f$ for all $f \in C_c^\infty(\mathbb{R})$.* A feature of this theorem is that continuity of L is not required a priori to obtain (2.6.6).

2.6.6 Example. Translation, Convolution, and $Lf = h * f$

A fundamental theorem in the *theory of multipliers* is that *the translation-invariant continuous linear operators $L : L^1(\mathbb{R}) \to L^1(\mathbb{R})$ are precisely of the form*

$$\forall f \in L^1(\mathbb{R}), \qquad Lf = \mu * f, \qquad (2.6.7)$$

where $\mu \in M_b(\mathbb{R})$, e.g., [Lar71]. It is a routine calculation to prove that if $\mu \in M_b(\mathbb{R})$ and L is defined by (2.6.7), then $L : L^1(\mathbb{R}) \rightarrow L^1(\mathbb{R})$ is a translation-invariant continuous linear operator, e.g., Exercise 2.58 *a*. The converse is more difficult, and the initial steps of its proof are the contents of Exercise 2.58 *b,c*.

2.6.7 Example. Bandpass and Lowpass Filters—The Hilbert Transform

In Example 1.2.2 *b* we discussed modulated signals and their carrier frequencies. This example continues that discussion in terms of lowpass and bandpass filters, and the fact that they are related by the Hilbert transform. To fix ideas, we say that a filter $\hat{h}$ is *bandpass* if supp $\hat{h} = [-\beta, -\alpha] \bigcup [\alpha, \beta]$, where $\beta > \alpha > 0$, and it is *lowpass* if supp $\hat{h} = [-\alpha, \alpha]$.

   a.   Let L be an LTI system with a real impulse response $h \in PW_\Omega$; in particular, h is a lowpass filter. Then the modulated signal $h_b(t) \equiv h(t) \cos 2\pi t \gamma_0$, with carrier frequency $\gamma_0 > \Omega$, is the impulse response for the bandpass filter $\hat{h}_b = A_b e^{i\varphi_b}$ with *bands* $[\gamma_0 - \Omega, \gamma_0 + \Omega]$, $[-\gamma_0 - \Omega, -\gamma_0 + \Omega]$; and A_b and φ_b satisfy the properties

$$\forall \gamma \in [0, \Omega], \; A_b(\gamma + \gamma_0) = A_b(-\gamma + \gamma_0) \text{ and}$$
$$\varphi_b(\gamma + \gamma_0) = -\varphi_b(-\gamma + \gamma_0), \tag{2.6.8}$$

i.e., in the band $[\gamma_0 - \Omega, \gamma_0 + \Omega]$, A_b satisfies mirror symmetry, and φ_b satisfies "point" symmetry about γ_0, e.g., Exercise 2.59c.

   b.   Conversely, let L_b be an LTI system with a real impulse response h_b and bandpass filter $\hat{h}_b = A_b e^{i\varphi_b}$. Assume the bands of $\hat{h}_b$ are $[\gamma_0 - \Omega, \gamma_0 + \Omega]$, $[-\gamma_0 - \Omega, -\gamma_0 + \Omega]$, where $\gamma_0 > \Omega$, and that the symmetry condition (2.6.8) is satisfied. We shall verify that h_b is a modulated signal $h_b(t) = h(t) \cos 2\pi t \gamma_0$, with carrier frequency γ_0, where h is the impulse response for a lowpass filter $\hat{h}$ for which supp $\hat{h} = [-\Omega, \Omega]$.

   Let $\hat{h} = A e^{i\varphi}$ be defined as

$$\hat{h}(\gamma) = 2(\hat{h}_b H)(\gamma + \gamma_0), \tag{2.6.9}$$

where H is the Heaviside function. By definition, it is easy to see that supp $\hat{h} = [-\Omega, \Omega]$, e.g., Figure 2.1. Exercise 2.57a, where it is required to verify

$$f + i\mathcal{H}f \longleftrightarrow 2\hat{f}H,$$

and (2.6.9) allow us to write

$$h(t) = (h_b + i\mathcal{H}h_b)(t)e^{-2\pi i t \gamma_0}. \tag{2.6.10}$$

   It turns out that (2.6.8) allows us to conclude that h is real-valued. In fact, (2.6.8) implies $\hat{h}_b(\gamma + \gamma_0) = \overline{\hat{h}_b(-\gamma + \gamma_0)}$ for $\gamma \in [-\Omega, \Omega]$; and hence, by (2.6.8), $\hat{h}(\gamma) = \overline{\hat{h}(-\gamma)}$ on $\hat{\mathbb{R}}$, i.e., h is real-valued.

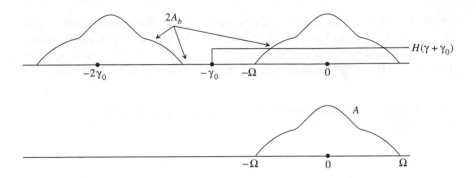

FIGURE 2.1.

Thus, by (2.6.10), we have $h_b(t) = h(t) \cos 2\pi t \gamma_0$ and our calculation is complete.

c. Parts *a* and *b* combine for a result of both theoretical and practical value. Theoretically, we establish isomorphisms between PW_Ω and spaces of $L^2(\mathbb{R})$ functions whose Fourier transforms are supported by more unusual sets than $[-\Omega, \Omega]$. Practically, ideas based on the above calculations are used in bandpass sampling, in the design of equivalent systems where one may be more efficient than another in some desired way, or in narrow band communications theory, e.g., [OS75, Chapters 7 and 10].

2.7. Measure Theory

We shall address two profound ideas from modern analysis: the Riesz Representation Theorem (RRT) and the Herglotz–Bochner Theorem. In the process we hope to tie together our discussions of Radon measures in Definition 2.3.6 and of real analysis in Appendix A. Integration and measure, truly a "keystone combination" for analysis, began with ancient efforts to *define* areas of nonrectilinear regions, e.g., [Ben77] on Archimedes and integration, reached a new level of precision and generality in the mid-twentieth century to cope with emerging requirements of trigonometric series, e.g., [Rie1873] and Section 3.2, generalized around 1900 to measure bizarre sets and to grapple with subjects such as statistical mechanics, e.g., [Wie81, Volume II, page 801] on sets of measure 0, and assumed a sophisticated twentieth century functional analytic identity by means of the RRT, e.g., [Ries49] for a readable history for the first half of this century by the master, cf., [Gra84].

2.7.1 Definition. Dual Spaces Associated with RRT

a. A linear functional $T : C_c(\mathbb{R}) \longrightarrow \mathbb{C}$ is an element of the *dual space* $C_c(\mathbb{R})'$ (of the vector space $C_c(\mathbb{R})$) if $\lim_{n\to\infty} T(f_n) = 0$ for every sequence $\{f_n\} \subseteq C_c(\mathbb{R})$ satisfying the properties:

i. $\exists K \subseteq \mathbb{R}$, compact, such that $\forall n$, $\operatorname{supp} f_n \subseteq K$;

ii. $\lim_{n\to\infty} \|f_n\|_{L^\infty(\mathbb{R})} = 0$.

These two properties define a topology on $C_c(\mathbb{R})$, and $C_c(\mathbb{R})'$ is the space of *continuous linear functionals* on $C_c(\mathbb{R})$.

b. A linear functional $T : C_0(\mathbb{R}) \longrightarrow \mathbb{C}$ is an element of the *dual space* $C_0(\mathbb{R})'$ (of the vector space $C_0(\mathbb{R})$) if $\lim T(f_n) = 0$ for every sequence $\{f_n\} \subseteq C_0(\mathbb{R})$ for which $\lim_{n\to\infty} \|f_n\|_{L^\infty(\mathbb{R})} = 0$. The sup norm, $\| \ldots \|_{L^\infty(\mathbb{R})}$, defines a topology on $C_0(\mathbb{R})$, by which it becomes a Banach space; and $C_0(\mathbb{R})'$ is the space of *continuous linear functionals* on $C_0(\mathbb{R})$, cf., Example B.17.

2.7.2 Remark. Preliminaries for RRT

a. If $F \in BV_{\mathrm{loc}}(\mathbb{R})$ there are two Radon measures, T_F and S_F, defined in terms of F, that are fundamental for RRT. Recalling that $BV_{\mathrm{loc}}(\mathbb{R}) \subseteq L^1_{\mathrm{loc}}(\mathbb{R})$ (Exercise 2.23), T_F is defined by (2.2.4) and (2.2.5) as

$$\forall f \in C_c^\infty(\mathbb{R}), \qquad T_F(f) = \int F(t) f(t) \, dt. \tag{2.7.1}$$

S_F is the *Riemann–Stieltjes Radon measure* defined in terms of the Riemann–Stieltjes integral [Apo57] as

$$\forall f \in C_c^\infty(\mathbb{R}), \qquad S_F(f) = \int f(t) dF(t). \tag{2.7.2}$$

It is easy to check that T_F and S_F are, in fact, Radon measures.

b. Equation (2.7.2) defines a mapping,

$$\begin{aligned} L_S : BV_{\mathrm{loc}}(\mathbb{R}) &\longrightarrow M(\mathbb{R}) \\ F &\longmapsto S_F, \end{aligned} \tag{2.7.3}$$

that is *not* injective. In fact, if $F \in BV_{\mathrm{loc}}(\mathbb{R})$ and $C \in \mathbb{C}$, then $F + C \in BV_{\mathrm{loc}}(\mathbb{R})$ and $S_F = S_{F+C}$. This should be compared to the mapping L of (2.2.5), restricted to $BV_{\mathrm{loc}}(\mathbb{R})$, which is injective.

c. Let $F \in BV_{\mathrm{loc}}(\mathbb{R})$. Clearly, the integration by parts formula for Riemann–Stieltjes integrals [Apo57],

$$\forall f \in C_c^\infty(\mathbb{R}), \qquad \int f(t)dF(t) = -\int f'(t)F(t)\,dt,$$

can be rewritten as $S_F = T_F'(= F')$.

The space $M(\mathbb{R})$, resp., $M_b(\mathbb{R})$, of Radon measures, resp., bounded Radon measures, was defined in Definition 2.3.6 in terms of distributional derivatives.

2.7.3 Theorem. Riesz Representation Theorem

a. $C_c(\mathbb{R})' = M(\mathbb{R})$, *the space of Radon measures. In fact, for every* $\mu \in M(\mathbb{R})$ *there is* $F \in BV_{\text{loc}}(\mathbb{R})$ *such that* $\mu = S_F = F'$.

b. $C_0(\mathbb{R})' = M_b(\mathbb{R})$, *the space of bounded Radon measures. In fact, for every* $\mu \in M_b(\mathbb{R})$ *there is* $F \in BV(\mathbb{R})$ *such that* $\mu = S_F = F'$.

2.7.4 Remark. RRT: Theorem to Definition

a. RRT asserts that the mapping (2.7.3) is a surjection, and that its restriction to $BV(\mathbb{R})$ is a surjection onto $M_b(\mathbb{R})$. An important part of the proof of RRT utilizes the theory of integration from Appendix A to extend $\mu \in C_0(\mathbb{R})'$ to a functional on the vector space generated by linear combinations of characteristic functions of intervals.

Frigyes (Frederick) Riesz proved RRT in 1909. Although Riesz did not deal with distributions, it is most efficient to write RRT as we did in Theorem 2.7.3, e.g., [Sch66, pages 53–54], cf., [Ries14], [RN55], [Ben76, pages 255–257] for classical and readable proofs.

RRT has evolved from a *theorem*, associating certain linear functionals with elements of $BV_{\text{loc}}(\mathbb{R})$ by means of the Riemann–Stieltjes integral, to the *definition of integral* in terms of such functionals, e.g., [Bou65]. Notation reflects this meta-morphisis, e.g., (2.3.10). If $\mu \in M(\mathbb{R})$, then μ is a distribution $\mu = F'$ for some $F \in BV_{\text{loc}}(\mathbb{R})$, $\mu \in C_c(\mathbb{R})'$ where $\mu(f)$ is denoted by $\int f(t)d\mu(t)$ but is really the Riemann–Stieltjes integral $\int f(t)dF(t)$, *and there is a space* $L_\mu^1(\mathbb{R})$ *of functions integrable with respect to* μ. This last concept is developed in integration theory by extending the functional $\mu : C_c(\mathbb{R}) \to \mathbb{C}$ to a large space, viz., $L_\mu^1(\mathbb{R})$, e.g., [Ben76], [Bou65], [Mal82], [Rud66], cf., part *b*. In this setting $L^1(\mathbb{R})$ is the space of functions integrable with respect to Lebesgue measure.

b. In integration theory we define the *Borel algebra* $\mathcal{B}(\mathbb{R})$ to be the smallest σ-algebra of subsets of $\mathbb{R}$ that contains all the open sets. A function $\mu : \mathcal{B}(\mathbb{R}) \to \mathbb{R}^+ \cup \{\infty\}$ is a *locally finite Borel measure* if the following conditions are satisfied:

i. $\mu(\cup B_n) = \sum \mu(B_n)$ for every disjoint sequence $\{B_n\} \subseteq \mathcal{B}(\mathbb{R})$;

ii. $\mu(K) < \infty$ for every compact set $K \subseteq \mathbb{R}$.

Let $\mathcal{M}_+(\mathbb{R})$ be the space of locally finite Borel measures.

An integral (sic) part of the technique and philosophy associated with RRT is the following result. Define the mapping

$$I : M_+(\mathbb{R}) \longrightarrow M_+(\mathbb{R})$$
$$\mu \longmapsto I_\mu,$$

where, for all $f \in C_c(\mathbb{R})$, $I_\mu(f)$ is the integral of f with respect to μ, as defined in integration theory. F. Riesz and Radon proved the existence and uniqueness theorem that *I is a bijection*, e.g., [Mal82, pages 61–76].

2.7.5 Definition. Continuous and Discrete Measures

a. A Radon measure $\mu \in M(\mathbb{R})$ is *discrete* if μ is of the form $\mu = \sum_{x\in S} a_x \delta_x$, where each $a_x \in \mathbb{C}\backslash\{0\}$ and $\sum_{x\in S} |a_x| < \infty$. Thus, each such index set S is countable, cf., Exercise 2.33. $M_d(\mathbb{R})$ denotes the space of discrete measures. Clearly, $M_d(\mathbb{R}) \subseteq M_b(\mathbb{R})$.

b. A Radon measure $\mu \in M(\mathbb{R})$ is *continuous* if $F \in BV_{\mathrm{loc}}(\mathbb{R})$ can be chosen as a continuous function on $\mathbb{R}$ in the representation $F' = \mu$.

The Cantor measure μ_C defined in Example 2.3.9b is a continuous measure, as are the elements of $L^1(\mathbb{R})$.

c. The *problem of spectral estimation* can be viewed in terms of determining the discrete part of Radon measures, e.g., Definition 2.8.6.

The following result is one of the basic decompositions of real analysis, e.g., [Ben76], [Bou65], [Rud66].

2.7.6 Theorem. Decomposition of Measures

Let $F \in BV(\mathbb{R})$ and $\mu \in M_b(\mathbb{R})$, and assume $F' = \mu$.

a. $F = F_{ac} + F_{sc} + \sum a_x \tau_x H$, where H is the Heaviside function, $F_{ac} \in BV(\mathbb{R}) \cap AC_{\mathrm{loc}}(\mathbb{R})$, $\sum |a_x| < \infty$, F_{sc} is continuous on $\mathbb{R}$, and $DF_{sc} = 0$ a.e., where D is ordinary pointwise differentiation.

b. $\mu = g + \mu_{sc} + \sum a_x \delta_x$, where $g \in L^1(\mathbb{R})$, $\sum |a_x| < \infty$, and $\mu_{sc} \in M_b(\mathbb{R})$ is designated the continuous singular part of μ.

c. $g = F_{ac}'$, $\mu_{sc} = F_{sc}'$, and $\sum a_x \delta_x = (\sum a_x \tau_x H)'$.

d. For all $f \in C_c(\mathbb{R})$,

$$\mu(f) = \int g(t)f(t)\,dt + \int f(t)\,dF_{sc}(t) + \sum a_x f(x),$$

where the integrals on the right side are Lebesgue and Riemann–Stieltjes integrals, respectively.

e. $\|\mu\|_1 = \|g\|_{L^1(\mathbb{R})} + \|\mu_{sc}\|_1 + \sum |a_x|$.

2.7.7   Theorem. Positive Distributions

If $T \in D'(\mathbb{R})$ is positive, then $T \in M_+(\mathbb{R})$.

PROOF.   Let $\{f_n\} \subseteq C_c^\infty(\mathbb{R})$ satisfy the properties that there is a compact set $K \subseteq \mathbb{R}$ such that supp $f_n \subseteq K$ for each n and that $\lim_{n \to \infty} \|f_n\|_{L^\infty(\mathbb{R})} = 0$. We shall prove that $\lim_{n \to \infty} T(f_n) = 0$, from which we can conclude that $T \in M(\mathbb{R})$ by RRT.

Let nonnegative $f \in C_c^\infty(\mathbb{R})$ equal 1 on K. By the uniform convergence of $\{f_n\}$ there is $\{\epsilon_n\} \subseteq (0, \infty)$ decreasing to 0 such that

$$\forall t \text{ and } \forall n, \qquad |f_n(t)| \leq \epsilon_n f(t).$$

A straightforward calculation yields the fact that the real and imaginary parts of each f_n are in $C_c^\infty(\mathbb{R})$, and so we assume that each f_n is real-valued. Thus,

$$\forall t \text{ and } \forall n, \qquad -\epsilon_n f(t) \leq f_n(t) \leq \epsilon_n f(t),$$

so that, by the positivity assumption, $\lim_{n \to \infty} T(f_n) = 0$, and, hence, $T \in M(\mathbb{R})$.   ∎

2.7.8   Definition. Positive Definite Functions

a.   A function $P : \hat{\mathbb{R}} \to \mathbb{C}$ is *positive definite* if

$$\forall n, \forall c_1, \ldots, c_n \in \mathbb{C}, \text{ and } \forall \gamma_1, \ldots, \gamma_n \in \hat{\mathbb{R}},$$

$$\sum_{j,k} c_j \bar{c}_k P(\gamma_j - \gamma_k) \geq 0. \tag{2.7.4}$$

In this case we write $P >> 0$.

b.   If $P >> 0$, then:

i. $\forall \gamma \in \hat{\mathbb{R}}, P(\gamma) = \overline{P(-\gamma)}$;

ii. $\forall \gamma \in \hat{\mathbb{R}}, |P(\gamma)| \leq P(0)$, and hence $P(0) \geq 0$ and $P \in L^\infty(\hat{\mathbb{R}})$;

iii. $\forall \lambda, \gamma \in \hat{\mathbb{R}}$,

$$|P(\gamma) - P(\lambda)|^2 \leq 2P(0) \operatorname{Re}(P(0) - P(\gamma - \lambda));$$

iv. P continuous at 0 implies P is continuous on $\hat{\mathbb{R}}$ (by part *iii*), e.g., Exercise 2.52 for verification of these facts and the relationship to positive definite matrices.

c.   Positive definite functions arise in *moment problems*, e.g., [RN55], [Wid41]; and, since (2.7.4) is a pure quadratic form, such functions also arise in minimization problems (in economics, for example) and minimum principles (in physics, for example), e.g., [CH53], [Str88].

2.7.9 Example. Positive Definite Functions

a. If $F \in L^2(\mathbb{R})$, then the L^2-*autocorrelation* of F, viz., $P \equiv F * \widetilde{F}$, is positive definite. In fact,

$$
\sum_{j,k} c_j \bar{c}_k P(\gamma_j - \gamma_k) = \int \sum_{j,k} c_j \bar{c}_k F(\gamma_j - \lambda) \overline{F(\gamma_k - \lambda)} \, d\lambda
$$

$$
= \int \left| \sum_j c_j F(\gamma_j - \lambda) \right|^2 d\lambda \geq 0.
$$

Clearly, $P \in A(\hat{\mathbb{R}})$ since $(F * \widetilde{F})^{\vee} = |F^{\vee}|^2 \in L^1(\mathbb{R})$.

b. If $\mu \in M_{b+}(\mathbb{R})$, then $P \equiv \hat{\mu} >> 0$ and P is continuous.

For the continuity, if $\mu \in M_b(\mathbb{R})$, then it is easy to check that $\mu \in \mathcal{S}'(\mathbb{R})$ and

$$
\forall \gamma \in \hat{\mathbb{R}}, \qquad \hat{\mu}(\gamma) = \int e^{-2\pi i t \gamma} \, d\mu(t),
$$

e.g., Remark B.16. This formulation in terms of an integral allows us to conclude that $\hat{\mu}$ is continuous on $\hat{\mathbb{R}}$.

To prove that $P >> 0$, it is only necessary to make the computation

$$
\sum_{j,k} c_j c_k \hat{\mu}(\gamma_j - \gamma_k) = \int \left| \sum_j c_j e^{-2\pi i t \gamma_j} \right|^2 d\mu(t) \geq 0.
$$

c. The set of positive definite functions is not a vector space. However, if $P, Q >> 0$, then $P + Q >> 0$ and $PQ >> 0$. The latter fact follows from a theorem of I. Schur, and is an easy calculation such as those of parts *a, b* in the case $Q \in M_{b+}(\mathbb{R})^{\hat{}}$.

In Example 2.7.9*b* we showed that if $\mu \in M_{b+}(\mathbb{R})$, then $\hat{\mu}$ is a continuous positive definite function on $\hat{\mathbb{R}}$. The converse of this fact is the *Herglotz–Bochner Theorem* (Theorem 2.7.10). Herglotz (1911) proved it for positive definite functions on $\mathbb{Z}$. The proofs for positive definite functions on $\hat{\mathbb{R}}$ and locally compact abelian groups are due to Bochner (1933) and Weil (1938), respectively. There are several conceptually different proofs, e.g., [Don69], [Kat76], [Rud62], cf., [Sch66].

2.7.10 Theorem. Herglotz–Bochner Theorem

Let $P : \hat{\mathbb{R}} \to \mathbb{C}$ be a continuous positive definite function on $\hat{\mathbb{R}}$. Then there is a unique positive bounded Radon measure $\mu \in M_b(\mathbb{R})$ for which $\hat{\mu} = P$ on $\hat{\mathbb{R}}$.

PROOF.   Let $F \in \mathcal{S}(\hat{\mathbb{R}})$. Since $P \in C_b(\hat{\mathbb{R}})$, the integral

$$\iint P(\gamma - \lambda) F(\gamma) \overline{F(\lambda)} \, d\gamma \, d\lambda \tag{2.7.5}$$

is the limit of the Riemann sums

$$\sum_j \sum_k P(\gamma_j - \lambda_k) F(\gamma_j) \overline{F(\lambda_k)} \Delta \gamma_j \Delta \lambda_k. \tag{2.7.6}$$

$P >> 0$ and (2.7.6) allow us to conclude that the integral of (2.7.5) is nonnegative. Rewriting (2.7.5), we have

$$0 \le \iint P(\lambda) F(\gamma) \overline{F(\gamma - \lambda)} \, d\gamma \, d\lambda = \int P(\lambda) F * \tilde{F}(\lambda) \, d\lambda.$$

Since $C_b(\hat{\mathbb{R}}) \subseteq \mathcal{S}'(\hat{\mathbb{R}})$, we can compute

$$P(F * \tilde{F}) = P^{\vee}(|\check{F}|^2);$$

and so $P^{\vee} \in \mathcal{S}'(\mathbb{R})$ is nonnegative on all $k \in \mathcal{S}(\mathbb{R})$ of the form $k = |h|^2, h \in \mathcal{S}(\mathbb{R})$.

Now let $f \in C_c^{\infty}(\mathbb{R})$ be nonnegative. For each $\epsilon > 0$, $f + \epsilon^2 g \in \mathcal{S}(\mathbb{R})$ is positive on $\mathbb{R}$, where g is the normalized Gaussian; and $h \equiv (f + \epsilon^2 g)^{1/2} \in \mathcal{S}(\mathbb{R})$. Thus, $P^{\vee}(f + \epsilon^2 g) = P^{\vee}(h^2) = P^{\vee}(|h|^2) \ge 0$. By linearity and the fact that $\epsilon > 0$ is arbitrary, we see that $P^{\vee}(f) \ge 0$; and so $\mu \equiv P^{\vee} \in M(\mathbb{R})$ by Theorem 2.7.7.

Finally, we must show that $\mu \in M_b(\mathbb{R})$. First note that

$$\forall \alpha > 0, \qquad P * g_\alpha(0) \le P(0),$$

where g_α is the dilation of g. This follows since $|P(\gamma)| \le P(0)$ and

$$\forall \gamma \in \hat{\mathbb{R}},$$

$$|P * g_\alpha(\gamma)| \le \int |P(\gamma - \lambda)| g_\alpha(\lambda) \, d\lambda \le P(0) \int g_\alpha(\lambda) \, d\lambda = P(0).$$

Next note that the approximate identity $\{g_\alpha\}$ satisfies the following properties: $g_\alpha \ge 0$, $(g_\alpha)^{\vee} \ge 0$, and g_α is even. Consequently, we have the computation

$$P(0) \ge |P * g_\alpha(0)| = \left| \int P(\gamma) g_\alpha(\gamma) d\gamma \right|$$

$$= |\alpha P^{\vee}((g_{1/\alpha})^{\vee})| = \int g^{\vee}(\alpha t) d\mu(t) \ge 0.$$

We can apply the Beppo Levi Theorem (Theorem A.8) since $0 \le g^{\vee}(\alpha t) \le g^{\vee}(\beta t)$ for $\alpha \le \beta$ and since $\lim_{\alpha \to \infty} g^{\vee}(\alpha t) = 1$ by Proposition 1.6.11. Thus

$$0 \le \int d\mu(t) \le P(0)$$

for the positive measure μ, and this allows us to conclude that $\mu \in M_b(\mathbb{R})$. ∎

2.7.11 Definition. Fourier–Stieltjes Transforms

a. The space of Fourier transforms of bounded Radon measures is denoted by $B(\hat{\mathbb{R}})$, i.e.,

$$B(\hat{\mathbb{R}}) = \{F : \hat{\mathbb{R}} \to \mathbb{C} : \exists \mu \in M_b(\mathbb{R}) \text{ such that } \hat{\mu} = F\}.$$

An element of $B(\hat{\mathbb{R}})$ is a *Fourier–Stieltjes transform*, e.g., [Rud62] for an elegant, incisive, authoritative exposition.

b. It is elementary to check that the elements of $B(\hat{\mathbb{R}})$ are uniformly continuous members of $C_b(\hat{\mathbb{R}})$. It is natural to ask for an intrinsic characterization of $B(\hat{\mathbb{R}})$ as we did for $A(\hat{\mathbb{R}})$ in Example 1.4.4, i.e., to seek a theorem of the form, "a uniformly continuous element $F \in C_b(\hat{\mathbb{R}})$ is an element of $B(\hat{\mathbb{R}})$ if and only if ...", where "..." is a statement about the behavior of F on $\hat{\mathbb{R}}$. In spite of Theorem 2.7.10, and some wonderful contributions by A. C. Berry (1931) for $A(\hat{\mathbb{R}})$, Bochner (1934), Schoenberg (1934), Kreĭn (1940), Yosida (1944), R. S. Phillips (1950), and Doss (1971), the problem remains unsolved, cf., Exercise 2.60.

Even though $B(\hat{\mathbb{R}})$ is much larger than $A(\hat{\mathbb{R}})$ there are still elements from $C_0(\hat{\mathbb{R}})$, and even $C_c(\hat{\mathbb{R}})$, that are not in $B(\hat{\mathbb{R}})$. In fact, membership in $B(\hat{\mathbb{R}})$ is a predominantly local property, and, in that sense, is closely related to membership in $A(\hat{\mathbb{R}})$ except at infinity, e.g., [Ben75, Definition 2.4.2] for an explanation of this opaque remark as well as further references, cf., Definition 3.5.6.

c. The hypotheses for the inversion formula in Theorem 1.7.8 can be weakened. In fact, if $f \in L^1(\mathbb{R}) \cap B(\mathbb{R})$, then $f \in A(\mathbb{R})$. To see this, let $\mu \in A(\hat{\mathbb{R}}) \cap M_b(\hat{\mathbb{R}})$ have the property that $\mu^{\vee} = f$. Since $\mu \in A(\hat{\mathbb{R}})$, we have $\mu \in L^1_{\text{loc}}(\hat{\mathbb{R}})$, and so μ has no discrete or continuous singular part. Thus, $\mu \in L^1(\hat{\mathbb{R}})$ by Theorem 2.7.6.

2.8. Definitions from Probability Theory

After our statement of the *Central Limit Theorem* in Example 1.6.8, we noted that the hypotheses involved the notion of a *probability density function (of a random variable) having mean* 0 *and variance* 1. We shall now define these and some other probabilistic terms that arise in Fourier analysis; but we do not make any pretense about explaining probabilistic ideas, see [Lam66], [Lam77], [Pri81] for such explanations.

2.8.1 Definition. Probabilistic Setup

a. A *probability measure* on $\mathbb{R}$ is an element $p \in M_{b+}(\mathbb{R})$ for which $\|p\|_1 = 1$.
The pair $(\mathbb{R}, p)$ is a *probability space*. The *distribution function* $F \in BV(\mathbb{R})$
associated with p is the increasing function

$$F : \mathbb{R} \longrightarrow [0, 1]$$

having the properties:

i. $\forall \alpha \in \mathbb{R}, \quad \lim_{\beta \to \alpha+} F(\beta) = F(\alpha)$;

ii. $F' = p$,

e.g., let p be the Dirac measure δ. Caveat: The (Schwartz) distributional derivative
in part *ii* has *nothing* to do with the probabilistic designation *distribution function*
that is given to F.

b. A *random variable*,

$$X : (\mathbb{R}, p) \longrightarrow \mathbb{R},$$

is a measurable function $X : \mathbb{R} \longrightarrow \mathbb{R}$. Measurable functions are defined as
limits of sequences of continuous functions in Definition A.10 *a*; there is also a
primordial measure-theoretic definition. The *cumulative distribution function* F_X
defined on the range of X is defined as

$$\forall t \in \mathbb{R}, \qquad F_X(t) = \int \mathbf{1}_{S(t)}(\alpha) \, dp(\alpha),$$

where integration is over the probability space $(\mathbb{R}, p)$ and

$$S(t) = \{\alpha \in \mathbb{R} : X(\alpha) \le t\}.$$

$F_X \in BV(\mathbb{R})$ is an increasing function

$$F_X : \mathbb{R} \longrightarrow [0, 1].$$

We have Figure 2.2.

$$\mathbb{R} \quad \xrightarrow{F} \quad [0, 1]$$

$$X \downarrow$$

$$\mathbb{R} \quad \xrightarrow{F_X} \quad [0, 1]$$

FIGURE 2.2.

$F_X{}'$ is a probability measure on the range of X. X has a *probability density function* (pdf) f_X if

$$F_X{}' = f_X \in L^1(\mathbb{R}),$$

i.e., f is the *pdf* of a random variable X if $f \geq 0$, $\int f(t)\,dt = 1$, and F_X can be written as

$$F_X(t) = \int_{-\infty}^{t} f(u)\,du.$$

2.8.2 Definition. Mean and Variance

a.   The *mean* or *expectation* of a random variable $X \in L^1_p(\mathbb{R})$ defined on a probability space $(\mathbb{R}, p)$ is

$$m_X = E\{X\} = \int X(\alpha)\,dp(\alpha),$$

where the value of the integral, which defines the notation m_X and $E\{X\}$, is $p(X)$ in our distributional notation, e.g., Remark 2.7.4. $X \in L^1_p(\mathbb{R})$ means that $p(|X|) \equiv \int |X(\alpha)|\,dp(\alpha) < \infty$.

b.   We can prove by straightforward calculations that

$$E\{X\} = \int X(\alpha)\,dF(\alpha) = \int t\,dF_X(t), \qquad (2.8.1)$$

where the integrals are Riemann–Stieltjes integrals over $\mathbb{R}$, the domain $\mathbb{R}$ of the first integral being the probability space $(\mathbb{R}, p)$, and the domain $\mathbb{R}$ of the second integral being the range space of X. In the case X has a pdf f_X, then (2.8.1) allows us to write

$$E\{X\} = \int t f_X(t)\,dt.$$

c.   Let $X \in L^2_p(\mathbb{R})$, i.e.,

$$\|X\|_{2,p} = \left(\int |X(\alpha)|^2\,dp(\alpha) \right)^{1/2} < \infty.$$

The *variance* σ_X^2 of the random variable X is

$$\sigma_X^2 = E\{(X - m_X)^2\}.$$

Therefore,

$$\sigma_X^2 = E\{X^2\} - (E\{X\})^2 = E\{X^2\} - m_X^2.$$

σ_X is the *standard deviation* of X. In the case X has a pdf f_X, then

$$\sigma_X^2 = \int (t - m_X)^2 \, f_X(t) \, dt.$$

In fact, a formal calculation shows that

$$\int g(X)(\alpha) \, dp(\alpha) = \int g(t) f_X(t) \, dt.$$

2.8.3 Example. Variance and Dispersion

The notion of variance can be viewed as providing more relevant information than that provided by knowing a mean value. For example, the complacency of knowing that the mean or average family income in a nation can provide a good standard of living would be offset by a revolution for justice if such a nation had some destitute and homeless in its population. Thus, it is important to know how a nation's wealth is *dispersed* among the population. Variance is a measure of dispersion.

To fix ideas, let P be a finite set, for example, a class of students; and let $X : P \longrightarrow \mathbb{R}$ be a random variable, for example, a test score, i.e., $X(\alpha)$ is the score that student $\alpha \in P$ received. (Technically, we should first put a measure p on P.) The *mean* (score) of X is

$$m_X = \frac{1}{\operatorname{card} P} \sum_{\alpha \in P} X(\alpha),$$

which is a discrete way of writing $m_X = \int_P X(\alpha) \, dp(\alpha)$, where p is a "probability measure" defined by

$$\forall A \subseteq P, \qquad p(A) = \frac{\operatorname{card} A}{\operatorname{card} P}.$$

The *average squared distance of X from the average* is the variance, viz.,

$$\sigma_X^2 = \frac{1}{\operatorname{card} P} \sum_{\alpha \in P} (X(\alpha) - m_X)^2.$$

Consequently, the standard deviation is

$$\sigma_X = \text{``} \|X - m_X\|_{L_p^2(P)} \text{''};$$

and it makes sense to discuss *dispersion* in terms of how many standard deviations a given value (or score) $X(\alpha)$ is from the mean.

2.8.4 Definition. Stochastic Processes

a. Let $(\mathbb{R}, p)$ be a probability space with elements $\alpha \in \mathbb{R}$. The mapping

$$
\begin{aligned}
X : \mathbb{R} \times \mathbb{R} &\longrightarrow \mathbb{C} \\
(t, \alpha) &\longmapsto X(t, \alpha)
\end{aligned}
$$

is a *weakly stationary stochastic process* (WSSP) or, equivalently, a *wide-sense stationary stochastic* process, if the following properties are valid:

i. $\forall t \in \mathbb{R}$, $\quad X(t, \cdot) \in L_p^2(\mathbb{R})$;

ii. $\exists m \in \mathbb{C}$ such that $\quad \forall t$, $E\{X(t)\} = m$, where

$$
E\{X(t)\} = \int X(t, \alpha) \, dp(\alpha);
$$

iii. $\forall t, u, h \in \mathbb{R}$, $\quad E\{X(t + u)\overline{X(t + h)}\} = E\{X(u)\overline{X(h)}\}$;

iv. $\lim_{t \to 0} E\{|X(t) - X(0)|^2\} = 0$.

This definition makes sense for mappings $X : \mathbb{R} \times P \to \mathbb{C}$, whether (P, p) is a quite general probability space (which we haven't defined) or something as specific as the set P of Example 2.8.3.

b. Part *iii* of the above definition implies $E\{X(t + u)\overline{X(t)}\} = E\{X(u)\overline{X(0)}\}$ for all $t, u \in \mathbb{R}$; and the function R, defined by

$$
\forall t \in \mathbb{R}, \qquad R(t) = E\{X(t + u)\overline{X(u)}\},
$$

is the *stochastic autocorrelation* of the WSSP X. The *autocovariance* of X is the function C, defined by

$$
\forall t \in \mathbb{R}, \qquad C(t) = E\{(X(t + u) - m)(\overline{X(u) - m})\} = R(t) - |m|^2.
$$

Thus, the variance of the WSSP X is

$$
\sigma_X^2 = C(0) = R(0) - |m|^2.
$$

2.8.5 Proposition.

The stochastic autocorrelation R of a WSSP X is a continuous positive definite function.

This result is proved by an elementary calculation, and the continuity of R results from part *iv* of the definition of a WSSP. If X does not satisfy part *iv*, then R is still positive definite; and it has the decomposition $R = R_C + R_0$, where

R_C, $R_0 \gg 0$, R_C is continuous, and $R_0 = 0$ a.e. This decomposition is due mostly to F. Riesz (*Acta Sci. Math.*, 1933).

2.8.6 Definition. Power Spectrum and Spectral Estimation

a. Let X be a WSSP with stochastic autocorrelation R. Because of Proposition 2.8.5 and the Herglotz–Bochner Theorem, there is $\mu \in M_{b+}(\hat{\mathbb{R}})$ for which $\hat{\mu} = R$ on $\mathbb{R}$. By notational tradition we set $S = \mu$, and, by definition, S is the *power spectrum* of X.

b. The *spectral estimation problem* is to clarify and quantify the statement: find periodicities in a signal X recorded over a fixed time interval. In more picturesque language, we want to filter the noise from the incoming signal X in order to determine the intelligent message (periodicities) therein, e.g., [Chi78], [IEEE82].

Such signals can sometimes be modeled as WSSPs [Bar78], [BTu59], [Bri81], [Pri81], and, then, the spectral estimation problem is one of power spectrum computation [Ben83, Part IV] or approximation.

2.8.7 Definition. Periodogram

Let X be a WSSP, for which $X(\cdot, \alpha) \in L^\infty(\mathbb{R})$ for each $\alpha \in \mathbb{R}$, and let $b \in L^1(\mathbb{R})$. The function

$$
\begin{aligned}
S_b : \hat{\mathbb{R}} \times \mathbb{R} &\longrightarrow \mathbb{R} \\
(\gamma, \alpha) &\longmapsto S_b(\gamma, \alpha),
\end{aligned}
$$

defined as

$$
S_b(\gamma, \alpha) = \left| \int b(t) X(t, \alpha) e^{-2\pi i t \gamma} \, dt \right|^2,
$$

is the *periodogram* associated with the process X and the *data window b*.

Schuster initiated periodogram analysis, and his work was one of the major influences on Wiener's Generalized Harmonic Analysis (1930), e.g., [Wie81, Volume II, pages 183–324] and Section 2.9. The following calculation shows the role of periodograms in spectral estimation.

2.8.8 Proposition.

Let X be a real-valued WSSP, for which $X(\cdot, \alpha) \in L^\infty(\mathbb{R})$ for each $\alpha \in \mathbb{R}$, and let S_b be the periodogram associated with X for the real and even data window $b \in L^1(\mathbb{R})$. Assume $B^2 \equiv (\hat{b})^2 \in L^1(\hat{\mathbb{R}})$ and $\int B^2(\gamma) \, d\gamma = 1$. If S is the power spectrum of X, then

$$E\{S_b(\gamma)\} = S * B^2(\gamma), \tag{2.8.2}$$

and

$$\lim_{\lambda \to \infty} E\{S_{(1/\sqrt{\lambda})\, b(t/\lambda)}(\gamma)\} = \lim_{\lambda \to \infty} S * (B^2)_\lambda = S \tag{2.8.3}$$

in the sense that

$$\forall F \in C_b(\hat{\mathbb{R}}), \qquad \lim_{\lambda \to \infty} (S * (B^2)_\lambda)(F) = S(F), \tag{2.8.4}$$

where the subscript "λ" of B^2 designates dilation.

PROOF.   Our hypotheses allow us to verify (2.8.2) as follows:

$$E\{S_b(\gamma)\} = \iint b(t)\overline{b(u)} e^{-2\pi i(t-u)\gamma}\, \hat{S}(t-u)\, dt\, du$$

$$= \int |B(\gamma + \omega)|^2\, dS(\omega) = S * B^2(\omega).$$

The last step is a consequence of the following: since b is real and even, $|B|^2 = B^2$; and S is even since X real allows us to conclude that R is real and even.

Equation (2.8.3) is a consequence of the facts that $\{(B^2)_\lambda\}$ is an approximate identity and that $(1/\sqrt{\lambda})b(t/\lambda) \longleftrightarrow \sqrt{\lambda}B(\lambda\gamma)$. In fact,

$$G(\alpha) = \int F(\alpha + \beta)\, dS(\beta) \in C_b(\hat{\mathbb{R}}),$$

so that by the definition of convolution and the evenness of B^2 we can apply (2.1.2), which in turn yields (2.8.3), cf., Exercise 2.38.   ∎

2.8.9 Remark. Asymptotically Unbiased Estimator

For any real and even data window $b \in L^1(\mathbb{R})$, Proposition 2.8.8 permits us to refer to the periodograms $\{S_{(1/\sqrt{\lambda})\, b(t/\lambda)}\}$ as an *asymptotically unbiased estimator* of the power spectrum S.

The relatively weak type of convergence in (2.8.3) and (2.8.4) allows for a great deal of mischief on the part of the raw periodogram S_b if one thinks of it as an approximant to S. There are results due to Beurling, Herz, and Pollard, which are similar to (2.8.4), but for which the convergence is much "stronger". An example of such a theorem for closed intervals I, disjoint from supp S, is

$$\lim_{T \to \infty} \int_{-T}^{T} R(t)e^{2\pi i t \gamma}\, e^{|t|/T}\, dt = 0 \quad \text{uniformly on} \quad I,$$

cf., [Ben75, Section 2.1]. These results provide quantitative estimates for the support of S in terms of the support of the approximants.

2.8.10 Example. Michelson Interferometer and Spectral Estimation

a. In Example 2.7.9 we defined a "deterministic" version of autocorrelation P for $f \in L^2(\mathbb{R})$, viz., the L^2-*autocorrelation*

$$P(t) = f * \tilde{f}(t) = \int f(t+u)\overline{f(u)}\,du;$$

and we noted that P is positive definite, just as the stochastic autocorrelation R of a WSSP is positive definite, cf., Exercise 1.33. We mention this to illustrate the following specific, but in some sense typical, use of autocorrelation.

b. It is often difficult to measure a signal directly, whereas one can experimentally measure its power. To be more precise, the spectral analysis of a beam of light f, a real-valued signal, can be made by a Michelson interferometer in the following way. The power or intensity of the beam is the energy flow per unit time (assuming area normalization) and is measured by a power-sensitive photometer. The interferometer allows the beam to take different paths of different length to the photometer. As such the intensity of $f(u) + f(u+t)$ can be measured for various lags t. Thus, the left side of the equation,

$$\int |f(u) + f(u+t)|^2\,du - 2\int |f(u)|^2\,du = 2\int f(u+t)f(u)\,du,$$

can be measured, noting that $\int |f(u+t)|^2\,du = \int |f(u)|^2\,du$. Consequently, the L^2-autocorrelation P is computable even though f may not be.

c. Summarizing, suppose we have a complicated signal f, considered deterministically on $\mathbb{R}$ or modeled by a WSSP X. Suppose, further, that f or X cannot be analyzed directly, but that there are power-measuring devices that allow us to quantify the autocorrelations $P >> 0$ or $R >> 0$, as we did in part *b*. Then the computation of the power spectra $P^{\vee}$ or $R^{\vee}$ allows us to determine significant frequencies of $P^{\vee}$ or $R^{\vee}$; and these frequencies are also significant in the behavior of f or X since, in the case of $P^{\vee}$, $P^{\vee} = |\hat{f}|^2$. This theme is further developed in Section 2.9.

2.8.11 Example. The Wave Function and the Uncertainty Principle

The wave theory in quantum mechanics arose since electron beams diffracted through crystals produced an effect analogous to Newton's spectral theory of white light diffracted through a prism. (Wiener's spectral theory explaining the polychromatic nature of sunlight, i.e., white light, is the Generalized Harmonic Analysis of Section 2.9 [Wie81, Volume II, pages 183–324].)

For a fixed time t_0, the wave function $\Psi(\gamma)$, normalized so that $\|\Psi\|_{L^2(\hat{\mathbb{R}})} = 1$, is a solution of Schrödinger's equation for a freely moving particle X (Schrödinger, 1926); and an important aspect of its physical significance is that the "probability" that X is in a given subset $A \subseteq \hat{\mathbb{R}}$ is $\int_A |\Psi(\gamma)|^2 \, d\gamma$, e.g., [Schi68]. This assertion defines a probability measure $p \in M_{b+}(\mathbb{R})$, and allows us to think of X as a random variable $X : (\mathbb{R}, p) \longrightarrow \hat{\mathbb{R}}$. In fact, X is considered as a "measure of subsets" of $(\mathbb{R}, p)$ as in Remark 2.7.4 b, and p is defined as

$$p\{\alpha : X(\alpha) \in A\} = \int_A |\Psi(\gamma)|^2 \, d\gamma.$$

The associated pdf is $f_X \equiv |\Psi|^2$. For the case of 0-mean, the *classical* (or *Heisenberg) uncertainty principle inequality* associated with the wave function Ψ and $\psi \equiv \Psi^{\vee}$ is

$$1 = \|\Psi\|^2_{L^2(\hat{\mathbb{R}})} \leq 4\pi \|t\psi(t)\|_{L^2(\mathbb{R})} \|\gamma\Psi(\gamma)\|_{L^2(\hat{\mathbb{R}})}, \tag{2.8.5}$$

cf., [BF94, Chapter 7] as well as the "intuitive calculation" of the Heisenberg inequality in [Ben75, pages 77–79].

2.9. Wiener's Generalized Harmonic Analysis (GHA)

In 1930, Norbert Wiener [Wie81, Volume II, pages 183–324] proved an analogue of the Parseval–Plancherel formula, $\|f\|_{L^2(\mathbb{R})} = \|\hat{f}\|_{L^2(\hat{\mathbb{R}})}$, for functions that are *not* elements of $L^2(\mathbb{R})$. We refer to his formula as the *Wiener–Plancherel* formula, see (2.9.2). It became a beacon in his perception and formulation of the statistical theory of communication, e.g., [Wie49], [Lee60]. Wiener even chose to have the formula appear on the cover of his autobiography, *I Am a Mathematician*. (This is a twentieth century analogue of Archimedes' tombstone, which had a carving of a sphere inscribed in a cylinder to commemorate his "1:2:3" theorem, e.g., [Ben77] for details concerning the mathematical results, Cicero's role, and a recent update.)

Besides the motivation for GHA mentioned in Example 2.8.11, Wiener discussed the background for GHA in [Wie81, Volume II, pages 183–324]; and this background has been explained scientifically and historically in a virtuoso display of scholarship by Masani, e.g., Masani's remarkable commentaries in [Wie81, Volume II, pages 333–379], as well as [Mas90]. Two precursors, whose work Wiener studied and who should be mentioned vis á vis GHA, were Sir Arthur Schuster, cf., Definition 2.8.7, and Sir Geoffrey I. Taylor. Schuster pointed out analogies between the harmonic analysis of light and the statistical analysis of hidden periods associated with meteorological and astronomical data. Taylor conducted experiments in fluid mechanics dealing with the *onset to turbulence*, and formulated a special case of correlation. A third scientist, whose work (1914) vis á vis GHA was not known to Wiener, was Albert Einstein. Einstein writes: "Suppose the quantity y (for example, the number of sun spots) is determined empirically as a function of time, for a very large interval, T. How can one represent the statistical behavior of y?" In his heuristic answer to this question he came close to the notions of autocorrelation and power spectrum, e.g., Section 2.8 and Definition 2.9.5, cf., [Mas90, pages 112–113], Einstein's paper (in *Archive des Sciences Physiques et Naturelles* 37 (1914), 254–255), and commentaries by Masani and Yaglom.

The Fourier analysis of $L^1(\mathbb{R})$ or $L^2(\mathbb{R})$ (Chapter 1) or the theory of Fourier series (Chapter 3) were inadequate tools to analyze the issues confronting Schuster, Taylor, and Einstein. On the other hand, GHA became a successful device to gain some insight into their problems, as well as other problems where the data and/or noises cannot be modeled by the Fourier transform decay, finite energy, or periodicity inherent in the above classical theories, e.g., [Ars66, Chapter II], [Bas84], [Ric54].

The material in Sections 2.9.1–2.9.10 outlines GHA and is due to Wiener [Wie81, Volume II, pages 183–324 and 519–619], [Wie33], cf., [Ben75, Chapter 2], [Ber87]. The higher dimensional theory, with its geometrical ramifications, is found in [BBE89], [Ben91a], cf., [AKM80].

2.9.1 Definition. Bounded Quadratic Means

The space $BQM(\mathbb{R})$ of functions having *bounded quadratic means* is the set of all functions $f \in L^2_{\text{loc}}(\mathbb{R})$ for which

$$\sup_{T>0} \frac{1}{2T} \int_{-T}^{T} |f(t)|^2 \, dt < \infty.$$

The *Wiener space* $W(\mathbb{R})$ is the set of all functions $f \in L^2_{\text{loc}}(\mathbb{R})$ for which

$$\int \frac{|f(t)|^2}{1 + t^2} \, dt < \infty.$$

2.9.2 Theorem. Inclusions for GHA

$$L^\infty(\mathbb{R}) \subseteq BQM(\mathbb{R}) \subseteq W(\mathbb{R}) \subseteq \mathcal{S}'(\mathbb{R}),$$

and the inclusions are proper.

2.9.3 Definition. The Wiener s-Function

The *Wiener s-function* associated with $f \in BQM(\mathbb{R})$ is defined as the sum $s = s_1 + s_2$ where

$$s_1(\gamma) = \int_{-1}^{1} f(t) \frac{e^{-2\pi i t \gamma} - 1}{-2\pi i t} \, dt$$

and

$$s_2(\gamma) = \int_{|t| \geq 1} f(t) \frac{e^{-2\pi i t \gamma}}{-2\pi i t} \, dt.$$

Since $f \in L^1[-1, 1]$, we have $s_1 \in C(\mathbb{R})$ and $|s_1(\gamma)| \leq 2|\gamma| \|f\|_{L^1[-1,1]}$. Since $f \in BQM(\mathbb{R})$, Theorem 2.9.2 and the Parseval–Plancherel Theorem allow us to conclude that $s_2 \in L^2(\mathbb{R})$. In particular, $s \in L^2_{loc}(\hat{\mathbb{R}}) \cap S'(\hat{\mathbb{R}})$. The Wiener s-function is also called the *Wiener transform* or integrated Fourier transform of f.

2.9.4 Theorem. The Derivative of the Wiener s-Function

Let $f \in BQM(\mathbb{R})$. Then $f \in S'(\mathbb{R})$ and

$$s' = \hat{f},$$

where $s \in L^2_{loc}(\hat{\mathbb{R}}) \cap S'(\hat{\mathbb{R}})$ is the Wiener s-function associated with f (Exercise 2.61).

2.9.5 Definition. Deterministic Autocorrelation

The *deterministic autocorrelation R* of a function $f : \mathbb{R} \to \mathbb{C}$ is formally defined as

$$R(t) = \lim_{T \to \infty} \frac{1}{2T} \int_{-T}^{T} f(u+t) \overline{f(u)} \, du.$$

To fix ideas, suppose R exists for each $t \in \mathbb{R}$. It is easy to prove that $R >> 0$, and so $R = \hat{S}$ for some $S \in M_{b+}(\hat{\mathbb{R}})$. We have used the same notation, viz., R, to denote both deterministic and stochastic autocorrelation since they are often the same, e.g., Theorem 2.9.11. As in the stochastic case, S is called the *power spectrum* of f.

The *Wiener–Plancherel formula* is (2.9.2) in the following result.

2.9.6 Theorem. Wiener–Plancherel Formula

Let $f \in BQM(\mathbb{R})$, and suppose its deterministic autocorrelation $R = \hat{S}$ exists for each $t \in \mathbb{R}$.

a. *Then*

$$\forall t \in \mathbb{R}, \qquad R(t) = \lim_{\epsilon \to 0} \frac{2}{\epsilon} \int |\Delta_\epsilon s(\gamma)|^2 e^{-2\pi i t \gamma} \, d\gamma, \qquad (2.9.1)$$

where $\Delta_\epsilon s(\gamma) \equiv \frac{1}{2}(s(\gamma + \epsilon) - s(\gamma - \epsilon))$.

b.   *In particular,*

$$\lim_{T\to\infty}\frac{1}{2T}\int_{-T}^{T}|f(t)|^2\,dt = \lim_{\epsilon\to0}\frac{2}{\epsilon}\int|\Delta_\epsilon s(\gamma)|^2\,d\gamma. \tag{2.9.2}$$

$$f \quad\longleftrightarrow\quad \hat{f}=s' \qquad s$$

$$\downarrow \qquad\qquad\qquad \downarrow$$

$$R=\hat{S} \quad\longleftrightarrow\quad S \qquad \{\tfrac{2}{\epsilon}|\Delta_\epsilon s|^2\}$$

FIGURE 2.3.

2.9.7 Example. Related Formulas and Spectral Estimation

a.   Because of (2.9.1) and assuming the setup of Theorem 2.9.6, the following formulas are true under the proper hypotheses, e.g., [Ben75, page 90], [Ben91b, page 847]:

$$\lim_{\epsilon\to0}\frac{2}{\epsilon}|\Delta_\epsilon s(\gamma)|^2 = S \tag{2.9.3}$$

and

$$\int|\hat{k}(\gamma)|^2\,dS(\gamma) = \lim_{T\to\infty}\frac{1}{2T}\int_{-T}^{T}|k*f(t)|^2\,dt$$
$$= \lim_{\epsilon\to0}\frac{2}{\epsilon}\int|\hat{k}(\gamma)\Delta_\epsilon s(\gamma)|^2\,d\gamma. \tag{2.9.4}$$

b.   Formally, (2.9.4) is (2.9.2) for the case $k=\delta$. For $k\in C_c(\mathbb{R})$ the first equality of (2.9.4) is not difficult, e.g., [Ben91b, pages 847–848]. The second equality or, equivalently, Theorem 2.9.6, requires *Wiener's Tauberian Theorem*, e.g., Theorem 2.9.12.

c.   Figure 2.3 illustrates the action and "levels" of the functions and measure in Theorem 2.9.6 for a given signal f.

d.   Since S is the "power" spectrum, (2.9.2) and (2.9.3) allow us to assert that

$$\lim_{T\to\infty}\frac{1}{2T}\int_{-T}^{T}|f(t)|^2\,dt$$

is a measure of the total power of f, cf., Wiener's comparison of energy and power in [Wie49, pages 39–40 and 42]. In light of the spectral estimation problem of

Definition 2.8.6, the middle term of (2.9.4) is a measure of the power in a frequency band $[\alpha, \beta]$ if $\hat{k} = \mathbf{1}_{[\alpha,\beta]}$ in the first term of (2.9.4), cf., [Ben91b, Theorem 5.2].

2.9.8 Remark. Wiener–Plancherel Formula

The Parseval–Plancherel formula, $\|f\|_{L^2(\mathbb{R})} = \|\hat{f}\|_{L^2(\hat{\mathbb{R}})}$, allowed us to define the Fourier transform of a square integrable function (Theorem 1.10.2), and, at certain levels of abstraction, it is considered to characterize what is meant by an harmonic analysis of f. On the other hand, for most applications in $\mathbb{R}$, the formula assumes the workaday role of an effective tool used to obtain quantitative results. It is this latter role that was envisaged for the Wiener–Plancherel formula in dealing with the non-square-integrable case. After all, distribution theory gives the proper definition of the Fourier transform of tempered distributions. The real issue is to obtain quantitative results for problems where an harmonic analysis of a non-square-integrable function is desired. As mentioned above, a host of such problems comes under the heading of an harmonic (spectral) analysis of signals containing non-square-integrable noise and/or random components, whether it be speech recognition, image processing, geophysical modeling, or turbulence in fluid mechanics. Such problems can be attacked by Beurling's profound theory of spectral synthesis, e.g., Example 2.4.6f, as well as by the extensive multifaceted theory of time series, e.g., Section 2.8. Beurling's spectral synthesis does not deal with energy and power considerations, i.e., quadratic criteria, and time series relies on a stochastic point of view. The Wiener–Plancherel formula deals with these problems deterministically and, hence, with potential for real implementation, e.g., Example 2.9.7d.

2.9.9 Example. Elementary Power Spectra

a. The value of the autocorrelation R is that it can be measured in many cases where the underlying signal f cannot be quantified, e.g., Example 2.8.10. Also, the discrete part of the power spectrum S characterizes periodicities in f, e.g., [Wie48, Chapter X]. This fact is illustrated by taking $f(t) = \sum_{k=1}^{n} r_k e^{-2\pi i t \lambda_k}$, $r_k \in \mathbb{C}$, $\lambda_k \in \hat{\mathbb{R}}$. The L^2-autocorrelation is not defined, but the deterministic autocorrelation is $R(t) = \sum_{k=1}^{n} |r_k|^2 e^{-2\pi i t \lambda_k}$ (by direct calculation); and hence the power spectrum is

$$S = \sum_{k=1}^{n} |r_k|^2 \delta_{\lambda_k}.$$

b. If $f : \mathbb{R} \to \mathbb{C}$ has the property that $\lim_{|t| \to \pm\infty} f(t) = 0$, then $S = 0$. It is elementary to construct examples f for which $S = 0$ whereas $\overline{\lim}_{|t| \to \pm\infty} |f(t)| > 0$, cf., [Wie33, pages 151–154], [Bas84, pages 99–100], [Ben75, pages 84 and 87], [Ben83, Section IV].

2.9.10 Definition. Correlation Ergodicity

Let X be a WSSP with stochastic autocorrelation R. X is a *correlation ergodic process* if

$$\forall t \in \mathbb{R}, \qquad \lim_{T \to \infty} \frac{1}{2T} \int_{-T}^{T} X(t+u, \alpha)\overline{X(u, \alpha)}\, du = R(t) \qquad (2.9.5)$$

in measure. (Convergence in measure is defined in Example A.11c.)

Because of (2.9.5), the following result establishes the relationship between the notions of deterministic and stochastic autocorrelation, cf., [Pap77, pages 354–360].

2.9.11 Theorem. Criterion for Correlation Ergodicity

Let X be a WSSP with stochastic autocorrelation R. X is a correlation ergodic process if

$$\forall t \in \mathbb{R}, \qquad \lim_{T \to \infty} \frac{1}{2T} \int_{-T}^{T} C(t, v) \left(1 - \frac{|v|}{2T} \right) dv = 0,$$

where $C(t, v) = E\{X(t + u + v)\overline{X(u + v)}X(t + u)\overline{X(u)}\} - |R(t)|^2.$

As mentioned in Example 2.9.7, the following result is required to prove the Wiener–Plancherel formula.

2.9.12 Theorem. Wiener Tauberian Theorem

Let $g \in L^1(\mathbb{R})$ have a nonvanishing Fourier transform and let $\varphi \in L^\infty(\mathbb{R})$. If

$$\lim_{t \to \infty} g * \varphi(t) = r \int g(u)\, du, \qquad (2.9.6)$$

then

$$\forall f \in L^1(\mathbb{R}), \qquad \lim_{t \to \infty} f * \varphi(t) = r \int f(u)\, du. \qquad (2.9.7)$$

2.9.13 Remark. Wiener Tauberian Theorem

a. Theorem 2.9.12 has the format of *classical* Tauberian theorems: a boundedness (or some other) condition and "summability" by a certain method yield "summability" by other methods. In Theorem 2.9.12, the boundedness or "Tauberian" condition is the hypothesis that $\varphi \in L^\infty(\mathbb{R})$. The given summability is (2.9.6), where g represents a so-called "summability method". The conclusion (2.9.7) of the theorem is summability for a whole class of summability methods, viz., for all $f \in L^1(\mathbb{R})$. A classical and masterful treatment of summability methods is due to Hardy [Har49].

If g is the Gaussian defined in Example 1.3.3, then $\hat{g}$ never vanishes. Thus, in this case, if $\varphi \in L^\infty(\mathbb{R})$ has the property that

$$\lim_{t \to \infty} g * \varphi(t) = r,$$

then

$$\forall \lambda, \qquad \lim_{t \to \infty} w_\lambda * \varphi(t) = r,$$

where $\{w_\lambda\}$ is the Fejér kernel.

The particular functions used by Wiener to prove his Wiener Tauberian formulas are found in [Wie33], [Ben75, pages 91–92].

b. *Modern* Tauberian theorems have a more algebraic and/or functional analytic flavor to them. For example, the Wiener Tauberian Theorem is a special case of the fact that *if* $\hat{g} \in A(\hat{\mathbb{R}})$, $T \in A'(\hat{\mathbb{R}})$, *and* $T\hat{g} = 0$, *then* $\hat{g} = 0$ *on* supp T. In fact, the generalizations of Theorem 2.9.12 are much more far reaching than this. [Ben75] gives an extensive treatment of both classical and modern Tauberian theory, as well as the history of the subject, and applications to spectral synthesis and analytic number theory.

Because of the importance of translation invariant systems and the theory of multipliers, e.g., Definition 2.6.5 and Example 2.6.6, we define the *closed translation-invariant subspace* V_g *generated by* $g \in X$, where X is $L^1(\mathbb{R})$ or $L^2(\mathbb{R})$, to be the closure in X of the linear span of $\{\tau_t g : t \in \mathbb{R}\}$. We write

$$V_g = \overline{\text{span}}\{\tau_t g : t \in \mathbb{R}\}. \tag{2.9.8}$$

2.9.14 Theorem. Zero Sets and Dense Subspaces

a. If $g \in L^1(\mathbb{R})$ *and* $\hat{g}$ *never vanishes, then* $V_g = L^1(\mathbb{R})$.

b. If $g \in L^2(\mathbb{R})$ *and* $|\hat{g}| > 0$ *a.e., then* $V_g = L^2(\mathbb{R})$.

PROOF. Part *a* is the Wiener Tauberian Theorem, and we refer to [Wie33], [Ben75, pages 25–26, 49–50, and 94–95, and Section 2.3] for proofs.

The proof of part *b* is much simpler than that of part *a*, and so we shall give it here. Suppose $V_g \neq L^2(\mathbb{R})$. Then there is $h \in L^2(\mathbb{R}) \setminus \{0\}$ such that

$$\forall t \in \mathbb{R}, \qquad \int (\tau_t g)(u) \overline{h(u)} \, du = 0. \tag{2.9.9}$$

Equation (2.9.9) is a consequence of the Hahn–Banach Theorem and the fact that $L^2(\mathbb{R})' = L^2(\mathbb{R})$, e.g., Theorem B.14. By the Parseval–Plancherel Theorem,

$$\forall t \in \mathbb{R}, \qquad \int \hat{g}(\gamma)\overline{\hat{h}(\gamma)}e^{-2\pi i t\gamma}\,d\gamma = 0.$$

$\hat{g}\overline{\hat{h}} \in L^1(\hat{\mathbb{R}})$ by Hölder's inequality, and so, by the L^1-uniqueness theorem (Theorem 1.6.9c), $\hat{g}\overline{\hat{h}} = 0$ a.e. Since $|\hat{g}| > 0$ a.e., we conclude that $\hat{h} = 0$ a.e., and this contradicts the hypothesis on h. Thus, $V_g = L^2(\mathbb{R})$.   ∎

Subspaces such as V_g in (2.9.8) play an important role in Gabor and wavelet decompositions in the case that the set of translates $\tau_t g$ is reduced to $\{\tau_r g : r \in D\}$ where D is a discrete subset of $\mathbb{R}$, e.g., [Mey90], [Dau92], [BF94].

2.9.15 Remark. GHA and the Khinchin Theorem

a.   In GHA, a function f is analyzed for its frequency information by computing its autocorrelation R and its power spectrum $S = R^{\vee} \in M_{b+}(\hat{\mathbb{R}})$. Mathematically, this is a mapping between a class of functions f and a class of measures $S \in M_{b+}(\hat{\mathbb{R}})$. A natural question to ask is the following: for any $\mu \in M_{b+}(\hat{\mathbb{R}})$, does there exist f whose autocorrelation R exists, and for which $R^{\vee} = \mu$?

b.   The question of part *a* is answered affirmatively in the WSSP case by the Khinchin Theorem: *a necessary and sufficient condition for R to be the stochastic autocorrelation of some WSSP X is that there exist $S \in M_{b+}(\hat{\mathbb{R}})$ for which $\hat{S} = R$.* In one direction, if R is the stochastic autocorrelation of a WSSP X, then $S \equiv R^{\vee} \in M_{b+}(\hat{\mathbb{R}})$ by the Herglotz–Bochner Theorem. The question in part *a* deals with the opposite direction, and the positive answer is not difficult to prove, e.g., [Pri81, pages 221–222], [DM76, pages 62–63 and 72–73]. Khinchin's proof dates from 1934, and there were further probabilistic contributions by Wold (1938), Cramér (1940), and Kolmogorov [Kol41], cf., [Ben92a]. Khinchin's Theorem is sometimes called the Wiener–Khinchin Theorem, but this is a misnomer. In fact, Wiener began using Khinchin's Theorem as such only after 1950.

c.   The deterministic and constructive affirmative answer to the question in part *a* is the *Wiener–Wintner Theorem* (1939) [Wie81]. Bass and Bertrandias made significant contributions to this result, e.g., [Bas84]; and the multidimensional version is found in [Ben91b], [Ker90].

2.9.16 Theorem. Wiener–Wintner Theorem

Let $\mu \in M_{b+}(\hat{\mathbb{R}})$. There is a constructible function $f \in L^{\infty}_{\mathrm{loc}}(\mathbb{R})$ such that its deterministic autocorrelation R exists for all $t \in \mathbb{R}$, and $R^{\vee} = \mu$.

2.10. exp{it²}

The function

$$s(t) = \frac{1}{\sqrt{\pi i}} e^{it^2}$$

has properties that serve as a paradigm for the *method of stationary phase* (Example 2.10.4), as well as being an underlying kernel for the *oscillatory integrals* that arise in areas as diverse as analytic number theory, e.g., [Tit51, pages 61–70 and 83–84], and partial differential equations, e.g., [Hör83, Sections 7.7 and 7.8], cf., [Ste93]. In optics, *s* is the convolution kernel in *Fresnel's approximation* to the Huygens–Fresnel principle. (Fresnel's approximation allows for realistic diffraction-pattern calculations, e.g., [Good68].) In signal processing, *s* is the linear *chirp signal* whose frequency changes linearly with time, e.g., Figure 2.4. It is an example from the class of *frequency modulated (FM) signals* $e^{2\pi i \varphi(t)}$ that arise in subjects such as radar and sonar, e.g., [Rih85], [BMW91], and that are characterized by the property that φ' is not a constant.

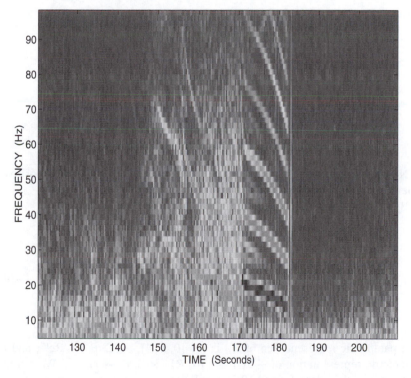

FIGURE 2.4. Spectrogram for electrocorticogram data [BeC95]. Chirp signal behavior is exhibited in the time interval from 170 to 182 seconds.

With regard to *Fourier optics*, the integrals $x(t) = \int_0^t \cos^2 u \, du$ and $y(t) = \int_0^t \sin^2 u \, du$ in Theorem 2.10.1 are called *Fresnel integrals*, cf., Example 2.10.7. Also, in light of (2.10.1), note that $s \notin L^1(\mathbb{R}) \cup L^2(\mathbb{R})$.

2.10.1 Theorem. Fresnel Integrals

a.   *Let s_λ be the dilation of s for $\lambda > 0$. Then*

$$\int s_\lambda(t) \, dt = 1 \tag{2.10.1}$$

in the sense that the improper Riemann integral $\int_0^\infty s_\lambda(t) \, dt$ equals $\frac{1}{2}$.
   b.

$$\int_0^\infty \cos t^2 \, dt = \int_0^\infty \sin t^2 \, dt = \frac{1}{2} \sqrt{\frac{\pi}{2}}. \tag{2.10.2}$$

PROOF.   To obtain (2.10.2) from (2.10.1) we compute

$$\sqrt{i\pi} = \int_{-\infty}^\infty \cos t^2 \, dt + i \int_{-\infty}^\infty \sin t^2 \, dt$$
$$= 2 \int_0^\infty \cos t^2 \, dt + 2i \int_0^\infty \sin t^2 \, dt.$$

Thus, since $i^{1/2} = (e^{i\pi/2})^{1/2} = e^{i\pi/4} = \frac{1}{\sqrt{2}}(1 + i)$, we have (2.10.2).

Now, let us verify (2.10.1). Let $s(z) = (1/\sqrt{i\pi})e^{iz^2}$, and consider the positively oriented wedge $C = [0, R] \cup C_R \cup L_R$ as illustrated in Figure 2.5.
By Cauchy's theorem, e.g., [Rud66], we compute

$$\sqrt{i\pi} \int_C s(z) \, dz = \int_0^R e^{ix^2} \, dx + \int_{C_R} e^{iz^2} \, dz$$
$$+ \int_{L_R} e^{iz^2} \, dz = 0. \tag{2.10.3}$$

The parametric representation of C_R is $z : [0, \pi/4] \to \mathbb{C}$, $\theta \mapsto Re^{i\theta}$; and the parametric representation of $-L_R$ is $z : [0, R] \to \mathbb{C}$, $r \mapsto re^{i\pi/4}$. We estimate and compute the integrals in (2.10.3) as follows:

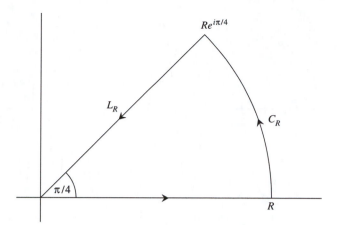

FIGURE 2.5.

$$\left| \int_{C_R} e^{iz^2}\, dz \right| = \left| iR \int_0^{\pi/4} e^{iR^2(\cos\theta + i\sin\theta)^2} e^{i\theta}\, d\theta \right|$$

$$\leq R \int_0^{\pi/4} e^{-(2R^2\cos\theta)\sin\theta}\, d\theta$$

$$\leq R \int_0^{\pi/4} e^{-(2R^2\cos(\pi/4))\sin\theta}\, d\theta \qquad (2.10.4)$$

$$\leq R \int_0^{\pi/2} e^{-(2R^2\cos(\pi/4))\sin\theta}\, d\theta < \frac{\pi}{2}\left(\frac{1}{2R^2\cos(\pi/4)} \right)$$

and

$$\int_{L_R} e^{iz^2}\, dz = -\int_0^R e^{ir^2 e^{i\pi/2}} e^{i\pi/4}\, dr$$

$$= -e^{i\pi/4}\left(\int_0^R e^{-r^2}\, dr \right). \qquad (2.10.5)$$

The last inequality in (2.10.4) follows from Jordan's inequality,

$$\int_0^{\pi/2} e^{-r\sin\theta}\, d\theta < \frac{\pi}{2r},$$

e.g., Exercise 2.62 *a*. The right side of (2.10.4) tends to 0 as $R \to \infty$ and the right side of (2.10.5) tend to $-\frac{\sqrt{\pi}}{2}\left(\frac{1}{\sqrt{2}} + i\frac{1}{\sqrt{2}} \right)$ as $R \to \infty$. Equation (2.10.1) then follows from (2.10.3). ∎

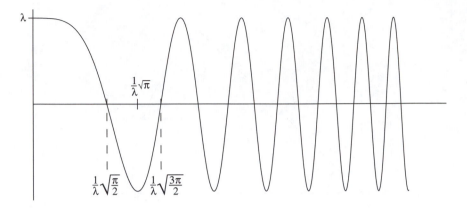

FIGURE 2.6.

2.10.2 Example. $\lambda \cos(\lambda t)^2$

Figure 2.6 is the graph of the dilation $\lambda \cos(\lambda t)^2$. Clearly, $\lambda \cos(\lambda t)^2$ is rapidly oscillating. In light of the fact that $\int s_\lambda(t)\,dt = 1$, it should be noted that, except for the main lobe about the origin, Figure 2.6 does not resemble the approximate identities of Chapter 1, cf., Theorem 2.10.3 *b*.

Note that the difference between consecutive "crests" of $\lambda \cos(\lambda t)^2$ has order of magnitude $1/(\lambda \sqrt{n})$, and so tends to 0 as $n \to \infty$.

In fact, we have

$$\frac{1}{\lambda}\left(\sqrt{2\pi n} - \sqrt{2\pi(n-1)}\right) = \frac{\sqrt{2\pi}}{\lambda}\left(\frac{\sqrt{n} - \sqrt{n-1}}{\sqrt{n} + \sqrt{n-1}}\right)\left(\sqrt{n} + \sqrt{n-1}\right)$$

$$= \frac{\sqrt{2\pi}}{\lambda}\frac{1}{\sqrt{n} + \sqrt{n-1}} \longrightarrow 0, \qquad n \longrightarrow \infty.$$

At this point we know that $\{s_\lambda\} \subseteq L^\infty(\mathbb{R}) \subseteq \mathcal{S}'(\mathbb{R})$, $s_\lambda \notin L^1(\mathbb{R}) \cup L^2(\mathbb{R})$, and $\int s_\lambda(t)\,dt = 1$. Thus, theoretically, $\hat{s}_\lambda \in A'(\hat{\mathbb{R}}) \subseteq \mathcal{S}'(\hat{\mathbb{R}})$ exists, and, tantalizingly, we'd like to know to what extent $\{s_\lambda\}$ is an "approximate identity", even though it isn't an approximate identity as defined in Chapter 1. The following result computes $\hat{s}_\lambda$ and answers the "approximate identity" query.

2.10.3 Theorem. $\hat{s}_\lambda$ and $s_\lambda \to \delta$

  a. The distributional Fourier transform $\hat{s}_\lambda$ of s_λ is

$$\forall \gamma \in \hat{\mathbb{R}}, \qquad \hat{s}_\lambda(\gamma) = e^{-i(\pi\gamma/\lambda)^2}, \tag{2.10.6}$$

and so $s_\lambda \in L^\infty(\mathbb{R}) \cap A'(\mathbb{R})$.

b. $\lim_{\lambda \to \infty} s_\lambda = \delta$ *in the sense that*

$$\forall f \in L^1(\mathbb{R}) \cap A(\mathbb{R}), \qquad \lim_{\lambda \to \infty} \int s_\lambda(t) f(t) \, dt = \delta(f). \qquad (2.10.7)$$

PROOF. ***i.*** Formally, by completing the square and invoking Theorem 2.10.1*a* we compute

$$\hat{s}_\lambda(\gamma) = \frac{\lambda}{\sqrt{\pi i}} \int \exp i \left\{ \left(\lambda t - \frac{\pi \gamma}{\lambda} \right)^2 - \left(\frac{\pi \gamma}{\lambda} \right)^2 \right\} dt = e^{-i(\pi \gamma/\lambda)^2}.$$

Similarly,

$$\int s_\lambda(t) \overline{f(t)} \, dt = \int e^{-i(\pi\gamma/\lambda)^2} \overline{\hat{f}(\gamma)} \, d\gamma. \qquad (2.10.8)$$

ii. The formal calculation to obtain equation (2.10.8) can be justified for $f \in L^1(\mathbb{R}) \cap A(\mathbb{R})$. In this case, the integrals of (2.10.8) are Lebesgue integrals, and the validity of the equality is a consequence of the inversion formula (Theorem 1.7.8), LDC, and the fact that

$$\exists M \text{ such that } \forall a, b, \qquad \left| \int_a^b e^{it^2} \, dt \right| \le M,$$

e.g., Exercises 2.64 and 2.68, cf., Exercise 2.63. Equation (2.10.6) follows from (2.10.8) since $S(\mathbb{R}) \subseteq L^1(\mathbb{R}) \cap A(\mathbb{R})$.

iii. Noting that

$$\forall \gamma \in \hat{\mathbb{R}}, \qquad \lim_{\lambda \to \infty} e^{-i(\pi\gamma/\lambda)} = 1,$$

we can use LDC and the inversion formula again to obtain (2.10.7). ∎

2.10.4 Example. Stationary Phase

a. For a given compactly supported function f and real-valued phase $\varphi \in C^2(\mathbb{R})$, there is the associated *oscillatory integral*

$$F_\varphi(\gamma) = \int f(t) e^{2\pi i \varphi(t) \gamma} \, dt. \qquad (2.10.9)$$

Obviously, the Fourier transform is an oscillatory integral. It is often important to investigate the behavior of F_φ for large values of γ. The *method of stationary phase* asserts that this behavior is determined by the so-called *stationary points* t for which $\varphi'(t) = 0$; and the method provides a means for quantifying this behavior, e.g., part *b*. Of course, in the case of the Fourier transform of $f \in L^1(\mathbb{R})$, where $\varphi' = 1$, we have $\lim_{\gamma \to \infty} F_\varphi(\gamma) = 0$, cf., Exercise 2.63 for more general phases φ without stationary points.

The investigation of integrals F_φ in this spirit goes back to Airy (1838) and Stokes (1850), and the method of stationary phase is just another brilliant chapter in Riemann's thesis on trigonometric series [Rie1873, Section XIII]. It should be mentioned that *Laplace's asymptotic method* (1820), e.g., part *c*, preceded stationary phase; and, although Laplace's method doesn't deal directly with oscillatory integrals, it has striking resemblances in both technique and result with stationary phase, e.g., [Olv74], [Wid41].

b.   Lord Kelvin (1887) made the following sort of observation about F_φ, most probably without knowledge of Riemann's work. If $f \in C_c(\mathbb{R})$ and γ is large, then the integral in (2.10.9) is very small because of the cancellation resulting from oscillation, *except* possibly near stationary points of φ since φ changes slowly near such points.

To quantify Lord Kelvin's point of view, assume supp $f = [a, b]$ for $f \in C_c(\mathbb{R})$, $\varphi \in C^2(\mathbb{R})$, $\varphi'(t_0) = 0$ for some $t_0 \in (a, b)$, $\varphi'(t) \neq 0$ for all $t \in [a, b] \setminus \{t_0\}$, and $\varphi^{(2)}(t_0) > 0$. Then, because of the oscillation, Lord Kelvin's observation asserts that

$$F_\varphi(\gamma) \approx \int_I f(t_0) \exp \left\{ 2\pi i \gamma \left[\varphi(t_0) + \frac{1}{2}(t - t_0)^2 \varphi^{(2)}(t_0) \right] \right\} \, dt \quad (2.10.10)$$

for large γ, where I is a small interval about t_0. By the oscillation again, (2.10.10) can be replaced by

$$F_\varphi(\gamma) \approx f(t_0) e^{2\pi i \varphi(t_0)\gamma} \int e^{\pi i t^2 \varphi^{(2)}(t_0)\gamma} \, dt$$

$$= e^{\pi i/4} f(t_0) e^{2\pi i \varphi(t_0)\gamma} \frac{1}{(\gamma \varphi^{(2)}(t_0))^{1/2}}, \quad (2.10.11)$$

where the right side is a consequence of Theorem 2.10.1. We shall say that F is *asymptotic* to G as $\gamma \to \infty$ to mean that $\lim_{\gamma \to \infty} F(\gamma)/G(\gamma) = 1$. The notation for this asymptotic behavior is $F(\gamma) \sim G(\gamma)$, $\gamma \to \infty$. As such, and noting that the heuristic argument leading to (2.10.11) can be made rigorous, the method of stationary phase allows us to assert that

$$F_\varphi(\gamma) \sim e^{\pi i/4} f(t_0) e^{2\pi i \varphi(t_0)\gamma} \frac{1}{(\gamma \varphi^{(2)}(t_0))^{1/2}}, \qquad \gamma \longrightarrow \infty, \quad (2.10.12)$$

given our hypotheses on f and φ, cf., [Hör83], [Olv74], [Ste93] for more advanced results.

c.   For comparison with (2.10.12), the asymptotic relation (2.10.13) is an elementary form of Laplace's asymptotic method. Let

$$G_\varphi(\gamma) = \int f(t) e^{2\pi \varphi(t)\gamma} \, dt.$$

Assume supp $f = [t_0, b]$, $f \in C[t_0, b]$, $f(t_0) \neq 0$, $\varphi \in C^2[t_0, b]$, $\varphi'(t_0) = 0$, $\varphi^{(2)}(t_0) < 0$, and φ nonincreasing on $[t_0, b]$. Then

$$G_\varphi(\gamma) \sim f(t_0)e^{2\pi\varphi(t_0)\gamma} \frac{1}{(-4\gamma\varphi^{(2)}(t_0))^{1/2}}, \qquad \gamma \longrightarrow \infty, \qquad (2.10.13)$$

cf., [Wid41, Chapter 7] for this and more advanced results.

We shall define the Stieltjes transform in Exercise 2.56, as well as noting its close relationship to the Hilbert and iterated Laplace transforms. It turns out that a version of Laplace's asymptotic method is central to establishing an inversion theory for the Stieltjes transforms of distributions, e.g., our theory of Stieltjes transforms in *Analytic representation of generalized functions, Math. Zeitschr.* **97** (1967), 303–319.

2.10.5 Example. Chirp Transform Algorithm

The *chirp transform algorithm* is the formula

$$\forall f \in L^1(\mathbb{R}), \qquad \hat{f} = \left[(f\bar{s}_{\sqrt{\pi}}) * s_{\sqrt{\pi}}\right](\overline{i^{1/2}s_{\sqrt{\pi}}}), \qquad (2.10.14)$$

where $s_{\sqrt{\pi}}$ is the $\sqrt{\pi}$-dilation of s. The verification of (2.10.14) is elementary:

$$(f\bar{s}_{\sqrt{\pi}}) * s_{\sqrt{\pi}}(\gamma) = \int f(u)(\overline{i^{-1/2}})e^{-i\pi u^2}(i^{-1/2})e^{i\pi(\gamma-u)^2}\, du = e^{i\pi\gamma^2}\hat{f}(\gamma).$$

Equation (2.10.14) can be "implemented" for a given function f by the following sequence of operations: multiply f by $\bar{s}_{\sqrt{\pi}}$, convolve $f\bar{s}_{\sqrt{\pi}}$ with $s_{\sqrt{\pi}}$, and multiply $(f\bar{s}_{\sqrt{\pi}}) * s_{\sqrt{\pi}}$ by $(\overline{i^{1/2}s_{\sqrt{\pi}}})$. In signal processing, there is a simple block diagram for these operations, cf., [OW83, pages 511–512] or the terminology in Definition 2.6.5. We have chosen the word "algorithm" to describe the "fact" (2.10.14), since both f and $\hat{f}$ have the same domain $\mathbb{R}$; whereas our point of view has been to consider f and $\hat{f}$ as being defined on different spaces, whether they be time and frequency axes or dual groups.

2.10.6 Example. Infinite Frequencies

a.   Let s_λ, $\lambda > 0$, be the λ-dilation of s. Then the deterministic autocorrelation R of s_λ is

$$R(t) = \begin{cases} 0, & \text{if } t \neq 0, \\ \dfrac{\lambda^2}{\pi}, & \text{if } t = 0. \end{cases} \qquad (2.10.15)$$

To verify (2.10.15), we need only substitute into the definition of R, and compute

$$R(t) = \lim_{T \to \infty} \frac{\lambda^2}{2\pi T} e^{i(\lambda t)^2} \int_{-T}^{T} e^{2i\lambda^2 ut} \, du.$$

$R \gg 0$ is discontinuous, and the power spectrum S of s_λ is the 0-measure, cf., Exercise 2.66.

b. Using the terminology of Example 2.9.7*d*, we see that the total power of s_λ is $R_\lambda(0)$, and that $R_\lambda(0) > \lim_{t \to 0} R_\lambda(t) = 0$. Since $S_\lambda = 0$, there is a portion of the power not represented by any finite frequencies; and Wiener reasoned that s_λ draws part of its power from so-called *infinite frequencies*, e.g., [Wie49, page 40].

c. For perspective with regard to the function s, if $f(t) = e^{2\pi i |t|}$, then its deterministic autocorrelation is $R(t) = \cos 2\pi t$ and its power spectrum is $S = \frac{1}{2}(\delta_1 + \delta_{-1})$. In fact, if $t > 0$, then $|t + u| - |u| = t$ for $u \geq 0$ and so

$$\lim_{T \to \infty} \frac{1}{2T} \int_{0}^{T} e^{2\pi i (|t+u|-|u|)} \, du = \frac{1}{2} e^{2\pi it}.$$

Similarly, for $t > 0$, we consider the intervals $[-T, -t)$ and $[-t, 0]$ separately, and note that $|u + t| - |u| = -t$ for $u \in [-T, -t)$; thus,

$$\lim_{T \to \infty} \frac{1}{2T} \int_{-T}^{0} e^{2\pi i (|t+u|-|u|)} \, du = \frac{1}{2} e^{-2\pi it}.$$

As another example, let

$$f(t) = e^{2\pi i |t|^{1/2}}.$$

Then, expanding

$$\exp 2\pi i \left(\frac{|t + u| - |u|}{|t + u|^{1/2} + |u|^{1/2}} \right),$$

we see that the deterministic autocorrelation of f is $R = 1$, and its power spectrum $S = \delta$.

2.10.7 Example. Curvature of the Cornu Spiral

The *Cornu spiral* C is the curve in the complex plane defined by the Fresnel integrals defined at the beginning of Section 2.10, i.e., $C = \{(x(t), y(t)) : t \geq 0\}$, where

$$x(t) = \int_{0}^{t} \cos u^2 \, du \quad \text{and} \quad y(t) = \int_{0}^{t} \sin u^2 \, du.$$

The length of C from the origin to $(x, y) \in C$ is easily computed. To compute the curvature $\kappa(t)$ of C at t, we define the vector

$$\mathbf{r}(t) = (x(t), y(t), 0),$$

as well as its velocity $\mathbf{r}^{(1)}(t) = (\cos t^2, \sin t^2, 0)$ and acceleration $\mathbf{r}^{(2)}(t) = (-2t \sin t^2, 2t \cos t^2, 0)$. Then

$$\mathbf{r}^{(1)}(t) \times \mathbf{r}^{(2)}(t) = (0, 0, 2t \cos^2 t^2 + 2t \sin^2 t^2) = (0, 0, 2t),$$

and, hence,

$$\kappa(t) = \frac{\|\mathbf{r}^{(1)}(t) \times \mathbf{r}^{(2)}(t)\|}{\|\mathbf{r}^{(1)}(t)\|^3} = 2t,$$

where $\| \ldots \|$ is the Euclidean norm in $\mathbb{R}^3$.

If the Cornu spiral is replaced by the curve

$$\int_0^t e^{i\varphi(u)} \, du,$$

then the curvature at t of the resulting curve is $\varphi'(t)$.

Exercises

Exercises 2.1–2.30 are appropriate for Course I.

2.1. Compute $\mathbf{1}'_{[-T,T)}$ (the distributional derivative).

2.2. Compute the following distributional derivatives, where H is the Heaviside function.

a) $\big(H(t) \cos t\big)'$.

b) $\big(H(t) \sin t\big)'$.

c) $\big(\mathbf{1}_{[-\pi/2,\pi/2)}(t) \cos t\big)^{(2)}$.

d) g' where $g(t) = \begin{cases} \sin t, & t < 0, \\ 3e^{-t}, & t > 0. \end{cases}$

e) g' where $g(t) = \begin{cases} 0, & \text{if } t \leq 0, \\ t, & \text{if } t > 0. \end{cases}$

2.3. Compute the nth distributional derivative of $g(t) = |t|$, $n = 1, 2, 3$.

2.4. Compute the nth distributional derivative of $g(t) = |\cos t|$, $n = 1, 2, 3$.

2.5. a) Prove that the function f defined in Example 2.2.2 is infinitely differentiable.

b) Prove that there are no analytic functions in $C_c^\infty(\mathbb{R}) \backslash \{0\}$.

2.6. Prove that if $f \in C_c^\infty(\mathbb{R})$ and $g \in L^1(\mathbb{R})$ has compact support, then $f * g \in C_c^\infty(\mathbb{R})$.

2.7.   Prove (2.3.4) and (2.3.5).

2.8.   From Example 2.2.6 *b* show that if $g \in L^1_{\text{loc}}(\mathbb{R}) \setminus \{0\}$, then there is $f \in C^\infty_c(\mathbb{R})$ such that

$$\int f(t)g(t)\, dt \neq 0.$$

2.9.   Prove that

$$\forall n \geq 0, \qquad t^n \delta^{(n)}(t) = (-1)^n n! \delta(t).$$

2.10. a) Show that $\delta(t) = \delta(-t)$.

   b) Consider the approximate identity $\{k_\lambda\}$, where $k = \mathbf{1}_{[-1/2, 1/2]}$. In the spirit of Example 2.1.6, evaluate $\delta(t^2)\big(f(t)\big)$, for $f \in C_b(\mathbb{R})$ and $f(0) \neq 0$, by computing

$$\lim_{\lambda \to \infty} \int k_\lambda(t^2) f(t)\, dt.$$

   c) Evaluate the limit in part *b* for $f \in C^\infty_c(\mathbb{R})$ for which $f(0) = f'(0) = 0$.

2.11. a) Let $f \in C^\infty_c(\mathbb{R})$ and assume $g \in L^1(\mathbb{R})$ has compact support. Prove that $\text{supp}(f * g) \subseteq \text{supp}\, f + \text{supp}\, g$, cf., Exercise 2.6.

   b) Prove the generalization of part *a* when g is replaced by $T \in D'(\mathbb{R})$.

2.12. Prove that

$$\sum \delta(t - n) + \sum \delta\left(t - n - \frac{1}{2}\right) = 2 \sum \delta(2t - n)$$

$$= \left(\sum \delta(t - n)\right) * \left(4 \left(M_{1/2}\delta\right) \left(2t - \frac{1}{2}\right)\right),$$

where $M_{1/2}\delta$ is the mean

$$\left(M_{1/2}\delta\right)(t) = \frac{1}{2}\left(\delta\left(t + \frac{1}{2}\right) + \delta\left(t - \frac{1}{2}\right)\right).$$

The middle term arises in wavelet theory.

2.13. a) Compute the "L^2-autocorrelation" of δ, i.e., compute

$$P_\delta(t) = \delta(t) * \overline{\delta(-t)},$$

cf., Exercises 1.33 and 2.10. $\overline{\delta}$ and $\overline{\delta(-t)}$ can be computed using the method of Example 2.1.6.

   b) Compute the L^2-autocorrelation P_Δ of the triangle function Δ defined in Example 1.3.4.

   c) Compute the L^2-autocorrelation $P_{\Delta_{2\pi/\lambda}}$ of the dilation $\Delta_{2\pi/\lambda}$.

   d) In what sense does $\lim_{\lambda \to 0} P_{\Delta_{2\pi/\lambda}} = P_\delta$?

2.14. Let

$$f - \frac{1}{16\pi^4} f^{(4)} = 4\pi^2 \delta + \delta^{(2)}$$

be a differential equation on $\mathbb{R}$. Solve for f using the method of Theorem 2.6.1.

2.15. a) Compute $\delta^{(n)} * H$.

   b) Compute the Fourier transform of $\delta^{(n)} * H$.

2.16. a) Compute the Fourier transform of $\delta^{(n)} * g$, where $g(t) = t H(t)$ and $n = 1, 2, 3$.

   b) Compute the Fourier transform of $\delta^{(n)} * g$, where $g(t) = t^2 H(t)$ and $n = 1, 2$.

2.17. Verify whether or not there are integrable or square integrable solutions of the following differential equations defined on $\mathbb{R}$.

   a) $f^{(5)} + f^{(1)} = \delta^{(1)}$.

   b) $f^{(5)} - i f^{(1)} = \delta^{(4)}$.

2.18. a) Compute $t^2 \delta^{(1)}(t)$, cf., Exercise 2.9.

   b) Compute $t \delta^{(2)}(t)$.

   c) Compute $t^{12} \delta^{(5)}(t)$.

   d) Compute $t^5 \delta^{(12)}(t)$.

   e) Compute $t^m \delta^{(n)}(t)$ for $m, n \geq 0$.

2.19. Verify that (2.4.17) is a consequence of (2.4.16). The calculation is elementary but involved.

2.20. Let $f_n = (1/n)f$, where $f \in C_c^\infty(\mathbb{R})$.

   a) Prove that $f_n \to 0$ in the sense that i and ii of Definition 2.2.3 are satisfied.

   b) Let T be any linear functional on $C_c^\infty(\mathbb{R})$. Prove that

$$\lim_{n\to\infty} T(f_n) = 0.$$

   c) Show that there is a linear functional on $C_c^\infty(\mathbb{R})$ that is not a distribution. [Hint. Let $g \in C_c^\infty(\mathbb{R}) \setminus \{0\}$ and let $g_n = (1/n)\tau_{1/n}g$. A Fourier transform argument shows that $\{g_n\}$ is linearly independent. There is a basis (in the algebraic sense) of $C_c^\infty(\mathbb{R})$ that contains $\{g_n\}$, e.g., [Tay58, pages 44–45] on Hamel bases. Define $T(g_n) = 1$ for each n and extend T linearly to $C_c^\infty(\mathbb{R})$.]

2.21. Verify that

$$\delta(t^3 + 3t) = \frac{1}{3}\delta(t)$$

   and

$$\delta'(t^3 + 3t) = \frac{1}{9}\delta'(t).$$

2.22. a) We evaluated $\int (\sin t / t)\,dt$ in Proposition 1.6.3. Now show that

$$\forall T > 0, \qquad \left| \int_{-T}^{T} \frac{\sin t}{t}\,dt - \pi \right| < \frac{\pi}{T}.$$

[Hint. Use Jordan's inequality, which is stated in Theorem 2.10.1 and Exercise 2.62 *a*.]

b) Refine part *a* by proving that

$$\forall T > 0, \qquad \int_{-T}^{T} \frac{\sin t}{t}\, dt - \pi = -\frac{2\cos T}{T} - \frac{2\sin T}{T^2} + \varepsilon_T,$$

where $|\varepsilon_T| \le 2\pi/T^2$. In particular, if $T_n = \pi/2 + \pi n$, then

$$\left| \int_{-T_n}^{T_n} \frac{\sin t}{t}\, dt - \pi \right| \le \frac{2(1 + \pi)}{T_n^2}.$$

2.23. Prove that $BV_{\text{loc}}(\mathbb{R}) \subseteq L^1_{\text{loc}}(\mathbb{R})$. Technically, if $F \in BV_{\text{loc}}(\mathbb{R})$, then it is an ordinary point function, whereas the elements of $L^1_{\text{loc}}(\mathbb{R})$ are sets of ordinary functions that are equal a.e. Show that this situation does not cause any problems in this exercise.

2.24. Verify that

$$L^1(\mathbb{R}) = \big\{ F' : F \in BV(\mathbb{R}) \cap AC_{\text{loc}}(\mathbb{R}) \big\},$$

where F' designates distributional differentiation.

2.25. Prove (2.2.7) and (2.2.8), cf., Exercise 2.33.

2.26. Let $f \longleftrightarrow F$, where $f, f' \in L^1(\mathbb{R})$. Verify that

$$\left(f'(u) * pv\left(\frac{1}{u} \right) \right)(t) \longleftrightarrow 2\pi |\gamma| F(\gamma).$$

2.27. a) Compute $\left(1 * \delta^{(n)} \right)\hat{}\,$.

   b) Compute $\left(\delta^{(m)} * \delta^{(n)} \right)\hat{}\,$.

2.28. Let $T \in D'(\mathbb{R})$ and let $g \in C^\infty(\mathbb{R})$. Prove that

$$(Tg)^{(n)} = \sum_{k=0}^{n} \frac{n!}{k!(n-k)!} T^{(k)} g^{(n-k)}.$$

2.29. Consider the distribution

$$A pv\left(\frac{1}{t} \right) + BH(t) + C\delta(t) + D\delta'(t) + E \log|t|. \qquad (\text{E} 2.1)$$

For each of the following distributions T, list the coefficients in (E 2.1) that must be 0. For example, if T is H', then the answer is A, B, D, and E.

   a) $\hat{H}$;   b) $(\log|t|)'$;   c) $\hat{1}$;   d) $[tH(t)]^{(3)}$.

2.30. Let $T \in D'(\mathbb{R})$ and assume $T' = 0$. Prove that T is a constant. [Hint. Let $f_0 \in C_c^\infty(\mathbb{R})$ have the property that $\int f_0(t)\, dt = 1$. First prove that each $f \in C_c^\infty(\mathbb{R})$ has a unique decomposition $f = c_f f_0 + g$ where $c_f = \int f(t)\, dt$ and $g = h'$ for some $h \in C_c^\infty(\mathbb{R})$, e.g., [Sch66, pages 51–52]. Then compute $T(f) = c_f T(f_0) - T'(h) = T(f_0) 1(f)$.]

2.31. Using MATLAB, graph the Cornu spiral C defined in Example 2.10.7.

2.32. Define

$$\forall \sigma \in (1/2, 1), \qquad f_\sigma(t) = \frac{e^{\sigma t}}{1 + \exp e^t}.$$

a) Compute $(f_\sigma)\hat{}$ in terms of the Riemann zeta function ζ, cf., [Ben75, pages 137–138]. ζ and the Riemann Hypothesis were defined in Example 2.4.6g.

b) Using the notation V_g defined in (2.9.8), prove that the Riemann Hypothesis is true if and only if $V_{f_\sigma} = L^1(\mathbb{R})$ for each $\sigma \in (\frac{1}{2}, 1)$. This elementary observation is due to Salem (1953) [Sal67].

2.33. a) Let $\mu = \sum_{n=1}^{\infty} (1/n^2)\, \delta_{1/n}$. Prove that $\mu \in M_d(\mathbb{R})$ and that $\operatorname{supp}\mu = \{0, \frac{1}{n} : n \in \mathbb{N}\}$.

b) Let $\{r_n\}$ be the subset of the $\frac{1}{3}$-Cantor set $C \subseteq [0, 1]$ where each r_n is of the form $k/3^m$, and let $\mu = \sum_{n=1}^{\infty} (1/n^2)\, \delta_{r_n}$. Prove that $\mu \in M_d(\mathbb{R})$ and that $\operatorname{supp}\mu = C$. Recall that C is a closed, uncountable set without isolated points.

c) Let $\{r_n\} = \mathbb{Q}$, the set of rational numbers, and let $\mu = \sum_{n=1}^{\infty} (1/n^2)\, \delta_{r_n}$. Prove that $\mu \in M_d(\mathbb{R})$ and that $\operatorname{supp}\mu = \mathbb{R}$.

2.34. Compute the Fourier transforms of $f(t) = \sin(\pi t) + \cos(\pi t)$ and $f(t) = \sin(\pi t)^2 + \cos(\pi t)^2$.

2.35. Let $n \in \mathbb{N} \cup \{0\}$ and define $T(t) = a\delta^{(n)}(t) + bt^n$ for $a, b \in \mathbb{C}\backslash\{0\}$. For which values of $a, b,$ and n do we have $\hat{T} = T$, cf., Example 1.10.12 c and Exercise 1.28?

2.36. Let $\mu_n = \frac{1}{2}(\delta_{1/n!} + \delta_{-1/n!})$, $n \geq 2$; and define $\mu = \mu_2 * \mu_3 * \cdots$. Prove that $\|\mu\|_1 = 1$ and $\operatorname{supp}\mu \subseteq [-1, 1]$. Compute $\hat{\mu}$, cf., Viète's formula in Exercise 1.36, where $\prod_{k=1}^{\infty} \cos(\gamma/2^k) \in L^2(\hat{\mathbb{R}})$.

The *Wiener–Pitt Theorem* asserts that *if* $\mu = g + \mu_{sc} + \sum a_x \delta_x$, *with notation as in Theorem 2.7.6, satisfies the properties that* $|\hat{\mu}| > 0$ *on* $\hat{\mathbb{R}}$ *and*

$$\|\mu_{sc}\|_1 < \inf \left\{ \left| \sum a_x e^{-2\pi i x y} \right| : \gamma \in \hat{\mathbb{R}} \right\},$$

then there is $\nu \in M_b(\mathbb{R})$ *for which* $\hat{\nu} = 1/\hat{\mu}$, e.g., [Ben75, pages 147–149] *where the proof uses Kronecker's Theorem (Exercises 3.40 and 3.41).* John Williamson (*International Congress of Mathematics*, 1958) *used the example of this exercise in analyzing the Wiener–Pitt phenomenon in another setting.*

2.37. Consider the convolution equation

$$g - k * g = f \quad \text{on } \mathbb{R},$$

where $k, f \in L^1(\mathbb{R})$ are a given kernel and forcing function, respectively. Assume $|1 - \hat{k}| > 0$ on $\hat{\mathbb{R}}$. Prove that there is a solution $g \in L^1(\mathbb{R})$. [Hint. Take the Fourier transform of the equation, invoke the Wiener–Pitt Theorem (stated in Exercise 2.36), and apply Exercise 2.49b.] There are other proofs.

The *Wiener–Hopf equation*,

$$g(t) - \int_0^\infty k(t - u)g(u)\, du = f(t), \qquad t \geq 0,$$

is more difficult, but has also been solved, e.g., [Wie49, Appendix C] for applications and solution, cf., *Constructive Methods of Wiener–Hopf Factorization* (I. Gohberg and M. A. Kaashoek, editors), Birkhauser-Verlag, Basel, 1986.

2.38. Let $\{k_{(\lambda)}\}$ be an approximate identity. Prove that

$$\lim_{\lambda \to \infty} k_{(\lambda)} = \delta$$

in the sense that

$$\forall f \in C_b(\mathbb{R}), \qquad \lim_{\lambda \to \infty} \int k_{(\lambda)}(t) f(t)\, dt = f(0). \qquad (\text{E}\,2.2)$$

[Hint. Add and subtract $f(0) \int k_{(\lambda)}(u)du$.] If we replace the condition, $\int k_{(\lambda)}(u)du = 1$, of an approximate identity by the condition $\hat{k}_{(\lambda)}(0) = \int k_{(\lambda)}(u)du = K$ for each λ (while retaining the other properties of an approximate identity), then the conclusion (E 2.2) is replaced by

$$\forall f \in C_b(\mathbb{R}), \qquad \lim_{\lambda \to \infty} \int k_{(\lambda)}(t) f(t)\, dt = K f(0).$$

2.39. a) Prove that $\delta \notin L^1_{\text{loc}}(\mathbb{R})$ and that $\delta' \notin M(\mathbb{R})$, cf., Remark 2.3.7.

  b) Prove that $L^1(\mathbb{R}) \subseteq M_b(\mathbb{R})$ and that $L^1_{\text{loc}}(\mathbb{R}) \subseteq M(\mathbb{R})$, e.g., Theorem 2.7.6.

2.40. Let $g \in L^1(\mathbb{R})$, and, for $s > 0$ and $(t, \gamma) \in \mathbb{R} \times \hat{\mathbb{R}}$, set $g_{s,t,\gamma}(u) = g_s(u - t)e^{2\pi i u \gamma}$ as in Exercise 1.49. Prove that

$$\lim_{s \to \infty} g_{s,t,\gamma} = \hat{g}(0)e^{2\pi i t \gamma} \delta_t$$

in the sense that

$$\forall f \in C_b(\mathbb{R}), \qquad \lim_{s \to \infty} \int g_{s,t,\gamma}(u) f(u)\, du = (\hat{g}(0)e^{2\pi i t \gamma} \tau_t \delta)(f)$$

for each fixed $(t, \gamma) \in \mathbb{R} \times \hat{\mathbb{R}}$, cf., Exercise 2.38.

2.41. We defined $\mathcal{E}'(\mathbb{R})$ in Example 2.4.6 b. With convolution as the multiplicative operation, $\mathcal{E}'(\mathbb{R})$ is a commutative, associative algebra with unit δ, cf., Example 2.5.2 c and Example 2.5.10. Note that if $h_s(t) \equiv e^{-st}$ for any fixed $s \in \mathbb{C}$, then $h_s : \mathcal{E}'(\mathbb{R}) \longrightarrow \mathbb{C}$ is a homomorphism. Prove that if $h \in C^\infty(\mathbb{R})$ is a homomorphism $\mathcal{E}'(\mathbb{R}) \longrightarrow \mathbb{C}$, then $h(t) = e^{-st}$ for some $s \in \mathbb{C}$.

The algebra $\mathcal{E}'(\mathbb{R})$ and its space of (continuous) homomorphisms $\{h_s : s \in \mathbb{C}\}$ lead to the definition of the bilateral Laplace transform $T(t)(e^{-st})$ of $T \in \mathcal{E}'(\mathbb{R})$, cf., Example 2.4.6g. In fact, as a general point of view, any *algebra* $\mathcal{A}$ and a set $\mathcal{M}$ (for *maximal ideal space*) of its *homomorphisms* $\mathcal{A} \longrightarrow \mathbb{C}$ gives rise to a *transform* T for which there is an *exchange formula* $T(A * B) = T(A)T(B)$, e.g., Theorem 2.5.9. For example, if $\mathcal{A} = L^1(\mathbb{R})$ and $\mathcal{M} = \{e^{-2\pi i t \gamma} : \gamma \in \hat{\mathbb{R}}\}$, then we have Fourier analysis.

2.42. Prove that the Heaviside function H is an unbounded continuous measure.

2.43. Prove that the principal value distribution defined in Example 2.3.8 is, in fact, a distribution.

2.44. Let $\mu_C \in M_{b+}(\mathbb{R})$ be the Cantor measure corresponding to the $\frac{1}{3}$-Cantor set $C \subseteq [0, 1]$. Prove that

$$\int t \, d\mu_C(t) = \frac{1}{2},$$

noting that $\int d\mu_C(t) = 1$, cf., Example 2.3.9 b.

2.45. a) Prove that the mapping $S(\mathbb{R}) \to S(\hat{\mathbb{R}})$, $f \longmapsto \hat{f}$, is a linear bijection. [Hint. The linear injection follows by a straightforward calculation and the uniqueness theorem. For the surjectivity, let $F \in S(\hat{\mathbb{R}})$ and denote $F^{\vee}$ by g. $g \in S(\mathbb{R})$ by the first part of the calculation, and the goal is to show that $\hat{g} = F$. This follows by the inversion theorem for "˅" instead of "ˆ", i.e., $F(\gamma) = \int F^{\vee}(t)e^{-2\pi i t \gamma} dt$, which is the same as $\hat{g} = F$.]

   b) Prove that ρ defined by (2.4.7) is a metric on $S(\mathbb{R}) \times S(\mathbb{R})$, and that, as such, $S(\mathbb{R})$ is a complete metric space.

   c) Prove that the mapping of part a, where S is given the metrizable topology of part b, is bicontinuous.

   d) Prove that "$f_n \to 0$" in the sense of (2.4.8) if and only if $\lim_{n \to \infty} \rho(f_n, 0) = 0$.

2.46. Let $\{T_j\} \subseteq D'(\mathbb{R})$. Define $\lim_{j \to \infty} T_j = T$ for some $T \in D'(\mathbb{R})$ to mean that

$$\forall f \in C_c^\infty(\mathbb{R}), \qquad \lim_{j \to \infty} T_j(f) = T(f).$$

Similarly, define $\sum T_j = T$ for some $T \in D'(\mathbb{R})$ to mean that

$$\forall f \in C_c^\infty(\mathbb{R}), \qquad \sum T_j(f) = T(f).$$

   a) Prove that if $\sum T_j(f)$ exists for each $f \in C_c^\infty(\mathbb{R})$, then there is $T \in D'(\mathbb{R})$ for which $\sum T_j = T$.

   b) Prove that if $\sum T_j = T$ for some $T \in D'(\mathbb{R})$, then

$$\forall n \in \mathbb{N}, \qquad \left(\sum T_j\right)^{(n)} = \sum T_j^{(n)}.$$

2.47. Note that $S(\mathbb{R}) \subseteq X(\mathbb{R}) \equiv L^1(\mathbb{R}) \cap L^2(\mathbb{R}) \cap A(\mathbb{R})$. If the norm of $f \in X(\mathbb{R})$ is defined as $\|f\|_{X(\mathbb{R})} = \|f\|_{L^1(\mathbb{R})} + \|f\|_{L^2(\mathbb{R})} + \|f\|_{A(\mathbb{R})}$, where $\|f\|_{A(\mathbb{R})} \equiv \|\hat{f}\|_{L^1(\hat{\mathbb{R}})}$, then $\overline{S(\mathbb{R})} = X(\mathbb{R})$ and $X(\mathbb{R})\hat{} = X(\hat{\mathbb{R}})$. Thus, since $X'(\mathbb{R}) \subseteq S'(\mathbb{R})$, the distributional Fourier transform is a Banach space isomorphism of $X'(\mathbb{R})$ onto $X'(\hat{\mathbb{R}})$. Describe the elements of $X'(\mathbb{R})$.

2.48. Consider the n-fold convolution $g = 1_{[-\Omega,\Omega]} * \cdots * 1_{[-\Omega,\Omega]}$, introduced in terms of splines in Exercise 1.17. Compute $g^{(n)}$.

   Motivated by the results of this calculation, we introduce the following "usual" definition. A *spline of order r on $\mathbb{R}$ with knots at the integers* is a function $g \in L^2(\mathbb{R})$, whose restriction to each interval $[n, n+1)$ is a polynomial of degree at most $r-1$, and which is in $C^{r-2}(\mathbb{R})$, i.e., $g \in L^2(\mathbb{R}) \cap C^{r-2}(\mathbb{R})$ and $g^{(r)} = \sum a_n \delta_n$.

2.49. $M_b(\mathbb{R})$ is a Banach algebra with unit δ under convolution, and $L^1(\mathbb{R}) \subseteq M_b(\mathbb{R})$ is a closed ideal. This "algebraic" fact is a consequence of the following exercises.

a) Prove that, for all $\mu, \nu \in M_b(\mathbb{R})$, we have $\mu * \nu \in M_b(\mathbb{R})$, and, in fact,

$$\|\mu * \nu\|_1 \leq \|\mu\|_1 \|\nu\|_1.$$

b) We know $L^1(\mathbb{R}) \subseteq M_b(\mathbb{R})$, e.g., Exercise 2.39$b$. Prove that $\| \cdots \|_1$ reduces to $\| \cdots \|_{L^1(\mathbb{R})}$ on $L^1(\mathbb{R})$, and that, for all $\mu \in M_b(\mathbb{R})$, $f \in L^1(\mathbb{R})$, we have $f * \mu \in L^1(\mathbb{R})$, cf., Exercise 2.54.

   The characterization of the ideal structure of $L^1(\mathbb{R})$ is equivalent to solving the problems of spectral synthesis (for $L^1(\mathbb{R})$) mentioned in Example 2.4.6f, e.g., [Ben75]. The characterization of the ideal structure of $M_b(\mathbb{R})$, even just its maximal ideals, is also a deep topic, e.g., [DR71], [Kat76, Chapter 8].

2.50. Let $T \in D'(\mathbb{R})$, and let $h \in C^\infty(\mathbb{R})$ be a strictly monotonic surjection $\mathbb{R} \to \mathbb{R}$. In light of Example 2.1.6, Exercise 2.10, and Exercise 2.21, prove that the *composition* $T \circ h^{-1}$ is an element of $D'(\mathbb{R})$ if it is defined as $(T \circ h^{-1})(f) \equiv T((f \circ h)|h'|)$ for $f \in C_c^\infty(\mathbb{R})$.

2.51. Let $T \in D'(\mathbb{R})$ and let $f \in C_c^\infty(\mathbb{R})$. Prove that $T(\tau_t f) \in C^\infty(\mathbb{R})$ as a function of t.

2.52. An $N \times N$ matrix $A = (a_{jk})$ is *positive semidefinite* if $\bar{c}^T A c \geq 0$ for every $N \times 1$ matrix $c \in \mathbb{C}^N$, where $\bar{c}$ designates conjugation of each component and T designates transposition. (Unfortunately, "semidefinite" is like "maybe for sure".) Let $P : \hat{\mathbb{R}} \to \mathbb{C}$ be a function, and let $A_P(\gamma_1, \ldots, \gamma_N)$ be the $N \times N$ matrix $\big(P(\gamma_j - \gamma_k)\big)$.

a) Prove that $P >> 0$ if and only if $A_P(\gamma_1, \ldots, \gamma_N)$ is positive semidefinite for each $N \geq 1$ and each set $\{\gamma_1, \ldots, \gamma_N\} \subseteq \hat{\mathbb{R}}$.

b) Prove the four properties of positive definite functions listed in Definition 2.7.8b. [Hint. For part *iii*, let $\gamma_1 = 0$, $\gamma_2 = \lambda$, $\gamma_3 = \gamma$; assume $P(\lambda) \neq P(\gamma)$; and let $c_1 = 1$, $c_2 = -c_3$, and

$$c_2 = \frac{|P(\lambda) - P(\gamma)|}{P(\lambda) - P(\gamma)} x, \qquad x \in \mathbb{R}.$$

Consider the quadratic form (2.7.4) as a polynomial in x, and analyze its discriminant.]

c) Prove that a positive semidefinite matrix $A = (a_{jk})$ is Hermitian, i.e., $a_{jk} = \bar{a}_{kj}$ for all j, k, cf., [Don69, pages 181–182], [Str88] for further properties.

2.53. Suppose $f : \mathbb{R} \to \mathbb{R}$ has the properties that

$$\exists M \text{ such that } \forall a < b, \qquad \left| \int_a^b f(t)\, dt \right| \leq M,$$

f is differentiable, and $\|f'\|_{L^\infty(\mathbb{R})} = m$. Prove that $f, \mathcal{H}f \in L^\infty(\mathbb{R})$. This result was alluded to in Theorem 2.5.12 and is due to Logan [Log83]. [Hint. First show that

$$\lim_{\substack{\varepsilon \to 0 \\ R \to \infty}} \left(\int_{-R}^{-\varepsilon} \frac{f(t)}{t} \, dt + \int_{\varepsilon}^{R} \frac{f(t)}{t} \, dt \right)$$

$$= \int_{-T}^{T} \left\{ \log \left| \frac{T}{t} \right| - C(T - |t|) \right\} f'(t) \, dt \qquad \text{(E2.3)}$$

$$+ \int_{|t| \geq T} \frac{F(t)}{t^2} \, dt + \left(C - \frac{1}{T} \right) \{ F(T) + F(-T) \},$$

where $F(t) = \int_0^t f(x) dx$, $T > 0$, and $C \in \mathbb{R}$. We assume everywhere differentiability of f to ensure $f \in AC_{\text{loc}}(\mathbb{R})$, since we use FTC to verify (E2.3). Now estimate the right side of (E2.3), and correctly choose T and C.]

2.54. Apropros Example 2.5.2, prove that $S'(\mathbb{R}) * S(\mathbb{R})$ is not contained in $L^1(\mathbb{R})$. [Hint. Let $T(u) = u^2$ and let $g(u) = e^{-u^2}$.] It turns out that $S'(\mathbb{R}) * S(\mathbb{R}) \subseteq \mathcal{O}_C$, where $f \in \mathcal{O}_C$ means that $f \in C^\infty(\mathbb{R})$ and

$$\exists k = k(f) \in \mathbb{Z} \text{ such that } \forall n \in \mathbb{N},$$

$$\lim_{|t| \to \infty} \left(1 + |t|^2 \right)^k \left| f^{(n)}(t) \right| = 0,$$

e.g., [Hor66, pages 420–423].

2.55. We defined the even and odd parts of functions in Exercise 1.7. Verify (formally) that $\mathcal{H} f_e = (\mathcal{H} f)_e$ and $(\mathcal{H} f_o) = (\mathcal{H} f)_o$, where $\mathcal{H}$ is the Hilbert transform. Thus, if X is a space of functions, then, formally, $X = X_e \oplus X_o$ and $\mathcal{H} X = (\mathcal{H} X)_e \oplus (\mathcal{H} X)_o$.

2.56. Let f be a causal function, i.e., supp $f \subseteq [0, \infty)$. Formally define the *unilateral Stieltjes transform* of f as $S(f)(t) = \int_0^\infty (f(u)/(t + u)) du$ and the *(unilateral) Laplace transform* of f as $\mathcal{L}(f)(t) = \int_0^\infty f(u) e^{-tu} \, du$, e.g., [Wid41]. Assume that $\mathcal{L} f$ and $\mathcal{L}\mathcal{L} f$ exist on $(0, \infty)$. Show that

$$\mathcal{L}\mathcal{L} f(u) = S f(u) = -\pi \mathcal{H} f(-u).$$

2.57. a) Verify (formally) the Fourier transform pairings $2fH \longleftrightarrow \hat{f} - i\mathcal{H}\hat{f}$ and $f + i\mathcal{H}f \longleftrightarrow 2\hat{f}H$, where H is the Heaviside function and $\mathcal{H}$ is the Hilbert transform operator.

   b) Verify (formally) that if f is causal, then $\text{Re } \hat{f} = \mathcal{H} \text{Im } \hat{f}$ and $\text{Im } \hat{f} = -\mathcal{H} \text{Re } \hat{f}$.

   Parts *a* and *b* give elementary relations between the Fourier and Hilbert transforms. The situation can become more complex (sic). Recall the Paley–Wiener Logarithmic Integral Theorem [PW34] from Example 1.6.5. This result asserts the existence of a causal function $f \in L^2(\mathbb{R})$ for which $|\hat{f}| = \phi$ a.e. in the case

$$\int \frac{|\log \phi(\gamma)|}{1 + \gamma^2} \, d\gamma < \infty$$

for a given nonnegative function $\phi \in L^2(\mathbb{R})$. Formally

$$\hat{f} = \phi e^{-i\mathcal{H}\phi}$$

is a candidate for such a function f, cf., [BT93] for relevant calculations and a wavelet application to speech compression, and consider $\varphi(\gamma) = 1/(1+\gamma^2)$ for a dose of mathematical reality.

2.58. a) Prove that (2.6.7) defines a translation-invariant continuous linear operator $L : L^1(\mathbb{R}) \to L^1(\mathbb{R})$.

b) Let $L : L^1(\mathbb{R}) \to L^1(\mathbb{R})$ be a translation-invariant continuous linear operator, and let $g \in L^\infty(\mathbb{R})$. Prove that there is an $h \in L^\infty(\mathbb{R})$, depending on L and g, such that

$$\forall f \in L^1(\mathbb{R}), \qquad \int (Lf)(t)g(t)\,dt = \int f(t)h(t)\,dt.$$

c) Use part b to prove that

$$\forall f, k \in L^1(\mathbb{R}), \qquad (Lf) * k = L(f * k) = f * L(k).$$

[Hint. Calculate

$$\int (Lf) * k(t)g(t)\,dt = \int k(s) \int (\tau_s f)(t)h(t)\,dt\,ds$$

$$= \int h(t) f * k(t)\,dt = \int L(f * k)(t)g(t)\,dt.]$$

2.59. a) Prove that $\mathcal{H}(\cos 2\pi t \gamma_0) = \sin 2\pi t \gamma_0$ and $\mathcal{H}(\sin 2\pi t \gamma_0) = -\cos 2\pi t \gamma_0$, $\gamma_0 > 0$.

b) Let $f \in L^2(\mathbb{R})$ be real-valued. Prove that $\int \mathcal{H}f(t)\overline{f(t)}\,dt = 0$.

c) Prove the symmetry condition (2.6.7) of Example 2.6.7a. [Hint. The straightforward calculation depends on the fact that h, described in Example 2.6.7a, is real-valued.]

2.60. With regard to the problem of finding $F \in C_c(\hat{\mathbb{R}})$ for which $F \notin B(\hat{\mathbb{R}})$, prove that

$$\forall \mu \in M_b(\mathbb{R}), \qquad \exists \lim_{\substack{\varepsilon \to 0 \\ R \to \infty}} \int_{\varepsilon \le |\gamma| \le R} \frac{\hat{\mu}(\gamma)}{\gamma}\,d\gamma,$$

cf., Exercise 2.19.

2.61. Prove Theorem 2.9.4.

2.62. a) Prove Jordan's Inequality,

$$\forall r > 0, \qquad \int_0^{\pi/2} e^{-r\sin\theta}\,d\theta < \frac{\pi}{2r}.$$

b) In light of part a, prove that

$$\forall \theta \in \left(0, \frac{\pi}{4}\right), \qquad (\sin\theta)^{\sin\theta} < (\cos\theta)^{\cos\theta},$$

e.g., *Amer. Math. Monthly Problems*, **101** (1994), 690.

2.63. The following inequalities are van der Corput's Lemma.

   a) Let ϕ be a real, differentiable function on $[a, b]$ for which ϕ' is monotone and for which $|\phi'| \geq r > 0$. Prove that

$$\left| \int_a^b e^{i\phi(t)}\, dt \right| \leq \frac{4}{r}.$$

   [Hint. Write the integral in terms of its real and imaginary parts. For each case multiply and divide by ϕ', and use the first mean value theorem for integrals.]

   b) Let ϕ be a real, twice differentiable function on $[a, b]$ for which $|\phi^{(2)}| \geq r > 0$. Prove that

$$\left| \int_a^b e^{i\phi(t)}\, dt \right| \leq \frac{8}{\sqrt{r}}.$$

   [Hint. ϕ' vanishes at most once in (a, b). If this point is c, write $\int_a^b$ as $\int_a^{c-\epsilon} + \int_{c-\epsilon}^{c+\epsilon} + \int_{c+\epsilon}^b$. The proper choice of ϵ and an application of part a on the first and third integrals yield the result.]

   It is often important to compute or estimate exponential sums, e.g., the Gauss sums of Section 3.8. Finite trigonometric sums $\sum e^{2\pi i\varphi(n)}$ arise in analytic number theory in the process of estimating the growth of the Riemann zeta function. A classical method due to van der Corput is to write

$$\sum_{a < n \leq b} e^{2\pi i\varphi(n)} \approx \sum_{\varphi'(b)-\eta < m \leq \varphi'(a)+\eta} \int_a^b e^{2\pi i(\varphi(t)-mt)}\, dt, \qquad (\text{E}2.4)$$

   and to estimate the right side using refinements of parts a and b, e.g., [Ivi85, Chapter 2], [Tit51, Chapter 4]. Then it is possible to show that

$$\exists C > 0 \text{ such that } \left| \zeta\left(\frac{1}{2} + iy\right) \right| \leq C|y|^{1/6}$$

   for large $|y|$. Bombieri and Iwaniec proved a slightly better result using modular forms. In any case, the right side of (E2.4) is an approximate form of the Poisson Summation Formula, cf., Section 3.10.

2.64. Complete the details in the proof of Theorem 2.10.3.

2.65. The function $s(t) = (1/\sqrt{\pi i})e^{it^2}$ of Section 2.10 is in $L^\infty(\mathbb{R}) \cap A'(\mathbb{R}) \subseteq M(\mathbb{R})$, whereas $s \notin L^1(\mathbb{R})$. Verify whether or not s is a bounded Radon measure.

2.66. Compute the *deterministic cross-correlation*,

$$R_{\gamma,\lambda}(t) = \lim_{T \to \infty} \frac{1}{2T} \int_{-T}^{T} s_\gamma(t + u)\overline{s_\lambda(u)}\, du, \qquad \gamma, \lambda > 0,$$

   where s_γ is the dilation of the function s defined in Exercise 2.65, cf., Example 2.10.6 for the case $\gamma = \lambda$.

2.67. Let $X(\mathbb{R}) = L^\infty(\mathbb{R}) \cap A'(\mathbb{R}) \subseteq S'(\mathbb{R})$. Note that $X(\mathbb{R})\hat{} = X(\hat{\mathbb{R}})$, and that $X(\mathbb{R})$ is a Banach space, where the norm of $f \in X(\mathbb{R})$ is defined as $\|f\|_{X(\mathbb{R})} = \|f\|_{L^\infty(\mathbb{R})} + \|\hat{f}\|_{L^\infty(\hat{\mathbb{R}})}$, cf., Exercise 2.47. Prove that $\sin \pi t^2$ and $\cos \pi t^2$ are eigenfunctions of the Fourier transform mapping $\mathcal{F} : X(\mathbb{R}) \to X(\hat{\mathbb{R}})$.

2.68. a) Verify that if $c \in \mathbb{R} \setminus \{0\}$, then

$$\sup_{a < b} \left| \int_a^b e^{ict^2} \, dt \right| \le 2 + \frac{4}{|c|},$$

e.g., Exercise 2.63. The bound can be refined.

b) Let $f \in AC_{\text{loc}}([0, \infty))$ be positive and decreasing. (Recall that f is not necessarily in $AC[a, b]$ if it is decreasing on $[a, b]$.) Prove that if $c > 0$, then

$$\left| \int_0^\infty f(t) e^{\pi i (ct)^2} e^{-2\pi i t y} \, dt \right| \le \frac{1}{c} \left(2 + \frac{4}{\pi} \right) f(0).$$

Such estimates are used in the *Littlewood Flatness Problem* discussed in Remark 3.8.11.

3

Fourier Series

3.1. Fourier Series—Definitions and Convergence

3.1.1 Definition. Fourier Series

a. Let $\Omega > 0$, and let $F : \hat{\mathbb{R}} \to \mathbb{C}$ be a function. F is 2Ω-*periodic* with period 2Ω if $F(\gamma + 2\Omega) = F(\gamma)$ for all $\gamma \in \hat{\mathbb{R}}$. For example, $F(\gamma) = \sin \gamma$ is 2π-periodic. If F is defined a.e., then F is 2Ω-*periodic* if $F(\gamma + 2\Omega) = F(\gamma)$ a.e.

b. Let $F \in L^1_{\mathrm{loc}}(\hat{\mathbb{R}})$ be 2Ω-periodic. The *Fourier series* of F is the series

$$S(F)(\gamma) = \sum f[n] e^{-\pi i n \gamma / \Omega}, \tag{3.1.1}$$

where

$$\forall n \in \mathbb{Z}, \qquad f[n] = \frac{1}{2\Omega} \int_{-\Omega}^{\Omega} F(\gamma) e^{\pi i n \gamma / \Omega} \, d\gamma. \tag{3.1.2}$$

The numbers $f[n]$ are the *Fourier coefficients* of F. The symbol "$\sum$" denotes summation over all of $\mathbb{Z}$, i.e., "$\sum_{n=-\infty}^{\infty}$".

c. Formally, the right side of (3.1.1) can be thought of as defining the *Fourier transform* $\hat{f}$ or F of the sequence $f = \{f[n]\}$, cf., Remark 3.1.3d. In fact, a sequence $f = \{f[n]\}$ is a function

$$
\begin{aligned}
f : \mathbb{Z} &\longrightarrow \mathbb{C} \\
n &\longmapsto f[n].
\end{aligned}
$$

Letting $\ell^1(\mathbb{Z})$ be the space of all sequences $f = \{f[n]\}$ for which $\|f\|_{\ell^1(\mathbb{Z})} \equiv \sum |f[n]| < \infty$, the right side of (3.1.1) is well defined for $f \in \ell^1(\mathbb{Z})$. In this context, we shall write

$$f \longleftrightarrow F, \quad \hat{f} = F, \quad f = \check{F}, \tag{3.1.3}$$

just as we did in Definition 1.1.2 for the case of Fourier transforms. Thus, in the case of sequences we write $\check{F}[n] = f[n]$.

The notation (3.1.3) is based on the presumption that $S(F)$ should equal F, e.g., Remark 3.1.2 and Theorem 3.1.6; and that if the right side of (3.1.1) defines the Fourier transform of the sequence f, then (3.1.2) is the Fourier inversion formula on $\mathbb{Z}$ corresponding to the Fourier inversion formula (1.1.1) on $\mathbb{R}$.

In fact, $S(F)$ often does equal F in the sense that the partial sums of the series $S(F)$ will converge in some way to F. With this in mind, if $F \in L^1_{\text{loc}}(\hat{\mathbb{R}})$ is 2Ω-periodic we shall write

$$S_{M,N}(F)(\gamma) = \sum_{n=-M}^{N} f[n]e^{-\pi i n\gamma/\Omega},$$

where f is defined by (3.1.2). $S_N(F) \equiv S_{N,N}(F)$ is the *Nth partial sum* of $S(F)$.

   d.   The venerable subject of Fourier series has its share of venerable treatises, which include [Bary64], [Car30], [Edw67], [HR56], [KS63], [Kah70], [Kat76], [Kör88], [Rog59], and [Zyg59] (Zygmund's first edition is great, too).

   e.   We are using the notation "$f[n]$" for a Fourier coefficient of F to distinguish it from the notation "$f(n)$", which usually designates the value at $n \in \mathbb{Z}$ of a function f defined on $\mathbb{R}$. Similarly, we have chosen "$f[n]$" instead of "f_n", since "f_n" often indicates an element of a sequence of functions defined on $\mathbb{R}$. Also, we could use "c_n" instead of "$f[n]$", but then we lose contact with the letter "F".

   We can and shall consider Fourier series of periodic functions on $\mathbb{R}$ instead of $\hat{\mathbb{R}}$. Our choice of (3.1.1) to define Fourier series is based on the first part of c, the typical setting of spectral frequency information (in terms of Greek letters such as "γ") associated with digital signals (sequences), *and* whim!

   Finally, we could have defined Fourier series for 1-periodic or 2π-periodic functions; and then have developed the theory of Fourier series, unburdened by lots of Ω's. We have chosen the setting of 2Ω-periodic functions to give us the flexibility of dealing with different values of Ω that might arise in specific problems or applications. In more theoretical developments, we shall usually let $2\Omega = 1$.

3.1.2 Remark. Formal Calculation and Elementary Examples

   a.   The reason we deal with (3.1.1) and (3.1.2) as a *pair* is that the decomposition of F into its fundamental parts, viz., the formula $S(F) = F$, is only effective if there is quantitative knowledge of the coefficients $f[n]$ in (3.1.3). In the case $S(F) = F$ the following *formal calculation* allows us to obtain (3.1.2) from (3.1.1):

$$\int_{-\Omega}^{\Omega} F(\gamma) e^{\pi i m \gamma / \Omega} \, d\gamma = \int_{-\Omega}^{\Omega} \left(\sum f[n] e^{-\pi i n \gamma / \Omega} \right) e^{\pi i m \gamma / \Omega} \, d\gamma$$

$$= \sum f[n] \int_{-\Omega}^{\Omega} e^{\pi i (m-n) \gamma / \Omega} \, d\gamma = 2\Omega f[m].$$

b. Let $\Omega > 0$ and let $\alpha \in (0, \Omega)$. Define F as $F = \mathbf{1}_{[-\alpha, \alpha)}$ on $[-\Omega, \Omega)$, extended 2Ω-periodically on $\hat{\mathbb{R}}$, i.e., $F(\gamma + 2n\Omega) = F(\gamma)$ for all $\gamma \in \hat{\mathbb{R}}$ and all $n \in \mathbb{Z}$. The Fourier series of F is

$$S(F)(\gamma) = \sum d_{(\alpha)}[n] e^{-\pi i n \gamma / \Omega}$$

where $d_{(\alpha)}[0] = \alpha / \Omega$ and

$$\forall n \in \mathbb{Z} \setminus \{0\}, \qquad d_{(\alpha)}[n] = \frac{\alpha}{\Omega} \frac{\sin(\pi n \alpha / \Omega)}{(\pi n \alpha / \Omega)},$$

cf., the Dirichlet function in Example 1.3.1.

Next define F by $F(\gamma) = \max(1 - (|\gamma|/\alpha), 0)$ on $[-\Omega, \Omega)$, extended 2Ω-periodically on $\hat{\mathbb{R}}$. A straightforward calculation, similar to that in Example 1.3.4, shows that the Fourier series of F is

$$S(F)(\gamma) = \sum w_{(\alpha)}[n] e^{-\pi i n \gamma / \Omega},$$

where $w_{(\alpha)}[0] = \alpha / 2\Omega$ and

$$\forall n \in \mathbb{Z} \setminus \{0\}, \qquad w_{(\alpha)}[n] = \frac{\alpha}{2\Omega} \left(\frac{\sin(\pi n \alpha / 2\Omega)}{\pi n \alpha / 2\Omega} \right)^2,$$

cf., the Fejér function in Example 1.3.4.

The Fourier coefficients in this example define the *Dirichlet* and *Fejér kernels* on $\mathbb{Z}$, cf., Example 3.4.5.

3.1.3 Remark. Notation and Setting

a. If $\Omega > 0$ and $F \in L^1_{\text{loc}}(\hat{\mathbb{R}})$ is 2Ω-periodic, then we write $F \in L^1(\mathbb{T}_{2\Omega})$.

Mathematically, $\mathbb{T}_{2\Omega} = \hat{\mathbb{R}}/(2\Omega\mathbb{Z})$ is a quotient group, that is referred to as the *circle group* depending on Ω. We shall not be concerned with this terminology for the time being. The point is that, because of the periodicity of F, $F \in L^1(\mathbb{T}_{2\Omega})$ can be thought of as being defined on *any* fixed interval $I \subseteq \hat{\mathbb{R}}$ of length 2Ω; and that this periodicity, combined with knowledge of F on any such interval, completely determines F on $\hat{\mathbb{R}}$.

If $\lambda_0 \in \hat{\mathbb{R}}$, then the *notation* $\lambda_0 \in \mathbb{T}_{2\Omega}$ indicates any one of the points $\lambda_0 + 2n\Omega \in \hat{\mathbb{R}}$, where $n \in \mathbb{Z}$. Further, if $J \subseteq \hat{\mathbb{R}}$ is contained in an interval of length 2Ω, then the *notation* $J \subseteq \mathbb{T}_{2\Omega}$ indicates any one of the subsets

$$J + 2n\Omega = \{\gamma + 2n\Omega : \gamma \in J\}, \qquad n \in \mathbb{Z}.$$

If $\Omega = \frac{1}{2}$, we write $\mathbb{T} \equiv \mathbb{T}_1$.

This possibly cryptic exposition might be unraveled at this time by performing some of the calculations in Exercise 3.1, cf., part *c*.

In any case, if $F \in L^1(\mathbb{T}_{2\Omega})$, then $\int_I F(\gamma)\,d\gamma = \int_{-\Omega}^{\Omega} F(\gamma)\,d\gamma$ for any interval $I \subseteq \hat{\mathbb{R}}$ of length 2Ω. As such we introduce the notation

$$\int_{\mathbb{T}_{2\Omega}} F(\gamma)\,d\gamma = \frac{1}{2\Omega} \int_I F(\gamma)\,d\gamma.$$

The factor "$1/2\Omega$" is a normalization factor in the sense that if $F = 1$ on $\hat{\mathbb{R}}$, then

$$F = 1 \in L^1(\mathbb{T}_{2\Omega}) \quad \text{and} \quad \int_{\mathbb{T}_{2\Omega}} 1\,d\gamma = 1.$$

The *norm* of $F \in L^1(\mathbb{T}_{2\Omega})$ is

$$\|F\|_{L^1(\mathbb{T}_{2\Omega})} = \int_{\mathbb{T}_{2\Omega}} |F(\gamma)|\,d\gamma = \frac{1}{2\Omega} \int_{-\Omega}^{\Omega} |F(\gamma)|\,d\gamma.$$

b. If $\Omega > 0$, F is 2Ω-periodic, and $F^2 \in L^1_{\mathrm{loc}}(\hat{\mathbb{R}})$ is 2Ω-periodic, then we write $F \in L^2(\mathbb{T}_{2\Omega})$. The *norm* of $F \in L^2(\mathbb{T}_{2\Omega})$ is

$$\|F\|_{L^2(\mathbb{T}_{2\Omega})} = \left(\int_{\mathbb{T}_{2\Omega}} |F(\gamma)|^2\,d\gamma \right)^{1/2} = \left(\frac{1}{2\Omega} \int_{-\Omega}^{\Omega} |F(\gamma)|^2\,d\gamma \right)^{1/2}.$$

By Hölder's Inequality,

$$\int_\alpha^\beta |F(\gamma)G(\gamma)|\,d\gamma$$
$$\le \left(\int_\alpha^\beta |F(\gamma)|^2\,d\gamma \right)^{1/2} \left(\int_\alpha^\beta |G(\gamma)|^2\,d\gamma \right)^{1/2}, \tag{3.1.4}$$

e.g., Theorem A.15; and so we have the inclusion $L^2(\mathbb{T}_{2\Omega}) \subseteq L^1(\mathbb{T}_{2\Omega})$ and the inequality

$$\forall F \in L^2(\mathbb{T}_{2\Omega}), \qquad \|F\|_{L^1(\mathbb{T}_{2\Omega})} \le \|F\|_{L^2(\mathbb{T}_{2\Omega})}. \tag{3.1.5}$$

Recall the analogous (sic) situation on $\mathbb{R}$, i.e., Exercise 1.35. The inclusion and inequality (3.1.5) follow since, by taking $G = 1$ in (3.1.4), we obtain

$$\|F\|_{L^1(\mathbb{T}_{2\Omega})} = \frac{1}{2\Omega} \int_{-\Omega}^{\Omega} |F(\gamma)| \, d\gamma$$

$$\leq \frac{1}{2\Omega} \left(\int_{-\Omega}^{\Omega} |F(\gamma)|^2 \, d\gamma \right)^{1/2} (2\Omega)^{1/2} = \|F\|_{L^2(\mathbb{T}_{2\Omega})}.$$

The inclusion is also proper, e.g., Exercise 3.6.

c. Let $\Omega = \pi$ and let $F(\gamma) = \sin \gamma + \cos 2\gamma$ on $\hat{\mathbb{R}}$. Then

$$\int_{\mathbb{T}_{2\Omega}} F(\gamma) \, d\gamma = \frac{1}{2\pi} \int_0^{2\pi} F(\gamma) \, d\gamma$$

$$= \frac{1}{2\pi} \int_{-\pi}^{\pi} F(\gamma) \, d\gamma = \frac{1}{2\pi} \int_\alpha^{\alpha+2\pi} F(\gamma) \, d\gamma = 0$$

for any fixed $\alpha \in \hat{\mathbb{R}}$. Further, if $J_I = (\pi/4, \pi/2] \subseteq I = (0, 2\pi]$, then

$$\forall n \in \mathbb{Z}, \qquad \frac{1}{2\pi} \int_{J_I} F(\gamma) \, d\gamma = \frac{1}{2\pi} \int_{\pi/4}^{\pi/2} F(\gamma) \, d\gamma = \frac{1}{2\pi} \int_{(\pi/4)+2\pi n}^{(\pi/2)+2\pi n} F(\gamma) \, d\gamma.$$

Thus, if we let J be any one of the sets $J_I + 2\pi n \subseteq \hat{\mathbb{R}}$, then $J \subseteq \mathbb{T}_{2\Omega}$ and

$$\int_J F(\gamma) \, d\gamma = \frac{1}{2\pi} \int_{\pi/4}^{\pi/2} F(\gamma) \, d\gamma.$$

d. In Definition 3.1.1c we defined the Fourier transform of a sequence $f = \{f[n]\} \in \ell^1(\mathbb{Z})$. We now define the *Fourier transform* of $F \in L^1(\mathbb{T}_{2\Omega})$ as the sequence $f = \{f[n]\}$, where

$$\forall n \in \mathbb{Z}, \qquad f[n] = \int_{\mathbb{T}_{2\Omega}} F(\gamma) e^{-\pi i n \gamma/\Omega} \, d\gamma,$$

and where "$\int_{\mathbb{T}_{2\Omega}}$" is defined in part *a*. In this case, the formal inversion formula is

$$F(\gamma) = \sum f[n] e^{\pi i n \gamma/\Omega}.$$

e. In Chapter 1, for $f \in L^1(\mathbb{R})$, the Fourier transform of f was defined on $\hat{\mathbb{R}}(= \mathbb{R})$. In this chapter, we have two "dual" settings. First, for $f \in \ell^1(\mathbb{Z})$, the Fourier transform of f is defined on $\mathbb{T}_{2\Omega}$; and, second, for $F \in L^1(\mathbb{T}_{2\Omega})$, the Fourier transform of F is defined on $\mathbb{Z}$. Mathematically, $\mathbb{R}$ and $\hat{\mathbb{R}}$ are locally compact abelian groups (LCAGs) that are not compact and that are *dual*, in a technical sense, to each other; similarly, the discrete LCAG $\mathbb{Z}$ is the dual group of the compact LCAG $\mathbb{T}_{2\Omega}$, and vice versa, e.g., [Rud62], [Edw67], [Ben75].

3.1.4 Example. Fourier Series of Real-Valued Functions

Let $F \in L^1(\mathbb{T}_{2\Omega})$ be real-valued, and let $f = \{f[n]\}$ be the sequence of Fourier coefficients of F. Thus,

$$f[n] = \frac{1}{2\Omega} \int_{-\Omega}^{\Omega} F(\gamma) \cos\left(\frac{\pi n\gamma}{\Omega}\right) d\gamma + \frac{i}{2\Omega} \int_{-\Omega}^{\Omega} F(\gamma) \sin\left(\frac{\pi n\gamma}{\Omega}\right) d\gamma.$$

Formally, since $S(F)$ should be equal to F and therefore be real-valued, we have

$$S(F)(\gamma) = \frac{1}{2\Omega} \int_{-\Omega}^{\Omega} F(\lambda) d\lambda + \sum{}' \left(\frac{1}{2\Omega} \int_{-\Omega}^{\Omega} F(\lambda) \cos\left(\frac{\pi n\lambda}{\Omega}\right) d\lambda\right.$$

$$+ \left. \frac{i}{2\Omega} \int_{-\Omega}^{\Omega} F(\lambda) \sin\left(\frac{\pi n\lambda}{\Omega}\right) d\lambda\right) \left(\cos\left(\frac{\pi n\gamma}{\Omega}\right) - i \sin\left(\frac{\pi n\gamma}{\Omega}\right)\right)$$

$$= \frac{1}{2\Omega} \int_{-\Omega}^{\Omega} F(\lambda) d\lambda$$

$$+ \sum{}' \left(\frac{1}{2\Omega} \int_{-\Omega}^{\Omega} F(\lambda) \cos\left(\frac{\pi n\lambda}{\Omega}\right) d\lambda\right) \cos\left(\frac{\pi n\gamma}{\Omega}\right)$$

$$+ \sum{}' \left(\frac{1}{2\Omega} \int_{-\Omega}^{\Omega} F(\lambda) \sin\left(\frac{\pi n\lambda}{\Omega}\right) d\lambda\right) \sin\left(\frac{\pi n\gamma}{\Omega}\right)$$

$$= a_0 + \sum_{n=1}^{\infty} a_n \cos\left(\frac{\pi n\gamma}{\Omega}\right) + \sum_{n=1}^{\infty} b_n \sin\left(\frac{\pi n\gamma}{\Omega}\right),$$

where "$\sum'$" designates summation over $\mathbb{Z}\setminus\{0\}$ and where

$$a_0 = \frac{1}{2\Omega} \int_{-\Omega}^{\Omega} F(\lambda) d\lambda,$$

$$a_n = \frac{1}{\Omega} \int_{-\Omega}^{\Omega} F(\lambda) \cos\left(\frac{\pi n\lambda}{\Omega}\right) d\lambda,$$

and

$$b_n = \frac{1}{\Omega} \int_{-\Omega}^{\Omega} F(\lambda) \sin\left(\frac{\pi n\lambda}{\Omega}\right) d\lambda,$$

for $n > 0$. The coefficients a_n, b_n are obtained since, for example, if $n > 0$, then

$$\left(\frac{1}{2\Omega} \int_{-\Omega}^{\Omega} F(\lambda) \sin\left(\frac{-\pi n\lambda}{\Omega}\right) d\lambda\right) \sin\left(\frac{-\pi n\gamma}{\Omega}\right)$$

$$+ \left(\frac{1}{2\Omega} \int_{-\Omega}^{\Omega} F(\lambda) \sin\left(\frac{\pi n\lambda}{\Omega}\right) d\lambda\right) \sin\left(\frac{\pi n\gamma}{\Omega}\right)$$

$$= b_n \sin\left(\frac{\pi n\gamma}{\Omega}\right).$$

The Riemann–Lebesgue Lemma for $L^1(\mathbb{R})$ (Theorem 1.4.1c) has an analogue for $L^1(\mathbb{T}_{2\Omega})$.

3.1.5 Theorem. Riemann-Lebesgue Lemma

If $F \in L^1(\mathbb{T}_{2\Omega})$ then $\lim_{|n|\to\infty} f[n] = 0$, where $f = \{f[n]\}$ is the sequence of Fourier coefficients of F, i.e., $\hat{f} = F$.

PROOF. ***a.*** Assume $F \in C^1(\hat{\mathbb{R}})$. Then $G = F' \in L^1(\mathbb{T}_{2\Omega})$ has the properties that $\int_{-\Omega}^{\Omega} G(\gamma)\,d\gamma = 0$ and

$$\forall \gamma \in [-\Omega, \Omega), \qquad F(\gamma) = \int_{-\Omega}^{\gamma} G(\lambda)\,d\lambda + F(-\Omega), \tag{3.1.6}$$

cf., the approach in Theorem 1.4.1c.

We compute (for $n \neq 0$)

$$\begin{aligned}
f[n] &= \frac{1}{2\Omega} \int_{-\Omega}^{\Omega} F(\gamma) e^{\pi i n \gamma / \Omega}\, d\gamma \\
&= \frac{1}{2\Omega} \left[\frac{\Omega}{\pi i n} e^{\pi i n \gamma / \Omega} F(\gamma) \Big|_{-\Omega}^{\Omega} - \frac{\Omega}{\pi i n} \int_{-\Omega}^{\Omega} G(\gamma) e^{\pi i n \gamma / \Omega}\, d\gamma \right] \\
&= \frac{-1}{2\pi i n} \int_{-\Omega}^{\Omega} G(\gamma) e^{\pi i n \gamma / \Omega}\, d\gamma
\end{aligned}$$

and, hence,

$$|f[n]| \leq \frac{\Omega}{\pi |n|} \|G\|_{L^1(\mathbb{T}_{2\Omega})}.$$

Consequently, $\lim_{|n|\to\infty} f[n] = 0$.

b. Let $F \in L^1(\mathbb{T}_{2\Omega})$ and $\epsilon > 0$. There is $F_\epsilon \in C^1(\hat{\mathbb{R}})$ that is 2Ω-periodic and for which $\|F - F_\epsilon\|_{L^1(\mathbb{T}_{2\Omega})} < \epsilon$. Then (3.1.6) is valid with F_ϵ and $G_\epsilon = F_\epsilon' \in L^1(\mathbb{T}_{2\Omega})$ instead of F and G. (The existence of F_ϵ can be proven in many ways, including the convolution of F with an approximate identity, which we shall discuss in Section 3.4.)

We compute (for $n \neq 0$)

$$|f[n]| \leq |f[n] - f_\epsilon[n]| + |f_\epsilon[n]|$$

$$\leq \|F - F_\epsilon\|_{L^1(\mathbb{T}_{2\Omega})} + \frac{\Omega}{\pi |n|} \|G_\epsilon\|_{L^1(\mathbb{T}_{2\Omega})},$$

where $f_\epsilon = \{f_\epsilon[n]\}$ is the sequence of Fourier coefficients of F_ϵ and where we have invoked part *a* in the second inequality.

We know that

$$\overline{\lim_{|n|\to\infty}} \, a_n \le \overline{\lim_{|n|\to\infty}} \, b_n + \overline{\lim_{|n|\to\infty}} \, c_n$$

in case $a_n \le b_n + c_n$ and $a_n, b_n, c_n \ge 0$, e.g., Definition A.1. Consequently,

$$\overline{\lim_{|n|\to\infty}} \, |f[n]| \le \overline{\lim_{|n|\to\infty}} \, \|F - F_\epsilon\|_{L^1(\mathbb{T}_{2\Omega})} < \epsilon.$$

Since the left side is nonnegative and independent of ϵ, we conclude that $\lim_{|n|\to\infty} |f[n]| = 0$. ∎

We shall use the Riemann–Lebesgue Lemma to verify Dirichlet's fundamental theorem, which provides sufficient conditions on a function $F \in L^1(\mathbb{T}_{2\Omega})$ so that $S(F)(\gamma_0) = F(\gamma_0)$ for a given point γ_0. The following ingenious proof is due to P. Chernoff [Che80], cf., [Lio86] and the classical proof as found in [Zyg59]. Dirichlet's theorem for Fourier series naturally preceded the analogous inversion theorem for Fourier transforms, e.g., Sections 1.7 and 3.2.

3.1.6 Theorem. Dirichlet Theorem

If $F \in L^1(\mathbb{T}_{2\Omega})$ and F is differentiable at γ_0, then $S(F)(\gamma_0) = F(\gamma_0)$ in the sense that

$$\lim_{M,N\to\infty} \sum_{n=-M}^{N} f[n]e^{-\pi in\gamma_0/\Omega} = F(\gamma_0),$$

where $f = \{f[n]\}$ is the sequence of Fourier coefficients of F, i.e., $\hat{f} = F$.

PROOF. **a.** Without loss of generality, assume $\gamma_0 = 0$ and $F(\gamma_0) = 0$. In fact, if $F(\gamma_0) \neq 0$, then consider the function $F - F(\gamma_0)$ (instead of F), which is also an element of $L^1(\mathbb{T}_{2\Omega})$, and then translate this function to the origin.
b. Since $F(0) = 0$ and $F'(0)$ exists, we can verify that

$$G(\gamma) = \frac{F(\gamma)}{(e^{-\pi i\gamma/\Omega} - 1)}$$

is bounded in some interval centered at the origin. To see this, note that

$$G(\gamma) = \frac{F(\gamma)}{\gamma} \left(\frac{1}{(-\pi i/\Omega) + (-\pi i/\Omega)^2(1/2!)\gamma + (-\pi i/\Omega)^3(1/3!)\gamma^2 + \cdots} \right);$$

$$(3.1.7)$$

and, hence, $G(\gamma)$ is close to $-\Omega F'(0)/(\pi i)$ in a neighborhood of the origin, e.g., Exercise 3.4.

This boundedness near the origin, coupled with integrability of F on $\mathbb{T}_{2\Omega}$, yields the integrability of G on $\mathbb{T}_{2\Omega}$. Therefore, since $F(\gamma) = G(\gamma)(e^{-\pi i \gamma/\Omega} - 1)$, we compute $f[n] = g[n-1] - g[n]$, where $g = \{g[n]\}$ is the sequence of Fourier coefficients of G. Thus, the partial sum $S_{M,N}(F)(0)$ is the telescoping series

$$\sum_{n=-M}^{N} (g[n-1] - g[n]) = g[-M-1] - g[N].$$

Consequently, we can apply the Riemann–Lebesgue Lemma to obtain

$$\lim_{M,N\to\infty} S_{M,N}(F)(0) = 0. \quad \blacksquare$$

3.1.7 Remark. Fundamental Spaces and Elementary Convergence Results

a. Let $\Omega > 0$; and let $C^m(\hat{\mathbb{R}})$, $0 \le m \le \infty$, be the space of m-times continuously differentiable functions on $\hat{\mathbb{R}}$. It is convenient to define the following spaces:

$$C^m(\mathbb{T}_{2\Omega}) = \{F \in C^m(\hat{\mathbb{R}}) : F \text{ is } 2\Omega\text{-periodic on } \hat{\mathbb{R}}\}, \ 0 \le m \le \infty;$$
$$AC(\mathbb{T}_{2\Omega}) = \{F \in AC_{\text{loc}}(\hat{\mathbb{R}}) : F \text{ is } 2\Omega\text{-periodic on } \hat{\mathbb{R}}\};$$
$$BV(\mathbb{T}_{2\Omega}) = \{F \in BV_{\text{loc}}(\hat{\mathbb{R}}) : F \text{ is } 2\Omega\text{-periodic on } \hat{\mathbb{R}}\};$$
$$C(\mathbb{T}_{2\Omega}) = C^0(\mathbb{T}_{2\Omega}) = \{F \in C^0(\hat{\mathbb{R}}) : F \text{ is } 2\Omega\text{-periodic on } \hat{\mathbb{R}}\}.$$

Clearly,

$$C^\infty(\mathbb{T}_{2\Omega}) \subseteq \cdots \subseteq C^{m+1}(\mathbb{T}_{2\Omega}) \subseteq C^m(\mathbb{T}_{2\Omega}) \subseteq \cdots \subseteq C^1(\mathbb{T}_{2\Omega})$$

$$\subseteq AC(\mathbb{T}_{2\Omega}) \subseteq C(\mathbb{T}_{2\Omega}) \cap BV(\mathbb{T}_{2\Omega}),$$

cf., Remark 1.4.2.

b. The hypothesis in Theorem 3.1.6 that F is differentiable at γ_0 is strong; and the proof of Theorem 3.1.6 is still valid if the hypothesis on $F \in L^1(\mathbb{T}_{2\Omega})$ is weakened to the condition that $(F(\gamma) - F(\gamma_0))/(\gamma - \gamma_0)$ be integrable on some interval centered at γ_0, cf., Exercise 3.42. In particular, if $F \in BV(\mathbb{T}_{2\Omega})$, then

$$\forall \gamma \in \mathbb{T}_{2\Omega}, \qquad \lim_{N\to\infty} S_N(F)(\gamma) = \frac{F(\gamma+) + F(\gamma-)}{2}.$$

Further, Exercises 3.26 and 3.49 deal with properties of Fourier coefficients and rates of convergence of $\{S_N(F)\}$ for functions F belonging to the spaces defined in part *a*.

c. With regard to Theorem 3.1.6, we can further assert that *if $F \in BV(\mathbb{T}_{2\Omega})$ and if F is also continuous on a closed subinterval $I \subseteq \mathbb{T}_{2\Omega}$, then $\{S_N(F)\}$*

converges uniformly to F on I, cf., [Zyg59, Volume I, pages 57–58]. The *Dirichlet Theorem* and this version of it for intervals of continuity are often referred to as the *Dirichlet–Jordan Test*, cf., Section 3.2.3.

3.1.8 Definition. Relations Between Functions Defined on $\mathbb{Z}$ and $\mathbb{T}_{2\Omega}$

a. If $f \in \ell^1(\mathbb{Z})$ and $\Omega > 0$, then $F = \hat{f}$ is an *absolutely convergent Fourier series*, and the space of such series is denoted by $A(\mathbb{T}_{2\Omega})$. By definition, the norm of $F = \hat{f} \in A(\mathbb{T}_{2\Omega})$ is

$$\|F\|_{A(\mathbb{T}_{2\Omega})} = \|f\|_{\ell^1(\mathbb{Z})} = \sum |f[n]|,$$

cf., Definition 1.1.2 and Example 2.4.6f. We define

$$L^\infty(\mathbb{T}_{2\Omega}) = \{F \in L^\infty(\hat{\mathbb{R}}) : F \text{ is } 2\Omega \text{ -periodic on } \hat{\mathbb{R}}\},$$

and the *norm* of $F \in L^\infty(\mathbb{T}_{2\Omega})$ is $\|F\|_{L^\infty(\mathbb{T}_{2\Omega})} = \|F\|_{L^\infty(\hat{\mathbb{R}})}$, e.g., Definition A.10. $C(\mathbb{T}_{2\Omega})$ is a closed subspace of $L^\infty(\mathbb{T}_{2\Omega})$ if $C(\mathbb{T}_{2\Omega})$ is taken with the $\|\cdots\|_{L^\infty(\mathbb{T}_{2\Omega})}$ norm.

b. We have the proper inclusions

$$A(\mathbb{T}_{2\Omega}) \subseteq C(\mathbb{T}_{2\Omega}) \subseteq L^\infty(\mathbb{T}_{2\Omega}) \subseteq L^2(\mathbb{T}_{2\Omega}) \subseteq L^1(\mathbb{T}_{2\Omega}); \qquad (3.1.8)$$

and the identity map corresponding to any of these inclusions is continuous, e.g., Exercise 3.44. In fact,

$$\|F\|_{L^1(\mathbb{T}_{2\Omega})} \leq \|F\|_{L^2(\mathbb{T}_{2\Omega})} \leq \|F\|_{L^\infty(\mathbb{T}_{2\Omega})} \leq \|F\|_{A(\mathbb{T}_{2\Omega})}. \qquad (3.1.9)$$

We shall show in Example 3.5.3 that $C^1(\mathbb{T}_{2\Omega}) \subseteq A(\mathbb{T}_{2\Omega})$.

c. Let $\ell^2(\mathbb{Z}) = \{f : \mathbb{Z} \to \mathbb{C} : \|f\|_{\ell^2(\mathbb{Z})} = (\sum |f[n]|^2)^{1/2} < \infty\}$. Because of part *b*, we have the proper inclusions

$$\ell^1(\mathbb{Z}) \subseteq X(\mathbb{Z}) \subseteq A'(\mathbb{Z}) \subseteq \ell^2(\mathbb{Z}) \subseteq A(\mathbb{Z}), \qquad (3.1.10)$$

where the notation $X(\mathbb{Z})$, $A'(\mathbb{Z})$, $A(\mathbb{Z})$ is defined as follows:

$$X(\mathbb{Z}) = \{f : \mathbb{Z} \to \mathbb{C} : \hat{f} \in C(\mathbb{T}_{2\Omega})\},$$
$$A'(\mathbb{Z}) = \{f : \mathbb{Z} \to \mathbb{C} : \hat{f} \in L^\infty(\mathbb{T}_{2\Omega})\},$$
$$A(\mathbb{Z}) = \{f : \mathbb{Z} \to \mathbb{C} : \hat{f} \in L^1(\mathbb{T}_{2\Omega})\},$$

e.g., Exercise 3.44. If we *define* $\|f\|_{X(\mathbb{Z})} = \|\hat{f}\|_{L^\infty(\mathbb{T}_{2\Omega})}$ for $f \in X(\mathbb{Z})$, resp., $\|f\|_{A'(\mathbb{Z})} = \|\hat{f}\|_{L^\infty(\mathbb{T}_{2\Omega})}$ for $f \in A'(\mathbb{Z})$ and $\|f\|_{A(\mathbb{Z})} = \|\hat{f}\|_{L^1(\mathbb{T}_{2\Omega})}$ for $f \in A(\mathbb{Z})$, then the identity map corresponding to any of the inclusions in (3.1.10) is continuous. In fact,

$$\|f\|_{A(\mathbb{Z})} \le \|f\|_{\ell^2(\mathbb{Z})} \le \|f\|_{A'(\mathbb{Z})} = \|f\|_{X(\mathbb{Z})} \le \|f\|_{\ell^1(\mathbb{Z})}. \qquad (3.1.11)$$

From Theorem 3.1.5, we know that

$$A(\mathbb{Z}) \subseteq c_0(\mathbb{Z}),$$

where $c_0(\mathbb{Z}) = \{f : \mathbb{Z} \to \mathbb{C} : \lim_{|n| \to \infty} f[n] = 0\}$, cf., Example 3.3.4a.

d. $A'(\mathbb{Z})$ is the space of *pseudomeasures* on $\mathbb{Z}$, cf., Example 2.4.6f. In light of RRT, it is natural to define the "bounded Radon measures $M_b(\mathbb{Z})$" on $\mathbb{Z}$ as the dual space $c_0(\mathbb{Z})'$ of $c_0(\mathbb{Z})$, where the norm of $f \in c_0(\mathbb{Z})$ is defined as $\|f\|_{\ell^\infty(\mathbb{Z})}$, e.g., Appendix B. This handwaving in terms of plausible analogy fails in this case since $c_0(\mathbb{Z})' = \ell^1(\mathbb{Z})$, i.e., $M_b(\mathbb{Z}) = \ell^1(\mathbb{Z})$!

3.1.9 Definition. Measures on $\mathbb{T}$

a. A linear function $T : C(\mathbb{T}) \to \mathbb{C}$ is an element of the *dual space* $C(\mathbb{T})'$ (of the vector space $C(\mathbb{T})$) if $\lim_{n \to \infty} T(F_n) = 0$ for every sequence $\{F_n\} \subseteq C(\mathbb{T})$ for which $\lim_{n \to \infty} \|F_n\|_{L^\infty(\mathbb{T})} = 0$.

b. We denote $C(\mathbb{T})'$ by $M(\mathbb{T})$. $M(\mathbb{T})$ is the space of *Radon measures* on $\mathbb{T}$. The functionals $T \in M(\mathbb{T})$ are often denoted, for example, by μ and ν, and in this case we have the usual notation

$$T(F) = \mu(F) = \int_{\mathbb{T}} F(\gamma)\, d\mu(\gamma),$$

cf., Section 2.7. We also define $M_+(\mathbb{T}) = \{\mu \in M(\mathbb{T}) : \mu(F) \ge 0 \text{ if } F \ge 0\}$.

c. By the definition of $M(\mathbb{T})$, if $\mu \in M(\mathbb{T})$, then

$$\exists C_\mu > 0 \text{ such that } \forall F \in C(\mathbb{T}), \qquad |\mu(F)| \le C_\mu \|F\|_{L^\infty(\mathbb{T})}.$$

$\|\mu\|_1$ denotes the infimum over all such constants C_μ. As in the case of measures on $\hat{\mathbb{R}}$, $L^1(\mathbb{T})$ is naturally embedded in $M(\mathbb{T})$, and if the correspondence is denoted by $F \mapsto \mu_F$, then it is easy to show that $\|F\|_{L^1(\mathbb{T})} = \|\mu_F\|_1$.

d. The relationship between $M(\mathbb{T})$ and $M(\hat{\mathbb{R}})$ is established by the fact that $M(\mathbb{T})$ can be identified with

$$\{\mu \in M(\hat{\mathbb{R}}) : \tau_1 \mu = \mu\},$$

i.e., $M(\mathbb{T})$ can be considered as the subspace of 1-periodic elements of $M(\hat{\mathbb{R}})$. Of course, $\mu \in M(\mathbb{T})$ is *bounded* in the sense of the norm inequality in part *c*; *but if* $\mu \in M(\mathbb{T}) \backslash \{0\}$, then the 1-periodic measure on $\hat{\mathbb{R}}$ corresponding to μ is not in $M_b(\hat{\mathbb{R}})$. For example, if $F \in L^1(\mathbb{T}) \backslash \{0\}$, then the 1-periodic function F_1 on $\hat{\mathbb{R}}$, which equals F on $[0, 1)$, is not in $L^1(\hat{\mathbb{R}})$. As another example, define the *Dirac measure* δ_γ at $\gamma \in \mathbb{T}$ by the formula $\delta_\gamma(F) = F(\gamma)$, where $F \in C(\mathbb{T})$.

Then $\mu = \tau_\gamma(\sum \delta_n)$ is the 1-periodic measure on $\hat{\mathbb{R}}$ corresponding to δ_γ, and $\mu \notin M_b(\hat{\mathbb{R}})$. The verification of these assertions is left to Exercise 3.50.

3.1.10 Example. Periodicity: Potpourri and Titillation

a. *Periodicity.* We defined 2Ω-periodicity on $\hat{\mathbb{R}}$ in Definition 3.1.1. Let $p \in \mathbb{C}$, and let $F : \mathbb{C} \to \mathbb{C}$ be a function. F is *p-periodic* with *period* p if $F(z+p) = F(z)$ for all $z \in D$. If $D \subseteq \mathbb{C}$ is a domain, then F is *doubly periodic* with *periods* $p_1, p_2 \in \mathbb{C}$ if $\mathrm{Im}(p_2/p_1) > 0$ and

$$\forall z \in D, \qquad F(z + p_1) = F(z) \quad \text{and} \quad F(z + p_2) = F(z).$$

F is *quasi-periodic* if

$$\forall (x, \omega) \in \mathbb{R} \times \hat{\mathbb{R}},$$

$$F(x + 1, \omega) = e^{-2\pi i \omega} F(x, \omega) \text{ and } F(x, \omega + 1) = F(x, \omega).$$

b. *Jacobi theta function.* As examples, we first note that entire doubly periodic functions are constants. Also $F(z) = e^z$ is $2\pi i$-periodic on $\mathbb{C}$.

The *Jacobi theta function* ϑ_3 is a 1-periodic entire function, depending on a parameter $t \in \mathbb{C}$, and is defined as

$$\vartheta(z; t) = \vartheta(z) = \sum e^{-\pi n^2 t + 2\pi i n z},$$

where $\mathrm{Re}\, t > 0$. $\vartheta(z)$ is 1-periodic on $\mathbb{C}$.

c. *Elliptic functions.* An *elliptic function* is a meromorphic function in the plane that is doubly periodic in its domain of definition. If $p_1, p_3 \in \mathbb{C}, \mathrm{Im}(p_3/p_1) > 0$, and $p_{m,n} \equiv 2mp_1 + 2np_3$ for $m, n \in \mathbb{Z}$, then the *Weierstrass $\mathcal{P}$-function* is defined as

$$\mathcal{P}(z) = \frac{1}{z^2} + \sum \left\{ \frac{1}{(z - p_{m,n})^2} - \frac{1}{p_{m,n}^2} \right\},$$

where summation is over all $(m, n) \in \mathbb{Z} \times \mathbb{Z} \backslash \{(0, 0)\}$. $\mathcal{P}$ is an elliptic function.

d. *Zak transform.* The *Zak transform* Zf of $f : \mathbb{R} \to \mathbb{C}$ is formally defined as

$$\forall (x, \omega) \in \mathbb{R} \times \hat{\mathbb{R}},$$

$$Zf(x, \omega) = a^{1/2} \sum_k f(xa + ka) e^{2\pi i k \omega}, \tag{3.1.12}$$

where $a > 0$. Zf is a quasi-periodic function.

e. *Elliptic integrals.* Seventeenth and eighteenth century problems from astronomy for computing arc length of orbits, or from mechanics for computing the period of a simple pendulum, led to the problem of evaluating integrals of the form

$$s(y) = \int_0^y \left(\frac{1 - k^2 x^2}{1 - x^2} \right)^{1/2} dx \qquad (3.1.13)$$

or

$$s(y) = \int_0^y \frac{d\varphi}{(1 - k^2 \sin^2 \varphi)^{1/2}}, \qquad (3.1.14)$$

respectively. These are examples of *elliptic integrals*; and Liouville (1833) proved that such integrals cannot be evaluated in terms of algebraic, trigonometric, logarithmic, or exponential functions. The study of elliptic integrals was an important part of eighteenth and nineteenth century mathematics featuring the likes of Fagnano, Euler, and Legendre, and leading to the analyses of Gauss, Abel, and Jacobi. General elliptic integrals are of the form $\int R(x, \sqrt{P(x)}) \, dx$, where $P(x)$ is a third- or fourth-degree polynomial with distinct roots and $R(x, \omega)$ is a rational function of x and ω.

f. Quintics. Instead of studying the function $s(y)$ in (3.1.13) and (3.1.14), Abel (1802–1829) in 1826 and Jacobi in 1827 analyzed the inverse of elliptic integrals; and these are, in fact, the elliptic functions. The analogue in trigonometry is to study the sine function instead of the multiple-valued arcsine

$$s(y) = \int_0^y \frac{dx}{\sqrt{1 - x^2}}.$$

Jacobi's theta functions are relatively elementary functions from which elliptic functions can be constructed. Later, in the early 1860s, Weierstrass introduced $\mathcal{P}(z)$ as the elliptic function inverting a specific elliptic integral; and then he proved that every elliptic function can be expressed in terms of $\mathcal{P}(z)$ and $\mathcal{P}'(z)$ [Hil74, page 141].

Ruffini and Abel proved that quintic polynomial equations cannot necessarily be solved by algebraic operations. In 1858, Hermite used elliptic functions to obtain solutions of such equations, cf., [Kle56].

g. Shape of the sun. We can integrate the Newtonian equations of motion of a secondary body in the equatorial plane of a rotationally symmetric central body; the solution is in terms of the Weierstrass $\mathcal{P}$-function [SB65]. There are important applications of this technique. In particular, the motion of equatorial artificial earth satellites is characterized, and orbital apsidal line shifts of the secondary body can be computed. This latter point is interesting because of the apsidal line shift of mercury's orbit about the sun. This shift can be accounted for by Newtonian methods if the sun is sufficiently "flat", as an oblate spheroid. Robert Dicke and others, e.g., Hill and Stebbins in 1975, have provided an experimental tour de force by showing that the sun is too spherical, by an order of magnitude, to account for even 10% of mercury's apsidal line shift by Newtonian methods. Einstein's theory of general relativity does explain the shift.

h.   *Coherent states.* A *coherent state* is a family of functions of the form

$$\varphi(t) = g(t - a)e^{2\pi i t b}e^{2\pi i c x \omega}, \tag{3.1.15}$$

parameterized by $(a, b, c) \in \mathbb{R}^3$. In the quantum physics literature, g is often taken to be the Gaussian. There is a natural relation between coherent states and the Heisenberg group, e.g., our *Gabor representations and wavelets, Contemp. Math.*, **91** (1989), American Mathematical Society, Providence, RI, 9–27. Closure problems for coherent states have a history going back to von Neumann's classic from the early 1930s [vN55, page 407]. Zak's role in the evolution of the Zak transform and its use in quantum mechanics has been documented in [Jan88]. The Zak transform and knowledge of its zero set are relevant for solving a variety of closure problems for coherent states, e.g., [BF94, Chapter 3].

As we have seen, $\vartheta(z)$ is a 1-periodic entire (and therefore nonelliptic) function used in the construction of meromorphic doubly periodic elliptic functions. It turns out that $\vartheta(z)$ also plays a role in the nonanalytic quasi-periodic Zak transform of the Gaussian. In fact, it is easy to check that the Zak transform of a Gaussian is a product of $\vartheta(z; t)$, for certain t, and a Gaussian. Thus, the zeros of the Zak transforms of the Gaussian are determined by the zeros of ϑ, cf., [BF94, computation after Theorem 7.8].

3.2. History of Fourier Series

George Sarton, who founded the journal *Isis* in 1912, wrote that the "main duty of the historian of mathematics . . . is to explain the humanity of mathematics, to illustrate its greatness, beauty and dignity . . .". Alas, we can neither achieve such a noble goal with its accompanying deep scholarship nor even present a lapidary exposition of the history of Fourier series. Fortunately such expositions abound, e.g., the historical commentaries of Riemann [Rie1873], Gibson [Gib1893], Carslaw [Car30], Hobson [Hob26], Plancherel [Pla25], cf., [Zyg59, Preface], the masterful entries on *Fourier series* in the *Encyclopedia Britannica* of the last fifty years, and the relevant biographical entries in the *Dictionary of Scientific Biography*. There are also important historical contributions by Burkhardt, Plessner, and Tonelli referenced in these works.

Our treatment in this section is selective and perhaps idiosyncratic. We shall not discuss the history of Fourier series vis á vis its major applications to heat and light and celestial mechanics by Fourier and Fresnel and Hill, respectively. (Of course, there are brilliant, but perhaps curmudgeonly, thermodynamicists who assert that Fourier's theory of heat [Fou1822] did not really treat heat.) We shall mostly deal with the relation of Fourier series with real analysis and to some extent with number theory, cf., [Ben76] and [Mon94], respectively. We shall not discuss its relation with functional analysis, which begins with the profound work of Beurling [Beu89], or with complex analysis, which begins with the work of F. and M. Riesz, Lusin and Privalov, and Hardy and Littlewood .

3.2.1 d'Alembert (1717–1783), Euler (1707–1783), D. Bernoulli (1700–1782), and Lagrange (1736–1813)

In Section 1.8 we discussed some partial differential equations from mathematical physics, and it turns out that Fourier series originated in dealing with such equations.

In 1747, d'Alembert solved the *vibrating string problem*. This problem is to solve the equation

$$\frac{\partial^2 u}{\partial t^2} = c^2 \frac{\partial^2 u}{\partial x^2} \tag{3.2.1}$$

for a function $u(x, t)$, where $x \in [0, L]$, $c \in \mathbb{R}$, $t \geq 0$, $u(0, t) = u(L, t) = 0$ for all $t \geq 0$, and $u(x, 0) = f(x)$ is given on $[0, L]$. d'Alembert's solution u is in terms of f, and so f must be twice differentiable in this case. On the other hand, (3.2.1) was derived, after *significant* assumptions, to represent the motion of a taut string, such as a violin string, after it is released from a given initial position f on $[0, L]$. One can imagine an initial position f to have corners, so that f' need not exist everywhere.

In 1748, Euler made an important observation about the vibrating string problem. He noted that the motion of the string is completely determined for $x \in [0, L]$ and time $t \geq 0$ if the form of the string and its velocity at $t = 0$ are given. In particular, Euler was able to find the solution of (3.2.1) for a given initial position f and initial velocity g of the string. Euler's solution allowed for the initial positions f to have discontinuous derivatives. This led to a disagreement with d'Alembert on an issue that ultimately comes down to defining the notion of *function*, e.g., [Bir73, pages 16ff.] which is taken from [Rie1873], cf., [G-V92]. In any case, Euler felt he had solved the vibrating string problem for very general initial positions f.

Daniel Bernoulli entered the discussion in 1753 in the midst of the d'Alembert–Euler disagreement. Daniel Bernoulli had developed hydrodynamics from the principle of conservation of energy, and was a professor of anatomy and botany at Basel, before becoming a professor of physics. He approached the vibrating string problem with Brook Taylor's observation (1715) that if

$$u_n(x, t) = \sin\left(\frac{\pi n x}{L}\right) \cos\left(\frac{\pi n c t}{L}\right), \qquad n \in \mathbb{Z}, \tag{3.2.2}$$

then (3.2.1) is satisfied for $u = u_n$, and $u_n(0, t) = u_n(L, t) = 0$ for all $t \geq 0$.

Using infinite sums of terms of the form (3.2.2), Bernoulli wrote down expressions that he asserted were the most general solutions of the vibrating string problem (3.2.1), i.e., solutions for the most general initial position f. His argument was both formal and in terms of the physics of sound, cf., [BS82], [Pie83] for beautiful treatments of the fundamentals and harmonics used by Bernoulli. Later in 1753, Euler noted that Bernoulli's claim of general solutions could only be correct *if "arbitrary curves" f defined on* [0, L] *could be written as*, what were later called, *Fourier series*. Further, because of the periodicity of the individual terms in Bernoulli's series, Euler judged that Bernoulli was incorrect as far as generality of solution, cf., [Bra86, pages 462–464]. Once again, ill-defined terms such as "arbitrary curves", instead of a precise definition of *function*, were the root cause of these different opinions.

To add to the intellectual melee, Lagrange, at age 23, wrote in 1759 in support of Euler's solution being the most general. Amazingly, his "proof" used trigonometric series similar to Bernoulli's. For the case $L = 1$, Lagrange's solution was essentially of the form

$$
u(x, t) = \int_0^1 \sum_{n=1}^{\infty} (\sin \pi nx \cos \pi nct) f(y) \sin \pi ny \, dy
$$

$$
+ \frac{2}{c\pi} \int_0^1 \sum_{n=1}^{\infty} \frac{1}{n} (\sin \pi nx \sin \pi nct) g(y) \sin \pi ny \, dy. \tag{3.2.3}
$$

Note that if $t = 0$ and if there is an interchange of summation and integration, then (3.2.3) gives rise to the Fourier series expansion of f. *Lagrange series* were almost history! Lagrange seemed intent on verifying Euler's claims versus d'Alembert. In another bizarre twist, Euler did the formal calculation of Remark 3.1.2 to compute Fourier coefficients in 1777 when Fourier was a 9-year old, cf., [Car30, page 4] for a similar contribution by Clairaut.

3.2.2 Fourier (1768–1830)

Fourier submitted his 234-page manuscript, "Sur la propagation de la chaleur", to the Institut de France in 1807. At that time, Fourier was almost 40, and he had only three (unrelated) published papers. His work as Prefect of the Department of Isère in Grenoble dealt with drainage of marshlands, consultation on the achievements of Napoleon's Institut d'Egypte, modeled after the Institut de France, and with planning the first road between Grenoble and Torino, Lagrange's hometown.

The turbulent story of the evolution of this paper includes its critique by Lagrange et al., a prize competition that Fourier won in 1812 along with Lagrange's reservations, the publication of *the* book [Fou1822] in 1822, and the disappearance of the original paper, e.g., [Grat72] along with some of the other references listed at the beginning of Section 3.2. Darboux rediscovered the paper at the École Nationale des Ponts et Chaussées in 1890.

As we saw in Section 3.2.1, the technology was already in place for Fourier series long before Fourier came on the scene. What did Fourier do? He never claimed discovery of the Fourier coefficients (3.1.2) that he used. However, he had a point of view that introduced a "new epoch", to use Riemann's phrase. In the eighteenth century, Fourier coefficients were an integral part (sic) of trigonometric series that had already been derived by other means. Fourier asserted that an arbitrary function could be expanded in a trigonometric series whose coefficients could be computed as in Remark 3.1.2. Such an assertion led to questions of convergence of series and integration of arbitrary functions (in the definition of Fourier coefficients) and, of course, to questions about the meaning of *function*. Fourier's examples and applications in [Fou1822] are extraordinary, and were influential in establishing the field of *Fourier's series*.

There are also related subsequent contributions by Cauchy and Poisson; but we shall go directly to Dirichlet.

3.2.3   Dirichlet (1805–1859)

Who gave the first correct definition of *function*? Scholars of good will and excellent credentials disagree, cf., [Monn72, especially pages 57–65]. As mentioned in Remark 1.7.7*b*, it seems to us that Dirichlet has a valid claim.

In 1829 [Dir1829, page 121], he wrote that a *continuous function* f on $[0, h]$ is defined by the property that it "has a finite and well-determined value for each value of β between 0 and h, and moreover such that the difference $f(\beta+\epsilon) - f(\beta)$ decreases without limit (to 0) when ϵ becomes smaller and smaller".

In 1837 [Dir1837], he wrote the following, in which the parenthetical remark shows that continuity was not an intrinsic part of his definition. "Imagine a and b to be two fixed values and x a variable, which is supposed to assume one after the other all values between a and b. If to each x there corresponds a unique finite y in such a manner that while x runs continuously through the interval from a to b, $y = f(x)$ varies gradually also, then y is a continuous function of x for this interval. (Since in what follows, we shall only discuss continuous functions, this attribute can be omitted without loss.)" The lack of rigor in defining continuity in terms of the word "gradually" is compensated by his precision in 1829. "It is not at all necessary that y depends on x in this whole interval by the same law, and it is not even necessary to imagine a dependency expressible by mathematical operations. ...This definition does not prescribe a common law to the different parts of the curve; it can be thought of as being composed of parts of the most different kinds or completely without law." This last part addresses the confusion from the eighteenth century analysis, when a formula such as $f(x) = x^2$ on $[a, b]$ was often thought to characterize a function, instead of characterizations such as $f = 2$ on $[0, \frac{a+b}{2}]$ and $f(x) = x^2$ on $(\frac{a+b}{2}, b]$.

Of course, it was precisely in the papers [Dir1829] and [Dir1837], where Dirichlet defined the notion of function, that he also proved the fundamental Theorem 3.1.6. Dirichlet's theorem was generalized by Lipschitz in 1864, supposing so-called *Lipschitz conditions*; and generalized still further by Dini, who wrote an important book on Fourier series in 1880. In the spirit of Dini's analysis, there is the following *Dini–Lipschitz–Lebesgue test* for uniform convergence. *If $F \in L^2(\mathbb{T}_{2\Omega})$ and*

$$\lim_{\lambda \to 0} |F(\gamma + \lambda) - F(\gamma)| \log |\lambda| = 0$$

uniformly in an open interval I, then $S(F) = F$ on any closed subinterval $J \subseteq I$, and the convergence of the Fourier series $S(F)$ to F is uniform on J, e.g., [HR56, Theorem 59], cf., [Zyg59, page 52] for the original *Dini test*.

For perspective, recall that the Jordan Theorem, Theorem 1.7.6, is the analogue for Fourier transforms of the Dirichlet Theorem, Theorem 3.1.6. The second mean value theorem (Lemma 1.7.3) was used to prove Theorem 1.7.6, and, in fact, Bonnet (*Mem. Savant Étrangers of the Belgian Academy* **23** (1848–1850)) used Lemma 1.7.3 directly to prove Theorem 3.1.6. Dirichlet's original proof in 1829 used an argument similar to that required to prove Lemma 1.7.3.

Dirichlet made major contributions to number theory. It is not difficult to prove that the sequence $\{4n - 1 : n \in \mathbb{N}\}$ contains infinitely many primes. Dirichlet proved the general fact that *if $a \in \mathbb{N}$, $b \in \mathbb{Z}$, and a and b are relatively prime, then $\{an + b : n \in \mathbb{N}\}$ contains infinitely many primes*, cf., Remark 3.8.11a.

In this book we shall refer to two other number theoretic issues where Dirichlet had seminal ideas. The first concerns the Dirichlet Box Principle, related to rational approximation of irrationals, and leading to the Kronecker Theorem, which can be formulated in terms of trigonometric sums, e.g., Exercises 3.40 and 3.41. The second concerns a proof of Gauss' Law of Quadratic Reciprocity, e.g., Remark 3.8.11. The material of Section 2.10 plays a role, as well as subtle issues concerning Gauss sums and the so-called Littlewood Flatness Problem, e.g., Remark 3.8.11.

3.2.4   Riemann (1826–1866)

Bernhard Riemann's life was tragic in its briefness and transcendental in its brilliance. The excerpts in [Kli72] about Riemann's ideas barely scratch the surface of the depth and breadth and lustre of his creativity, cf., [Edwa74] for implications of just one of his gems.

Riemann's *Habilitationsschrift* [Rie1873] was presented in 1854 but was only published in 1867 after his death. It is the first part of this work that has provided us with some of the material in Sections 3.2.1–3.2.3. Next, Riemann developed the *Riemann integral*. His theory of integral was created to define Fourier coefficients and Fourier series expansions for a large class of functions. Finally, he developed the Riemann Localization Principle and several other important tools for dealing with trigonometric series, e.g., [Zyg59], cf., [Ben71] for the relation between the *Riemann Localization Principle* and the notion of *support*.

The Riemann Localization Principle is a key technique in the study of *sets of uniqueness (U-sets)*. A set $E \subseteq [0, 1)$ is a *U-set* if

$$\lim_{N \to \infty} \sum_{|n| \le N} c_n e^{-2\pi i n \gamma} = 0 \quad \text{off of } E \quad \text{implies} \quad c_n = 0 \quad \text{for all } n \in \mathbb{Z}.$$

Using Riemann's theory, Cantor proved that *the empty set $\varnothing$ is a U-set*, cf., Section 3.2.5. Cantor's theorem was apparently known to Riemann, e.g., [Leb06, page 110]. At the other extreme, *if $|E| > 0$, then E is not a U-set*, e.g., Exercise 3.37.

As is well known, *a bounded function $f : [a, b] \to \mathbb{C}$ is Riemann integrable if and only if f is continuous a.e.*, e.g., [Ben76, pages 94–96]. The concept of measure 0 and the notation "a.e." (Definition A.4) are now part of the Lebesgue theory (1902). Leading to Lebesgue, Vito Volterra (1881), at that time a student of Dini at the Scuola Normale Superiore in Pisa, constructed functions f whose derivative exists everywhere but for which f' is not Riemann integrable, e.g., [Ben76, pages 20–21], cf., the interesting examples in [Rie1873, Section 13]. Actually, H. J. Smith had solved the same problem in 1875; but Lebesgue was unaware of Smith's result in his thesis [Leb02], where he gives prominent mention of Volterra's example.

The point is that measure 0 was emerging in the late nineteenth century as an important idea. Norbert Wiener (1938) has made a case for formulating the notion of measure zero based on justifying Maxwell's and Gibbs' theories of statistical mechanics. He wrote that "the ideas of statistical randomness and phenomena of zero probability were current among the physicists and mathematicians in Paris around 1900, and it was in a medium, heavily ionized by these ideas, that Borel and Lebesgue solved the mathematical problem of measure" [Wie81, Volume II, pages 794–806], cf., [Carl80].

Besides non-(Riemann) integrability, the issue of nondifferentiability was prominent in this part of the nineteenth century. Weierstrass wrote, at least in a possibly edited version of a lecture he gave at the Royal Academy of Sciences on July 18, 1872, that "Riemann, as I learned from some of his students, stated decisively (in 1861, or perhaps even earlier)" that

$$f(x) = \sum_{n=1}^{\infty} \frac{\sin n^2 x}{n^2} \tag{3.2.4}$$

is an everywhere continuous nowhere differentiable function, e.g., *Weierstrass' Mathematische Werke* II, pages 71–74, cf., [BSt86] for a profound analysis of this area. Although (3.2.4) does have some points of differentiability, there are in fact many continuous nowhere differentiable functions including Weierstrass' lacunary Fourier series (1872),

$$f(x) = \sum_{n=1}^{\infty} b^n \cos(\pi a^n x),$$

where $a > 1$ is an odd integer, $b \in (0, 1)$, and $ab > 1 + 3\pi/2$, cf., [Ben76, pages 28–29] for other examples and [Dui91] for the relation to *selfsimilarity*.

Riemann convalesced and toured in Italy during the winter of 1862, arriving in Pisa in 1863. He became friendly with Betti and Beltrami. Betti, of "Betti number" fame, was Director of the Scuola Normale Superiore, and there is an interesting Betti–Riemann correspondence at the Scuola. (The Scuola Normale was started by Napoleon, and it is modeled after the École Normale Supériore in Paris.) Dini was a student at the Scuola at the time of Riemann's visit. He graduated in 1864, spent a year studying with Bertrand in Paris, and returned to the Scuola Normale where he spent the next 52 years. Besides Volterra, he counts Vitali as one of his students, e.g., [Ben76] for historical remarks and mathematical contributions of Vitali.

Riemann returned to Germany for the winter of 1864–1865, but then came back to Pisa. He died and was buried at Biganzolo in the northern part of Verbania (the Italian resort town on the western banks of Lago Maggiore just 15 miles south of the Swiss border). A mourner at the local cemetery will surely point you to the marker of "Il Tedesco".

3.2.5 Cantor (1845–1918)

Georg Cantor received his Ph.D. in 1867 at Berlin. His dissertation on quadratic Diophantine equations, related to some issues from Gauss' monumental *Disquisitiones Arithmeticae* [Gau66], was written under the direction of Kummer, cf., [Ben77] for remarks about Kummer, Fermat's Last Theorem, and ideals. Kronecker, who later became an intellectual adversary of Cantor's, was also a professor in Berlin at the time.

Cantor wrote several important papers on U-sets in the early 1870s, including his theorem quoted in Remark 3.2.4. In order to prove this result, that the empty set is a U-set, he first proved what is now known as the *Cantor–Lebesgue Lemma*: *if $X \subseteq [0, 1)$ is a Lebesgue measurable set of positive measure, and if*

$$\forall \gamma \in X, \qquad \lim_{N \to \infty} \sum_{|n| \leq N} c_n e^{-2\pi i n \gamma} \in \mathbb{C},$$

then $\lim_{|n| \to \infty} c_n = 0$. Cantor actually proved the result for the case that X is a nondegenerate interval. Fatou first investigated the converse. In this regard, Lusin found a trigonometric series that was a.e. divergent and for which $\lim_{|n| \to \infty} c_n = 0$. Steinhaus clinched the converse by constructing such a series that was everywhere divergent and for which $\lim_{|n| \to \infty} c_n = 0$, e.g., [Bary64, Volume I, pages 176–177].

After proving that the empty set was a U-set, Cantor showed that finite sets and certain countably infinite are also U-sets. This work certainly influenced his later research on set theory and the infinite.

It was in 1874 that he gave his famous, correct, and controversial proof of the fact that there are only countably many algebraic numbers. Recall that an *algebraic number* is a zero of a polynomial with integer coefficients.

In any case, Cantor tried to prove that all countable sets $E \subseteq [0, 1)$ were U-sets; and this was finally achieved by F. Bernstein (1908) and W. H. Young (1909), cf., [But95]. Actually, Bernstein proved somewhat more, cf., Sections 3.2.6 and 3.2.7.

The remainder of Cantor's life, from the mid-1870s, was devoted to the study of the infinite, not only in mathematics as in [Can55], but often delving into various philosophical notions of infinity due to the Greeks, the Scholastic philosophers, and his contemporaries, e.g., [Daub79]. Cantor certainly did not dote on philosophers. In a letter to Bertrand Russell, who was then at Trinity College, Cambridge, Cantor wrote (1911): "... and I am quite an adversary of Old Kant, who in my eyes has done much harm and mischief to philosophy, even to mankind; as you easily see by the perverted development of metaphysics in Germany in all that followed him, as in Fichte, Schelling, Hegel, Herbart, Schopenhauer, Hartman, Nietzche, etc., etc., on to this very day. I never could understand why ... reasonable ... peoples ... could follow yonder sophistical Philistine, who was so bad a mathematician."

3.2.6 Mensov (1892–1988)

Dmitrii Mensov proved a key result on U-sets in 1916 by finding a non-U-set X with Lebesgue measure $|X| = 0$. He did this just after graduating from Moscow University, where he wrote his thesis under N. Lusin. Mensov's example stimulated a great deal of study about sets of measure zero. Actually, on the basis of Mensov's example, Lusin and Bary defined the notion of U-sets as such. Earlier, de la Vallée–Poussin had proved that if a trigonometric series converged to $F \in L^1(\mathbb{T})$ off a countable set E, then the series is the Fourier series of F. It was generally felt that the same would be true for sets E with $|E| = 0$. Mensov changed that perception. Mensov showed that there exists a nontrivial trigonometric series that converges to $F = 0$ off a set of measure zero. Since $F \in L^1(\mathbb{T})$ is the 0-function a.e., and since the series has some nonzero coefficients, the series is not the Fourier series of F. Needless to say, Mensov's example had a certain amount of shock value, cf., [Men68], [Ben76, page 115] for other major results by Mensov.

Nina Bary (1923) asked for conditions on the coefficients $\{c_n\}$ of trigonometric series to ensure that $c_n = 0$ for all n whenever

$$\lim_{N \to \infty} \sum_{|n| \leq N} c_n e^{-2\pi i n \gamma} = 0 \quad \text{a.e. on } \mathbb{T}. \tag{3.2.5}$$

In light of de la Vallée–Poussin's and Mensov's results of the previous paragraph, it is interesting to observe that *if* $\sum |c_n|^2 < \infty$ *and* (3.2.5) *is true, then* $c_n = 0$ *for all* n, e.g., Exercise 3.36. There have been deep results in this problem area, in the case $\sum |c_n|^2$ diverges, by Littlewood (1936), Wiener and Wintner (1938), A. C. Schaeffer (1939), Salem (1942), Ivašev–Musatov (1957), Brown and Hewitt

(1980), and Körner [Kör87]. For example, Salem proved that *for each $\epsilon > 0$ there is a (bounded measure) $\mu \in M(\mathbb{T})\backslash\{0\}$ for which* $|\operatorname{supp}\mu| = 0$ *and*

$$\exists C, N \quad \text{such that} \quad \forall |n| \geq N, \ |\breve{\mu}[n]| \leq C\frac{1}{|n|^{1/2-\epsilon}}, \qquad (3.2.6)$$

e.g., [KS63, pages 106–112], cf., [Ben75, pages 96–97]. If the right side of (3.2.6) were $C(1/|n|^{(1/2)+\epsilon})$, then $\mu \in L^2(\mathbb{T})$, e.g., Theorem 3.4.13; this coupled with the condition $|\operatorname{supp}\mu| = 0$ implies μ is the 0-function.

With regard to (3.2.6) and the Lusin Conjecture of Section 3.2.8, deLeeuw, Kahane, and Katznelson proved the following result:

$$\forall \hat{f} = F \in L^2(\mathbb{T}), \ \exists \hat{g} = G \in C(\mathbb{T}) \quad \text{such that}$$

$$\forall n \in \mathbb{Z}, \qquad |f[n]| \leq |g[n]| \ \text{and} \ \|G\|_{L^\infty(\mathbb{T})} \leq 9\|F\|_{L^2(\mathbb{T})}$$

(*C. R. Acad. Sci. Paris* **285** (1977), 1001–1003).

3.2.7   Bary (1901–1961) and Rajchman (1890–1940)

What with Mensov's example, Aleksander Rajchman (who died at Sachsenhausen in 1940) "seems to have been the first to realize that for sets of measure zero that occur in the theory of trigonometric series it is not so much the metric as the *arithmetic* properties that matter" (from Zygmund's biography of Salem in [Sal67]). Rajchman (1922) proved the existence of some uncountable, closed U-sets including the $\frac{1}{3}$-Cantor set. He was motivated by some work of Hardy and Littlewood (*Acta Math.*, **37** (1914)) and Steinhaus (1920) on Diophantine approximation to introduce "H-sets"; and he proved that such sets are U-sets. In fact, the $\frac{1}{3}$-Cantor set is an H-set. In a letter to Lusin, he also expressed his considered opinion that any U-set is contained in a countable union of H-sets. Although this particular conjecture was proved false by Pyatetskii–Shapiro (1952), it was such questions that focused the direction of the subject.

Actually, Nina Bary had proved the existence of some uncountable, closed U-sets in 1921, and she had presented her results at Lusin's seminar at the University of Moscow. They were unpublished at the time of Rajchman's paper. This does not undermine the importance of Rajchman's theorems, since Rajchman's approach illustrated the need for number theoretic (Diophantine) properties in the construction of such sets.

Bary made significant contributions to the subject of U-sets throughout her life. One of her major results is that *the countable union of closed U-sets is a U-set*. The problem is open for the finite union of arbitrary U-sets. Another one of her theorems, which was proven in 1936–1937, asserts that *if α is rational and $E(\alpha)$ is the Cantor set with ratio of dissection α, then $E(\alpha)$ is a U-set if and only if $\beta = 1/\alpha$ is an integer*. This generalizes Rajchman's result about the $\frac{1}{3}$-Cantor set

C since $C = E(\frac{1}{3})$. In general, $E(\alpha)$ is constructed by "throwing away" centered open intervals of length $\alpha(b-a)$, where $[a, b]$ is any remaining closed interval at a given step and where the first step begins with $[a, b] = [0, 1]$, e.g., [KS63, Chapitre I].

A *Pisot–Vigayraghavan* (P-V) number is a real algebraic integer $\beta > 1$ with the property that all the other roots of its minimal polynomial have modulus less than 1. Bary's theorem on Cantor sets of uniqueness has the following spectacular sequel announced by Salem (1943): *if $\alpha \in (0, \frac{1}{2})$, then $E(\alpha)$ is a U-set if and only if $\beta = 1/\alpha$ is a P-V number*, e.g., [Bary64], [Ben76, pages 116–117], [Mey72], [Sal63] for the proof, a history of the proof, and recent developments.

3.2.8 The Lusin Conjecture

In his dissertation of 1915 (actually he published a *C. R. Acad. Sci. Paris* note on the relevant material), Lusin conjectured that the Fourier series of every $F \in L^2(\mathbb{T})$ is convergent a.e. This is the *Lusin Conjecture*.

As background for the Lusin Conjecture, du Bois–Reymond (1872) constructed functions $F \in C(\mathbb{T})$ whose Fourier series diverge at some points, cf., [Rog59, pages 75–77], [Zyg59, Volume I, Chapter VIII] and Exercise 3.45. Further, just prior to Lusin, there were contributions in this general area by Fatou (1906), Jerosch and Weyl (1908), Weyl (1909), Fejér (1911), W. H. Young (1912), Hobson (1913), Plancherel (1913), and Hardy (1913).

In 1926, Kolmogorov constructed functions $F \in L^1(\mathbb{T}) \backslash L^2(\mathbb{T})$ whose Fourier series diverge *everywhere*! His proof used Kronecker's Theorem, which we shall discuss in Section 3.2.10. There were subsequent relevant "log estimates" by Kolmogorov and Seliverstov (1925), Plessner (1926), and Littlewood and Paley (1931). Finally, Lennart Carleson (1966) proved that *if $F \in L^2(\mathbb{T})$, then $S(F) = F$ a.e.* [Carl66], cf., [Fef73] for a conceptually different proof and [Moz71] for a superb exposition. R. A. Hunt (1968) used the method of Carleson's proof and the theory of interpolation of operators to extend Carleson's result to $L^p(\mathbb{T})$, $p > 1$, cf., [Ash76, pages 20–37] for an elegant presentation by Hunt, and [Ben76, pages 208–210] for a connection between techniques used by Carleson and the FTC. Also, for perspective vis á vis du Bois–Reymond's example and Carleson's Theorem, we have Kahane and Katznelson's Theorem that *if $E \subseteq \mathbb{T}$ is a set of measure zero, then there is $F \in C(\mathbb{T})$ such that $S(F)(\gamma)$ diverges for all $\gamma \in E$*, e.g., [Kat76, Chapter 2].

We close this section with remarks by Carleson on the occasion of receiving the 1984 Steele Prize. They concern his proof and a remark about the FFT, cf., Section 3.9.

> "When I was a student at Harvard in 1950–1951, A. Zygmund and R. Salem were also in Cambridge and I learned very much from them. They also encouraged me to try to use Blaschke products as examples

of a Fourier series which diverges a.e. I worked hard at that then, and all through the years I tried different ideas. Then finally, in 1964 or so, I realized the basic reason why there should exist an example. Very briefly we can describe the main feature of the trigonometric system $\cos nx$, $n \leq 2^m$, by writing down a matrix of ± 1 giving the sequence of sign($\cos nx$) which can occur. This matrix is essentially $2^m \times 2^m$, i.e., very few sequences of signs occur which, of course, is very favorable for examples of divergence. (This is also the basic idea behind the fast Fourier transform.) To my great astonishment, it now turned out that for a random $2^m \times 2^m$ matrix there is no example and then a proof of the convergence theorem came naturally."

3.2.9   The Dirichlet Box Principle

a.   The *Dirichlet Box Principle* asserts that if Q boxes contain $Q + 1$ objects, then at least one of the boxes contains more than one object. This fact may not seem to be at the usual Dirichlet level of brilliance, but it has been a staple in the method of proof of many results since he first made use of it.

An adaptation of the Dirichlet Box Principle is even used in Wiles' proof of Fermat's Last Theorem. In this case the objects are *Hecke rings* and an infinite sequence of sets of boxes is created. The assertion, in the part of the proof due to Taylor and Wiles, is that there are Hecke rings in every set of boxes.

b.   Originally, Dirichlet used the box principle to give a new proof of the fact that *if $x \in (0, 1)$ is irrational, then*

$$\forall \epsilon > 0, \ \exists p, q \in \mathbb{N} \quad such \ that$$

$$\left| x - \frac{p}{q} \right| < \epsilon \quad and \quad \left| x - \frac{p}{q} \right| < \frac{1}{q^2}. \tag{3.2.7}$$

The pairs p, q can be chosen to be relatively prime. The assertion (3.2.7) is an elementary result in Diophantine approximation, and Dirichlet's proof (in part *c*) has the advantage of being applicable to d-dimensional problems, e.g., [HW65, Theorem 201]. The first inequality of (3.2.7) follows from basic properties of $\mathbb{R}$; and the second inequality gives insight into the rapidity of rational approximation to irrationals.

c.   To prove (3.2.7), let $Q > 1$, and consider the Q boxes $[\frac{n}{Q}, \frac{n+1}{Q}]$, $n = 0, \ldots, Q - 1$, and the $Q + 1$ numbers, $0, x - [x], 2x - [2x], \ldots, Qx - [Qx]$. By the Dirichlet Box Principle there are integers $0 \leq q_1 < q_2 \leq Q$ and $p \in \mathbb{N}$ for which $|qx - p| < 1/Q$, where $q = q_2 - q_1 \in (0, Q] \cap \mathbb{N}$ and $p = [q_2 x] - [q_1 x]$. Thus the second inequality of (3.2.7) is valid. This part of the proof also works for rational x.

Since $\epsilon > 0$ is given in (3.2.7), we choose $Q = Q(\epsilon) = [1/\epsilon] + 1$ for the above argument. In particular, $\epsilon \geq \frac{1}{Q}$ and so $|x - \frac{p}{q}| < \epsilon/q < \epsilon$.

3.2.10 Kronecker Sets and *U*-Sets

A significant refinement of (3.2.7), which is deeper than d-dimensional versions of (3.2.7), is *Kronecker's Theorem*. Kronecker (1884) proved that *if* $\{1, \gamma_1, \ldots, \gamma_d\} \subseteq \mathbb{R}$ *is linearly independent over the rationals, if* $\{\lambda_1, \ldots, \lambda_d\} \subseteq \mathbb{R}$, *and if* $\epsilon, N > 0$, *then there are integers* $q > N$ *and* $p_1, \ldots, p_d$ *such that*

$$\forall j = 1, \ldots, d, \qquad |q\gamma_j - p_j - \lambda_j| < \epsilon. \tag{3.2.8}$$

Dirichlet's analysis was for the case $\lambda_j = 0$. With the same hypotheses, the conclusion (3.2.8) can be reworded to assert the existence of q for which

$$\forall j = 1, \ldots, d, \qquad |e^{2\pi i q \gamma_j} - e^{2\pi i \lambda_j}| < \epsilon. \tag{3.2.9}$$

There are several different proofs, e.g., [Ben75, Theorem 3.2.7], [HW65, Chapter 23], [KK64], [Kat76, pages 181–183].

Because of (3.2.9) we say that a closed set $E \subseteq \mathbb{T}$ is a *Kronecker set* if for each $\epsilon > 0$ and continuous function $F : E \to \mathbb{C}$, for which $|F| = 1$ on E, there is $q \in \mathbb{Z}$ such that

$$\sup_{\gamma \in E} |e^{2\pi i q \gamma} - F(\gamma)| < \epsilon.$$

In 1962, Paul Malliavin proved that *if* $E \subseteq \mathbb{T}$ *is closed and if every closed subset of* E *is a set of spectral synthesis, then* E *is a* U-*set.* In 1965, Nicholas Varopoulos proved that Kronecker sets satisfy the hypothesis of Malliavin's result; and, hence, *Kronecker sets are* U-*sets*. This is the "tip of the iceberg", cf., [Rud62], [KS63], [Kah70], [Ben71], [LP71], [Mey72].

3.3. Integration and Differentiation of Fourier Series

3.3.1 Example. Integration of Series

If $\sum_{n=1}^{\infty} F_n$ is a uniformly convergent series of continuous functions F_n on $[0, 1]$, then

$$\int_0^1 \left(\sum_{n=1}^{\infty} F_n(\gamma) \right) d\gamma = \sum_{n=1}^{\infty} \int_0^1 F_n(\gamma) \, d\gamma, \tag{3.3.1}$$

e.g., [Apo57]. On the other hand, it is well known that the hypotheses, $\{F_n : n = 1, \ldots\} \subseteq C[0, 1]$ and $\sum_{n=1}^{\infty} F_n = F \in C[0, 1]$, where the convergence is pointwise on $[0, 1]$, are *not* sufficient to ensure (3.3.1), e.g., [Har49], [Ben76, Section 3.3]. For example, if we let $F_1(\gamma) = \gamma(1 - \gamma)$ and $F_n(\gamma) = n^2 \gamma (1 - \gamma)^n$ $-(n - 1)^2 \gamma (1 - \gamma)^{n-1}$ for $n > 1$, then it is easy to see that each $F_n \in C[0, 1]$

(and each $F_n \in C(\mathbb{T})$ since $F_n(0) = F_n(1) = 0$), $\sum_{n=1}^{\infty} F_n \equiv F = 0$ on $[0, 1]$, $\int_0^1 \left(\sum_{n=1}^{\infty} F_n(\gamma) \right) d\gamma = 0$, and

$$\sum_{n=1}^{\infty} \int_0^1 F_n(\gamma) \, d\gamma = \lim_{N \to \infty} \sum_{n=1}^{N} \int_0^1 F_n(\gamma) \, d\gamma$$

$$= \lim_{N \to \infty} \int_0^1 N^2 \gamma (1 - \gamma)^N \, d\gamma = \lim_{N \to \infty} \frac{N^2}{(N + 1)(N + 2)} = 1,$$

cf., Section 3.4 where we discuss approximate identities for $L^1(\mathbb{T})$.

The following theorem is a remarkable feature of Fourier series. It asserts that (3.3.1) is valid when the series $\sum_{n=1}^{\infty} F_n$ is replaced by the Fourier series of any function in $L^1(\mathbb{T})$. In particular, the Fourier series to be integrated can diverge everywhere, as in Kolmogorov's result mentioned in Section 3.2.8, cf., Exercise 3.45.

3.3.2 Theorem. Integration of Fourier Series

Let $F \in L^1(\mathbb{T}_{2\Omega})$. The Fourier series $S(F)$ of F, with Fourier coefficients $f = \{f[n]\}$, can be integrated term by term, i.e.,

$$\forall \alpha, \beta \in \hat{\mathbb{R}}, \qquad \int_\alpha^\beta S(F)(\gamma) \, d\gamma = \int_\alpha^\beta F(\gamma) \, d\gamma, \qquad (3.3.2)$$

where the left side of (3.3.2) denotes

$$\sum f[n] \int_\alpha^\beta e^{-\pi i n \gamma / \Omega} \, d\gamma.$$

PROOF.   Define

$$G(\gamma) = \frac{1}{2\Omega} \int_0^\gamma (F(\lambda) - f[0]) \, d\lambda$$

for $\gamma \in [0, 2\Omega)$. Consequently, $G(0) = G(2\Omega) = 0$, and G can be extended 2Ω-periodically to $\hat{\mathbb{R}}$ with the property that $G \in AC_{\text{loc}}(\hat{\mathbb{R}})$. Let $g = \{g[n]\}$ be the sequence of Fourier coefficients of G. As such, we can apply the Dirichlet Theorem, properly modified as in Remark 3.1.7b, to assert that $S(G) = G$ on $\mathbb{T}_{2\Omega}$, i.e.,

$$\forall \gamma \in \mathbb{T}_{2\Omega}, \qquad G(\gamma) = g[0] + \sum{}' g[n] e^{-\pi i n \gamma / \Omega}, \qquad (3.3.3)$$

where $\sum'$ denotes summation over $\mathbb{Z} \backslash \{0\}$. For $n \neq 0$, we compute

$$g[n] = \frac{1}{2\Omega} \int_0^{2\Omega} \left[\frac{1}{2\Omega} \int_0^{\gamma} (F(\lambda) - f[0]) \, d\lambda \right] e^{\pi i n \gamma / \Omega} \, d\lambda$$

$$= \frac{1}{(2\Omega)^2} \frac{\Omega}{\pi i n} e^{\pi i n \gamma / \Omega} \int_0^{\gamma} (F(\lambda) - f[0]) \, d\lambda \Big|_{\gamma=0}^{2\Omega}$$

$$- \frac{1}{(2\Omega)^2} \frac{\Omega}{\pi i n} \int_0^{2\Omega} (F(\gamma) - f[0]) \, e^{\pi i n \gamma / \Omega} \, d\gamma$$

$$= \frac{-f[n]}{2\pi i n}.$$

The last equation follows since $(1/2\Omega) \int_0^{2\Omega} F(\lambda) \, d\lambda = f[0]$ and $\int_0^{2\Omega} e^{\pi i n \gamma / \Omega} \, d\gamma = 0$. Integration by parts is allowable by the (local) absolute continuity. Combining the above computation for $g[n]$ with (3.3.3), we obtain

$$\forall \gamma \in \mathbb{T}_{2\Omega}, \qquad G(\gamma) = g[0] - \sum{}' \frac{f[n]}{2\pi i n} e^{-\pi i n \gamma / \Omega}, \tag{3.3.4}$$

so that, since $G(0) = 0$,

$$g[0] = \sum{}' \frac{f[n]}{2\pi i n}. \tag{3.3.5}$$

In particular, the series $\sum' f[n]/n$ converges, cf., Remark 3.1.7a about symmetric convergence.

By definition of G, (3.3.4) becomes

$$\frac{1}{2\Omega} \int_0^{\gamma} F(\lambda) \, d\lambda = \frac{\gamma}{2\Omega} f[0] + g[0] - \sum{}' \frac{f[n]}{2\pi i n} e^{-\pi i n \gamma / \Omega}. \tag{3.3.6}$$

By definition of $\int_0^{\gamma} S(F)(\lambda) \, d\lambda$ we have

$$\frac{1}{2\Omega} \int_0^{\gamma} S(F)(\lambda) \, d\lambda = \frac{1}{2\Omega} \sum f[n] \int_0^{\gamma} e^{-\pi i n \lambda / \Omega} \, d\lambda$$

$$= \frac{\gamma}{2\Omega} f[0] - \frac{1}{2\Omega} \sum{}' f[n] \frac{\Omega}{\pi i n} (e^{-\pi i n \gamma / \Omega} - 1)$$

$$= \frac{\gamma}{2\Omega} f[0] - \frac{1}{2\Omega} \sum{}' f[n] \frac{\Omega}{\pi i n} e^{-\pi i n \gamma / \Omega} + g[0].$$

Consequently, (3.3.2) is obtained for the interval $[0, \gamma] \subseteq [0, 2\Omega]$ by combining (3.3.6) with this last calculation.

The case for the interval $[\alpha, \beta] \subseteq [0, 2\Omega]$ is obtained by writing $\int_\alpha^\beta = \int_\alpha^0 + \int_0^\beta = \int_0^\beta - \int_0^\alpha$, and then making the natural adjustments for other values of α and β. ∎

The following result is a consequence of (3.3.5).

3.3.3 Corollary. A Property of Fourier Coefficients $f \in A(\mathbb{Z})$

Let $F \in L^1(\mathbb{T}_{2\Omega})$. The series

$$\sum{}' \frac{f[n]}{n}$$

converges, where $f = \{f[n]\}$ is the sequence of Fourier coefficients of F.

**3.3.4 Example. Necessary Conditions for Integrability
of Trigonometric Series**

   a.   The Riemann–Lebesgue Lemma asserts that if $F \in L^1(\mathbb{T}_{2\Omega})$, then

$$\lim_{|n|\to\infty} f[n] = 0,$$

where $f = \{f[n]\}$ is the sequence of Fourier coefficients of F. On the other hand, suppose we are given a trigonometric series $\sum c_n e^{-2\pi i n \gamma}$ for which $\lim_{|n|\to\infty} c_n = 0$. Is there any way we can determine if this series is or is not the Fourier series of some function $F \in L^1(\mathbb{T}_{2\Omega})$, cf., Section 3.2.6?

We *can* assert that the trigonometric series,

$$\sum_{n=2}^{\infty} \frac{\sin(\pi n \gamma / \Omega)}{\log n}, \tag{3.3.7}$$

is not the Fourier series of an element $F \in L^1(\mathbb{T}_{2\Omega})$ even though the coefficients tend to 0 at $\pm\infty$, à la the necessary conditions given by the Riemann–Lebesgue Lemma. To verify this claim we argue as follows. First, we write (3.3.7) as

$$-\frac{1}{2i} \sum_{|n|\geq 2} \frac{\operatorname{sgn} n}{\log |n|} e^{-\pi i n \gamma / \Omega},$$

where $\operatorname{sgn} n = n/|n|$. Then we apply Corollary 3.3.3 to observe that if (3.3.7) were a Fourier series (of an element $F \in L^1(\mathbb{T}_{2\Omega})$), then

$$\sum_{|n|\geq 2} \frac{1}{|n| \log |n|} < \infty,$$

which is false by the integral test, e.g., Exercise 3.10. Thus, we have constructed a sequence $f \in c_0(\mathbb{Z})\backslash A(\mathbb{Z})$, cf., Example 1.4.4 where we constructed the analogue of (3.3.7) for the case $C_0(\mathbb{R})\backslash A(\mathbb{R})$. An example $F \in C(\mathbb{T}_{2\Omega})\backslash A(\mathbb{T}_{2\Omega})$ is constructed in Exercise 3.43, cf., Exercise 3.45 and du Bois–Reymond's example mentioned in Section 3.2.8.

   b.   It does turn out, however, that the series (3.3.7) converges pointwise for each $\gamma \in \hat{\mathbb{R}}$, e.g., Exercise 3.29b.

3.3.5 Theorem. Differentiation of Fourier Series

Let $F \in AC(\mathbb{T}_{2\Omega})$. Then $F' \in L^1(\mathbb{T}_{2\Omega})$ (ordinary differentiation) and

$$S'(F) = S(F'),$$

where $S'(F)$ denotes the term-by-term differentiated series

$$-\sideset{}{'}\sum \frac{\pi i n}{\Omega} f[n] e^{-\pi i n \gamma / \Omega},$$

and where $f = \{f[n]\}$ is the sequence of Fourier coefficients of F.

PROOF. Clearly, $F' \in L^1(\mathbb{T}_{2\Omega})$, e.g., Remark A.21. By the absolute continuity and 2Ω-periodicity of F, and by the FTC, we compute

$$(F')^{\vee}[n] = -\frac{1}{2\Omega} \frac{\pi i n}{\Omega} \int_{-\Omega}^{\Omega} F(\gamma) e^{\pi i n \gamma / \Omega} \, d\gamma = -\frac{\pi i n}{\Omega} \check{F}[n]$$

for $n \neq 0$. If $n = 0$, then

$$(F')^{\vee}[0] = \frac{1}{2\Omega} \int_{-\Omega}^{\Omega} F'(\gamma) \, d\gamma = 0$$

since the absolute continuity again allows us to use FTC. The result is obtained. ∎

3.3.6 Example. $\zeta(2)$

a. Let $F(\gamma) = (\pi - \gamma)/2$ on $[0, 2\pi)$, and consider F as an element of $L^1(\mathbb{T}_{2\pi})$. The graph of F appears in Figure 3.1.

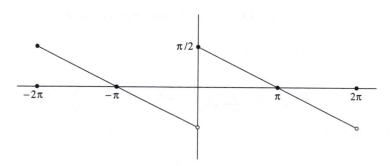

FIGURE 3.1.

We shall compute $S(F)$. F is odd on $(-\pi, \pi)$, and we have

$$S(F)(\gamma) = \sum_{n=1}^{\infty} b_n \sin n\gamma,$$

where

$$\forall n \geq 1, \qquad b_n = \frac{1}{\pi} \int_0^{2\pi} \frac{1}{2}(\pi - \gamma) \sin n\gamma \, d\gamma.$$

We calculate

$$b_n = -\frac{1}{2\pi} \left[-\frac{\gamma}{n} \cos n\gamma \Big|_0^{2\pi} + \frac{1}{n} \int_0^{2\pi} \cos n\gamma \, d\gamma \right] = \frac{1}{n}.$$

Thus,

$$S(F)(\gamma) = \sum_{n=1}^{\infty} \frac{\sin n\gamma}{n}.$$

By the Dirichlet Theorem, $S(F)(\gamma) = F(\gamma)$ on $\hat{\mathbb{R}} \setminus \cup \{2\pi n\}$ and $S(F)(2\pi n) = 0$.

b. Since $F \in L^1(\mathbb{T}_{2\pi})$, we have

$$\forall \gamma \in [0, 2\pi], \qquad \int_0^{\gamma} S(F)(\lambda) \, d\lambda = \int_0^{\gamma} F(\lambda) \, d\lambda$$

by Theorem 3.3.2. This becomes

$$\forall \gamma \in [0, 2\pi], \qquad \sum_{n=1}^{\infty} \frac{1}{n} \int_0^{\gamma} \sin n\lambda \, d\lambda = \frac{\gamma \pi}{2} - \frac{\gamma^2}{4}$$

and the left side is $-\sum_{n=1}^{\infty} (1/n^2)(\cos n\gamma - 1)$. Thus

$$\forall \gamma \in [0, 2\pi], \qquad \frac{\gamma}{4}(2\pi - \gamma) = -\sum_{n=1}^{\infty} \frac{1}{n^2}(\cos n\gamma - 1).$$

c. Integrating both sides of the last expression in part *b* we obtain

$$\int_0^{2\pi} \frac{\gamma}{4}(2\pi - \gamma) \, d\gamma$$

$$= 2\pi \sum_{n=1}^{\infty} \frac{1}{n^2} - \sum_{n=1}^{\infty} \frac{1}{n^2} \int_0^{2\pi} \cos n\gamma \, d\gamma = 2\pi \sum_{n=1}^{\infty} \frac{1}{n^2}.$$

Consequently,

$$\sum_{n=1}^{\infty} \frac{1}{n^2} = \frac{\pi^2}{6}. \qquad\qquad (3.3.8)$$

d. The Riemann ζ-function $\zeta(s) = \sum_{n=1}^{\infty} 1/n^s$, discussed in Example 2.4.6g, was defined by Pietro Mengoli (1625–1686) in 1650. He showed that the harmonic series $\zeta(1)$ diverges. The problem of evaluating $\zeta(k)$, for integers $k \geq 2$, attracted the attention of British mathematicians, including James Gregory (1638–1675). Henry Oldenburg (c. 1615–1677), first Secretary of the Royal Society in London, wrote to Gottfried Leibniz (1646–1716) asking him to evaluate $\zeta(2)$. This occurred during Leibniz's visit to London in 1673. In 1696, Leibniz admitted his inability to solve this problem. Earlier, in 1689, Jakob Bernoulli (1654-1705), Daniel's uncle, had also tried and apparently given-up summing $\zeta(2)$, cf., [Pól54, pages 17–22], [Kli72, pages 448–449], [Ebe83] for tantalizing historical remarks and incisive analysis.

Around 1736, Euler was able to state (3.3.8) by an ingenious argument outlined in Exercise 3.28, cf., Exercise 3.38. His *Introductio in Analysin Infinitorum* (1748) contains (3.3.8) and many similar results. In light of our discussion in Sections 3.2.1–3.2.3, it should also be pointed out that the *Introductio* also "defines" a *function* of a "variable quantity" as "any analytic expression whatsoever made up from that variable quantity and from numbers or constant quantities".

The following result can be used to prove the Classical Sampling Theorem (Theorem 3.10.10), e.g., [Ben92b, pages 447–449]. The proof of Theorem 3.3.7 is similar to that of Theorem 3.3.2, e.g., Exercise 3.15.

3.3.7 Theorem. Integration of Fourier Series—A Refinement

Let $F \in L^1(\mathbb{T}_{2\Omega})$ and $G \in BV(\mathbb{T}_{2\Omega})$. Then

$$\forall \alpha, \beta \in \hat{\mathbb{R}}, \qquad \int_{\alpha}^{\beta} S(F)(\gamma)G(\gamma)\,d\gamma = \int_{\alpha}^{\beta} F(\gamma)G(\gamma)\,d\gamma, \qquad (3.3.9)$$

where the left side of (3.3.9) denotes

$$\sum f[n] \int_{\alpha}^{\beta} G(\gamma)e^{-\pi i n \gamma / \Omega}\,d\gamma,$$

and where $f = \{f[n]\}$ is the sequence of Fourier coefficients of F.

3.3.8 Remark. Integration of Fourier Series—Evocations

a. If $\beta - \alpha = 2\Omega$ in Theorem 3.3.7, then (3.3.9) becomes *Parseval's formula*,

$$\lim_{N \to \infty} \sum_{|n| \leq N} f[n]g[-n] = \frac{1}{2\Omega} \int_{\alpha}^{\alpha + 2\Omega} F(\gamma)G(\gamma)\,d\gamma,$$

where $g = \check{G}$. With the adjustment as in Proposition 1.10.4, we obtain *Parseval's formula* in the form

$$\lim_{N \to \infty} \sum_{|n| \le N} f[n]\overline{g[n]} = \int_{\mathbb{T}_{2\Omega}} F(\gamma)\overline{G(\gamma)} \, d\gamma, \qquad (3.3.10)$$

where $F \in L^1(\mathbb{T}_{2\Omega})$, $G \in BV(\mathbb{T}_{2\Omega})$, and $\{f[n]\}$ and $\{g[n]\}$ are the Fourier coefficients of F and G, respectively, cf., Theorem 3.4.12*b* and Exercise 3.47 for other statements of *Parseval's formula*.

   b.   With respect to Corollary 3.3.3, we recall the following Hardy and Littlewood Theorem: *if $F \in L^1(\mathbb{T}_{2\Omega})$ and $f[n] = 0$ for all $n < 0$, where $f = \{f[n]\}$ is the sequence of Fourier coefficients of F, then*

$$\sum_{n=0}^{\infty} \frac{|f[n]|}{n+1} \le \pi \|F\|_{L^1(\mathbb{T}_{2\pi})} < \infty.$$

This result is difficult to prove. One proof involves a fundamental factorization theorem for the so-called *Hardy space* $H^1(\mathbb{T}_{2\Omega})$, as well as the following *double series theorem* due to Hilbert: *if $f, g \in \ell^2(\mathbb{N} \cup \{0\})$, then*

$$\sum_{j,k=0}^{\infty} \frac{|f[j]g[k]|}{j+k+1} \le \pi \|f\|_{\ell^2(\mathbb{N}\cup\{0\})} \|g\|_{\ell^2(\mathbb{N}\cup\{0\})},$$

where π is the best possible constant, as proved by Schur, e.g., [Hel83, pages 94–99], cf., Exercise 3.34 for related material on Hilbert transforms of sequences.

   c.   It is well known that the desirable statement,

$$\lim_{N \to \infty} \|S_N(F) - F\|_{L^1(\mathbb{T}_{2\Omega})} = 0, \qquad (3.3.11)$$

is *not* true for all $F \in L^1(\mathbb{T}_{2\Omega})$, e.g., Example 3.4.9.

   On the other hand, a sequence $\{F_N\} \subseteq L^1(\mathbb{T}_{2\Omega})$ converges to $F \in L^1(\mathbb{T}_{2\Omega})$ *weakly*, i.e.,

$$\forall G \in L^\infty(\mathbb{T}_{2\Omega}), \qquad \lim_{N \to \infty} \int_{\mathbb{T}_{2\Omega}} (F_N(\gamma) - F(\gamma))G(\gamma) \, d\gamma = 0,$$

if and only if

$$\lim_{N \to \infty} \int_A (F_N(\gamma) - F(\gamma)) \, d\gamma = 0 \qquad (3.3.12)$$

for every Lebesgue measurable set $A \subseteq \mathbb{T}_{2\Omega}$, e.g., [RN55, page 89]. If we have weak convergence, or, equivalently, (3.3.12), then (3.3.11) is true for $S_N(F) = F_N$ if $\{F_N\}$ converges to F in measure, e.g., [Ben76, page 226]. (Of course, norm convergence always implies convergence in measure.) Further, Dieudonné and Grothendieck proved that (3.3.12) is true for all Lebesgue measurable sets $A \subseteq \mathbb{T}_{2\Omega}$ if and only if it is true for all open sets $A \subseteq \mathbb{T}_{2\Omega}$, e.g., [Ben76, page 225].

Because of this general relationship between weak and norm convergence, and because we would like to have (3.3.11) (or at least know how close we are to it), we note the following reformulation of Theorem 3.3.2: Theorem 3.3.2 is (3.3.12) for all *intervals* A in the case $F_N = S_N(F)$. Further, Theorem 3.3.7 gives (3.3.12) in this case for all finite unions A of intervals.

3.4. The $L^1(\mathbb{T})$ and $L^2(\mathbb{T})$ Theories

We showed in Remark 3.1.3 that $L^2(\mathbb{T}) \subseteq L^1(\mathbb{T})$; and we have already noted in Section 3.2.8 that Fourier series of $L^2(\mathbb{T})$ functions converge a.e., whereas there are $L^1(\mathbb{T})$ functions whose Fourier series diverge at every point. Conceptually there are deeper differences between $L^1(\mathbb{T})$ and $L^2(\mathbb{T})$ than the fact that larger spaces may allow more instances of unusual behavior; briefly, $L^1(\mathbb{T})$ has *algebraic* properties and $L^2(\mathbb{T})$ has *geometric* properties that characterize their Fourier analysis.

3.4.1 Definition. Convolution

a. Let $F, G \in L^1(\mathbb{T}_{2\Omega})$. The *convolution* of F and G, denoted by $F * G$, is

$$F * G(\gamma) = \int_{\mathbb{T}_{2\Omega}} F(\gamma - \lambda)G(\lambda)\, d\lambda = \int_{\mathbb{T}_{2\Omega}} F(\lambda)G(\gamma - \lambda)\, d\lambda.$$

(Recall that "$\int_{\mathbb{T}_{2\Omega}}$" designates "$(1/2\Omega)\int_{\alpha}^{\alpha+2\Omega}$" for any fixed $\alpha \in \hat{\mathbb{R}}$.) As with $L^1(\mathbb{R})$ and Exercise 1.31, it is not difficult to prove that $F * G \in L^1(\mathbb{T}_{2\Omega})$ and

$$\forall F, G \in L^1(\mathbb{T}_{2\Omega}),$$

$$\|F * G\|_{L^1(\mathbb{T}_{2\Omega})} \leq \|F\|_{L^1(\mathbb{T}_{2\Omega})}\|G\|_{L^1(\mathbb{T}_{2\Omega})}. \tag{3.4.1}$$

b. $L^1(\mathbb{T}_{2\Omega})$ is a *commutative algebra* taken with the operations of addition and convolution, i.e., $L^1(\mathbb{T}_{2\Omega})$ is a complex vector space under addition, and convolution is distributive with respect to addition, as well as being associative, and commutative.

3.4.2 Proposition.

Let $F, G \in L^1(\mathbb{T}_{2\Omega})$, with corresponding sequences $f = \{f[n]\}, g = \{g[n]\} \in A(\mathbb{Z})$ of Fourier coefficients, i.e., $\hat{f} = F$ and $\hat{g} = G$. Then $fg = \{f[n]g[n]\} \in$

$A(\mathbb{Z})$ is the sequence of Fourier coefficients of $F * G \in L^1(\mathbb{T}_{2\Omega})$, i.e.,

$$\forall n \in \mathbb{Z}, \qquad f[n]g[n] = \int_{\mathbb{T}_{2\Omega}} F * G(\gamma) e^{\pi i n \gamma / \Omega} \, d\gamma.$$

The proof of Proposition 3.4.2 is the same as that of Proposition 1.5.2.

3.4.3 Definition. Approximate Identity

An *approximate identity* is a family $\{K_{(\lambda)} : \lambda > 0\} \subseteq L^1(\mathbb{T}_{2\Omega})$ of functions with the properties:

a) $\forall \lambda > 0, \quad \displaystyle\int_{\mathbb{T}_{2\Omega}} K_{(\lambda)}(\gamma) d\gamma = 1;$

b) $\exists C \geq 1$ such that $\forall \lambda > 0, \quad \|K_{(\lambda)}\|_{L^1(\mathbb{T}_{2\Omega})} \leq C;$

c) $\forall \eta \in (0, \Omega], \quad \lim_{\lambda \to \infty} \frac{1}{2\Omega} \displaystyle\int_{\eta \leq |\gamma| \leq \Omega} |K_{(\lambda)}(\gamma)| d\gamma = 0.$

3.4.4 Theorem. Approximate Identity Theorem

   a. Let $F \in C(\mathbb{T}_{2\Omega})$, and let $\{K_{(\lambda)}\} \subseteq L^1(\mathbb{T}_{2\Omega})$ be an approximate identity. Then

$$\lim_{\lambda \to \infty} \|F - F * K_{(\lambda)}\|_{L^\infty(\mathbb{T}_{2\Omega})} = 0. \tag{3.4.2}$$

   b. Let $F \in L^1(\mathbb{T}_{2\Omega})$, and let $\{K_{(\lambda)}\} \subseteq L^1(\mathbb{T}_{2\Omega})$ be an approximate identity. Then

$$\lim_{\lambda \to \infty} \|F - F * K_{(\lambda)}\|_{L^1(\mathbb{T}_{2\Omega})} = 0, \tag{3.4.3}$$

cf., Theorem 1.6.9a.

The proof of Theorem 3.4.4*a* follows the proof of Theorem 1.6.9*a*, but for the case of $\| \cdots \|_{L^\infty(\mathbb{T}_{2\Omega})}$ instead of $\| \cdots \|_{L^1(\mathbb{R})}$. In fact, by the uniform continuity of F, (1.6.6) can be replaced by the statement that for each $\epsilon > 0$ there is $\eta > 0$ for which

$$\forall |\lambda| < \eta, \qquad \|F - \tau_\lambda F\|_{L^\infty(\mathbb{T}_{2\Omega})} < \epsilon/C.$$

This allows us to prove

$$\overline{\lim_{\lambda \to \infty}} \|F - F * K_{(\lambda)}\|_{L^\infty(\mathbb{T}_{2\Omega})} \leq \epsilon$$

analogous to the proof of Theorem 1.6.9*a*.

The proof of Theorem 3.4.4*b* follows from part *a*, e.g., Exercise 3.16.

3.4.5 Example. The Dirichlet and Fejér Kernels

a. The *Dirichlet function* D_N on $\mathbb{T}_{2\Omega}$ is defined by

$$\forall \gamma \in \hat{\mathbb{R}}, \qquad D_N(\gamma) = \sum_{|n| \le N} e^{-\pi i n \gamma / \Omega}. \tag{3.4.4}$$

It is not difficult to show that the trigonometric polynomial D_N can be written as

$$D_N(\gamma) = \frac{\sin(N + 1/2)\pi \gamma / \Omega}{\sin(\pi \gamma / 2\Omega)}, \tag{3.4.5}$$

where $D_N(2k\Omega)$ is $2N + 1$ for $k \in \mathbb{Z}$, e.g., Exercise 3.11a. The family $\{D_N : N \in \mathbb{N} \cup \{0\}\}$ is the *Dirichlet kernel* on $\mathbb{T}_{2\Omega}$, cf., the Dirichlet kernels on $\mathbb{R}$ and $\mathbb{Z}$ in Remark 1.6.4 and Remark 3.1.2, respectively. Using the notation of Definition 3.4.3, we see that $K_{(\lambda)} \equiv D_N$ is not an approximate identity since $D_N \notin L^1(\mathbb{T}_{2\Omega})$.

b. The *Fejér function* W_N on $\mathbb{T}_{2\Omega}$ is defined by

$$\forall \gamma \in \hat{\mathbb{R}}, \qquad W_N(\gamma) = \sum_{|n| \le N} \left(1 - \frac{|n|}{N+1}\right) e^{-\pi i n \gamma / \Omega}. \tag{3.4.6}$$

It is not difficult to show that the trigonometric polynomial W_N can be written as

$$W_N(\gamma) = \frac{D_0(\gamma) + \cdots + D_N(\gamma)}{N+1} = \frac{1}{N+1}\left(\frac{\sin(N+1)\pi \gamma / 2\Omega}{\sin(\pi \gamma / 2\Omega)}\right)^2, \tag{3.4.7}$$

where $W_N(2k\Omega)$ is $N + 1$ for $k \in \mathbb{Z}$, e.g., Exercise 3.11b. The family $\{W_N : N \in \mathbb{N} \cup \{0\}\}$ is the *Fejér kernel* on $\mathbb{T}_{2\Omega}$, cf., the Fejér kernel on $\mathbb{R}$ and $\mathbb{Z}$ in Remark 1.6.4 and Remark 3.1.2, respectively.

c. *The Fejér kernel is an approximate identity.* To see this, first note that $W_N \ge 0$ by (3.4.7), and

$$\int_{\mathbb{T}_{2\Omega}} W_N(\gamma)\,d\gamma = 1$$

by the definition (3.4.6) and the fact that $\int_{\mathbb{T}_{2\Omega}} e^{-\pi i n \gamma / \Omega}\,d\gamma = 0$ for $n \ne 0$. Thus, parts a and b of Definition 3.4.3 are valid. Finally, we obtain part c of Definition 3.4.3 by (3.4.7) and the estimate

$$\frac{1}{2\Omega}\int_{\eta \le |\gamma| \le \Omega} |W_N(\gamma)|\,d\gamma \le \frac{2(\Omega - \eta)}{2\Omega(N+1)} \sup_{\eta \le |\gamma| \le \Omega} \frac{1}{\sin^2(\pi \gamma / 2\Omega)}$$

$$\le \frac{\Omega - \eta}{\Omega(N+1)} \frac{1}{\sin^2(\pi \eta / 2\Omega)}.$$

In Section 3.2.8, we mentioned du Bois–Reymond's example (1872) of a function $F \in C(\mathbb{T})$ for which $\{S_N(F)(0)\}$ diverges, cf., Exercise 3.45. At the risk of being an alarmist over 100 years after the fact, it is not an exaggeration to say that this example dampened some of the optimism for a comprehensive theory of the representation of functions by trigonometric series. Fejér's result (1904), stated below in Theorem 3.4.6a, came none too soon, e.g., [Bir73, pages 150–156] for a translation of the relevant parts of Fejér's original paper.

3.4.6 Theorem. Fejér Theorem

   a.   *Let $F \in C(\mathbb{T}_{2\Omega})$, and let $f = \{f[n]\} \in A'(\mathbb{Z}) \subseteq \ell^2(\mathbb{Z})$ be its sequence of Fourier coefficients. Then*

$$F * W_N(\gamma) = \sum_{|n| \leq N} \left(1 - \frac{|n|}{N+1} \right) f[n] e^{-\pi i n \gamma / \Omega} \qquad (3.4.8)$$

and

$$\lim_{N \to \infty} \| F - F * W_N \|_{L^\infty(\mathbb{T}_{2\Omega})} = 0. \qquad (3.4.9)$$

   b.   *Let $F \in L^1(\mathbb{T}_{2\Omega})$. Then*

$$\lim_{N \to \infty} \int_{\mathbb{T}_{2\Omega}} \left| F(\gamma) - \sum_{|n| \leq N} \left(1 - \frac{|n|}{N+1} \right) f[n] e^{-\pi i n \gamma / \Omega} \right| d\gamma = 0,$$

cf., Theorem 1.6.9b.

PROOF.   Equation (3.4.8) of part *a* follows from a direct computation of the left side, cf., Exercise 3.12*a*. Equation (3.4.9) follows by combining Example 3.4.5*c* with Theorem 3.4.4*a*. Part *b* follows from part *a*, e.g., Exercise 3.16.   ∎

3.4.7 Corollary. Uniqueness

Let $F \in L^1(\mathbb{T}_{2\Omega})$; and assume $f[n] = 0$ for each $n \in \mathbb{Z}$, where $f = \{f[n]\}$ is the sequence of Fourier coefficients of F. Then F is the 0-function, cf., Theorem 1.6.9c.

3.4.8 Remark. Weierstrass Approximation Theorem

   a.   The *Weierstrass Approximation Theorem* (1885) asserts that *if $F \in C[\alpha, \beta]$, then there is a sequence $\{P_N\}$ of polynomials for which*

$$\lim_{N \to \infty} \| F - P_N \|_{L^\infty[\alpha, \beta]} = 0. \qquad (3.4.10)$$

Equation (3.4.10) can be derived from (3.4.9) in the following way. By translation we can take $F \in C[-\Omega, \Omega]$ without loss of generality. Next choose c such that $G(-\Omega) = G(\Omega)$ where $G(\gamma) \equiv F(\gamma) - c\gamma$ for $\gamma \in [-\Omega, \Omega]$. In fact, let

$$c = \frac{F(\Omega) - F(-\Omega)}{2\Omega}.$$

Apply Theorem 3.4.6 to G considered as an element of $C(\mathbb{T}_{2\Omega})$. Finally, uniformly approximate the trigonometric polynomials $G * W_N$ on $[-\Omega, \Omega]$ by polynomial approximants of their Taylor series expansions.

b. There are extensive developments of the Weierstrass Theorem, many of which have evolved from Stone's celebrated Stone–Weierstrass Theorem (1937). We refer to [BD81] and [Bur84], replete with ingenuity and scholarship, for recent contributions to the Stone–Weierstrass Theorem in a functional analytic uniform algebra setting.

In Remark 3.3.8c we asserted that the partial sums $S_N(F)$ do not necessarily converge to F in $L^1(\mathbb{T}_{2\Omega})$. We shall now prove this assertion.

3.4.9 Example. Lebesgue Constants and an L^1-Convergence Problem

a. Analogous to (3.4.8), a direct computation shows that

$$\forall F \in L^1(\mathbb{T}_{2\Omega}), \qquad S_N(F)(\gamma) = F * D_N(\gamma),$$

cf., Exercise 3.12.

b. The *Lebesgue constants* are defined as $\|D_N\|_{L^1(\mathbb{T}_{2\Omega})}$ for each N. We shall prove that there is a *bounded sequence* $\{C(N) : N \geq 2\}$ such that

$$\forall N \geq 2, \quad \|D_N\|_{L^1(\mathbb{T}_{2\Omega})} = \frac{4}{\pi^2} \log N + C(N), \tag{3.4.11}$$

cf., Remark 3.3.8b.

To see this we use (3.4.5) as follows. First,

$$\|D_N\|_{L^1(\mathbb{T}_{2\Omega})} = \frac{2}{\pi} \int_0^{\pi/2} |\sin(2N+1)x| \left| \frac{1}{\sin x} - \frac{1}{x} + \frac{1}{x} \right| dx,$$

so that

$$\left| \|D_N\|_{L^1(\mathbb{T}_{2\Omega})} - \frac{2}{\pi} \int_0^{\pi/2} \frac{|\sin(2N+1)x|}{x} \, dx \right|$$
$$\leq \frac{2}{\pi} \int_0^{\pi/2} \left| \frac{1}{\sin x} - \frac{1}{x} \right| dx = C_1, \tag{3.4.12}$$

by the triangle inequality, where $C_1 < \infty$. In fact, it is easy to check that $\lim_{x \to 0}(\frac{1}{\sin x} - \frac{1}{x}) = 0$.

Next, letting $t = (2N + 1)x$ and dividing the domain of integration according to the sign of the sine, we have

$$\frac{2}{\pi} \int_0^{\pi/2} \frac{|\sin(2N+1)x|}{x}\, dx = \frac{2}{\pi} \sum_{k=0}^{N-1} \int_{k\pi}^{(k+1)\pi} \frac{|\sin t|}{t}\, dt$$

$$+ \frac{2}{\pi} \int_{N\pi}^{N\pi+\pi/2} \frac{|\sin t|}{t}\, dt$$

$$= \frac{2}{\pi} \int_0^{\pi} \sin u \left(\sum_{k=1}^{N-1} \frac{1}{u + k\pi} \right) du$$ (3.4.13)

$$+ \frac{2}{\pi} \int_0^{\pi} \frac{\sin u}{u}\, du + \frac{2}{\pi} \int_{N\pi}^{N\pi+\pi/2} \frac{|\sin u|}{u}\, du.$$

Since both $\sin u \geq 0$ and

$$\sum_{k=2}^{N} \frac{1}{k\pi} \leq \sum_{k=1}^{N-1} \frac{1}{u + k\pi} \leq \sum_{k=1}^{N-1} \frac{1}{k\pi}$$

on $[0, \pi]$, we can use the integral test to compute

$$\frac{4}{\pi^2}(-1 + \log(N+1)) \leq \frac{2}{\pi} \int_0^{\pi} \sin u \left(\sum_{k=1}^{N-1} \frac{1}{u+k\pi} \right) du$$

$$\leq \frac{4}{\pi^2}(1 + \log(N-1)).$$ (3.4.14)

We obtain (3.4.11) by combining (3.4.12), (3.4.13), and (3.4.14).

c.   Consider the linear mappings

$$L_N : L^1(\mathbb{T}_{2\Omega}) \longrightarrow L^1(\mathbb{T}_{2\Omega})$$
$$F \longmapsto S_N(F).$$

The norm of L_N, defined in Definition B.6, is

$$\|L_N\| = \sup_{\|F\|_{L^1(\mathbb{T}_{2\Omega})} \leq 1} \|L_N(F)\|_{L^1(\mathbb{T}_{2\Omega})} = \sup_{\|F\|_{L^1(\mathbb{T}_{2\Omega})} \leq 1} \|F * D_N\|_{L^1(\mathbb{T}_{2\Omega})},$$

and so $\|L_N\| \leq \|D_N\|_{L^1(\mathbb{T}_{2\Omega})}$ by (3.4.1) and part a. To prove the opposite inequality we first note that

$$\|L_N\| \geq \|L_N(W_n)\|_{L^1(\mathbb{T}_{2\Omega})} = \|D_N * W_n\|_{L^1(\mathbb{T}_{2\Omega})}.$$

Then, by Theorem 3.4.4*b* and Example 3.4.5*c*, we have

$$\lim_{n\to\infty} \|D_N * W_n\|_{L^1(\mathbb{T}_{2\Omega})} = \|D_N\|_{L^1(\mathbb{T}_{2\Omega})}.$$

Consequently,

$$\forall N \geq 1, \qquad \|L_N\| = \|D_N\|_{L^1(\mathbb{T}_{2\Omega})}. \tag{3.4.15}$$

Combining (3.4.15) and (3.4.11) with the Uniform Boundedness Principle (Theorem B.8), we can assert that there are functions $F \in L^1(\mathbb{T}_{2\Omega})$ for which $\sup_N \|S_N(F)\|_{L^1(\mathbb{T}_{2\Omega})} = \infty$. In particular, for such functions we do *not* have $\lim_{N\to\infty} \|S_N(F) - F\|_{L^1(\mathbb{T}_{2\Omega})} = 0$.

We begin our discussion of the $L^2(\mathbb{T})$ theory with the following definition (Definition 3.4.10) and background (Remark 3.4.11).

3.4.10 Definition. Orthonormal Basis

a. A sequence $\{E_n\} \subseteq L^2(\mathbb{T}_{2\Omega})$ is *orthonormal* in $L^2(\mathbb{T}_{2\Omega})$ if

$$\forall m, n \in \mathbb{Z}, \qquad \int_{\mathbb{T}_{2\Omega}} E_m(\gamma)\overline{E_n(\gamma)}\,d\gamma = \delta(m, n),$$

where

$$\delta(m, n) = \begin{cases} 1, & \text{if } m = n, \\ 0, & \text{if } m \neq n. \end{cases}$$

An orthonormal sequence $\{E_n\} \subseteq L^2(\mathbb{T}_{2\Omega})$ is an *orthonormal basis* (ONB) for $L^2(\mathbb{T}_{2\Omega})$ if

$$\forall F \in L^2(\mathbb{T}_{2\Omega}),\ \exists \{c_n\} \subseteq \mathbb{C} \quad \text{such that}$$

$$F = \sum c_n E_n \quad \text{in} \quad L^2(\mathbb{T}_{2\Omega}), \tag{3.4.16}$$

cf., (3.4.18).

b. Using Hölder's Inequality, it is an elementary calculation to show that

$$\lim_{n\to\infty} \int_{\mathbb{T}_{2\Omega}} F_n(\gamma)\overline{G_n(\gamma)}\,d\gamma = \int_{\mathbb{T}_{2\Omega}} F(\gamma)\overline{G(\gamma)}\,d\gamma \tag{3.4.17}$$

when $F, G, F_n, G_n \in L^2(\mathbb{T}_{2\Omega})$ and

$$\lim_{n\to\infty} \|F - F_n\|_{L^2(\mathbb{T}_{2\Omega})} = 0 \quad \text{and} \quad \lim_{n\to\infty} \|G - G_n\|_{L^2(\mathbb{T}_{2\Omega})} = 0,$$

e.g., Exercise 3.9.

c. In the case of an ONB $\{E_n\}$, the coefficients c_n in (3.4.16) are of the form

$$\forall n \in \mathbb{Z}, \qquad c_n = \int_{\mathbb{T}_{2\Omega}} F(\gamma)\overline{E_n(\gamma)}\,d\gamma. \qquad (3.4.18)$$

This follows from part *b*.

3.4.11 Remark. Integral Equations and the Riesz–Fischer Theorem

a. In Example 3.1.10*f* we mentioned Abel in conjunction with elliptic functions. In fact, this work was in the realm of *integral equations*, and Abel (1823) solved the "tautochrone" equation

$$\int_0^y \frac{f(x)}{\sqrt{y-x}}\,dx = g(y)$$

for a given forcing function g by computing

$$f(x) = \frac{1}{\pi}\int_0^x \frac{g'(y)}{\sqrt{x-y}}\,dy.$$

In the late nineteenth century it was realized that many problems in mathematical physics could be transformed into solving integral equations of the form

$$\int_{\mathbb{T}} F(\gamma)K(\gamma,\lambda)\,d\gamma = G(\lambda) \qquad (3.4.19)$$

e.g., [CH53], [Die81]. The Dirichlet problem in potential theory was solved in particular cases by Neumann. Vito Volterra (1896) used the Neumann method to solve a certain type of integral equation, and this led to Ivar Fredholm's (still) eminently readable and fundamental paper on integral equations (*Acta Mathematica* **27** (1903), 365–390).

b. Hilbert and (Erhard) Schmidt, in the periods 1904–1912 and 1905–1908, respectively, made great strides in solving (3.4.19), and, in the process, they established some of the fundamental ideas of functional analysis. They also set the stage for the Riesz–Fischer Theorem in the following way.

Suppose $\{E_n\} \subseteq L^2(\mathbb{T})$ is *orthonormal*. Assume we can write K, F, and G of (3.4.19) as $K = \sum_{m,n} k(m,n)\overline{E}_m E_n$, $F = \sum f(n)E_n$, $G = \sum g(n)E_n$. For a given kernel K and forcing function G, the goal is to find F. Formally,

$$\int_{\mathbb{T}} F(\gamma)K(\gamma,\lambda)\,d\gamma = \sum_m \sum_{p,q} f(m)k(p,q)E_q(\lambda)\int_{\mathbb{T}} E_m(\gamma)\overline{E_p(\gamma)}\,d\gamma$$

$$= \sum_q \left(\sum_m f(m)k(m,q)\right)E_q(\lambda),$$

and so (3.4.19) leads to the infinite system of linear equations

$$\forall q \in \mathbb{Z}, \qquad \sum_m f(m)k(m, q) = g(q). \tag{3.4.20}$$

Suppose $g \in \ell^2(\mathbb{Z})$, and assume $k : \mathbb{Z} \times \mathbb{Z} \to \mathbb{C}$ has the property that

$$\sum_{m,n} |k(m, n) - \delta(m, n)|^2 < \infty.$$

Then classical methods yield the construction of a unique solution $f \in \ell^2(\mathbb{Z})$ of (3.4.20), e.g., [GG81, pages 70–74].

    c.   Once the sequence $f \in \ell^2(\mathbb{Z})$ of part b is found, then the major problem in solving (3.4.19) in terms of $F = \sum f(n)E_n$ is accomplished by means of F. Riesz' Theorem: *if $\{E_n\} \subseteq L^2(\mathbb{T})$ is orthonormal and $f \in \ell^2(\mathbb{Z})$, then there is $F \in L^2(\mathbb{T})$ for which*

$$\forall n \in \mathbb{Z}, \qquad f(n) = \int_{\mathbb{T}} F(\gamma)\overline{E_n(\gamma)}\, d\gamma.$$

This is Riesz' formulation of the *Riesz–Fischer Theorem* (1906–1907). Fischer's formulation is that $L^2(\mathbb{T})$ *is complete*, i.e., *if $\{F_n\} \subseteq L^2(\mathbb{T})$ is a Cauchy sequence in the L^2-norm, then there is $F \in L^2(\mathbb{T})$ for which* $\lim \|F - F_n\|_{L^2(\mathbb{T})} = 0$. Zygmund refers to the Riesz–Fischer Theorem as "a great achievement of the Lebesgue theory".

    Fischer's formulation is a special case of Theorem A.18, which itself is a staple in a basic real variables course, e.g., [Ben76, pages 232–233], [HS65, pages 192–194], [Rud66, pages 66–67]. We shall use Theorem A.18 in the $L^2(\mathbb{T})$ theory, which follows.

3.4.12 Theorem. ONB, Parseval Formula, and Convergence

    a.   *ONB. $\{e^{-\pi i n\gamma/\Omega}\}$ is an ONB for $L^2(\mathbb{T}_{2\Omega})$.*
    b.   *Parseval formula. Let $F, G \in L^2(\mathbb{T}_{2\Omega})$, and consider the pairings $f \longleftrightarrow F$, $g \longleftrightarrow G$. Then*

$$\int_{\mathbb{T}_{2\Omega}} F(\gamma)\overline{G(\gamma)}\, d\gamma = \sum f[n]\overline{g[n]},$$

and, in particular,

$$\|F\|_{L^2(\mathbb{T}_{2\Omega})} = \left(\int_{\mathbb{T}_{2\Omega}} |F(\gamma)|^2 d\gamma\right)^{1/2} = \left(\sum |f[n]|^2\right)^{1/2} = \|f\|_{\ell^2(\mathbb{Z})}.$$

c. *Convergence.* For all $F \in L^2(\mathbb{T}_{2\Omega})$,

$$\lim_{N \to \infty} \|F - S_N(F)\|_{L^2(\mathbb{T}_{2\Omega})} = 0. \tag{3.4.21}$$

PROOF. ***i.*** $\{e^{-\pi i n\gamma/\Omega}\}$ is orthornormal in $L^2(\mathbb{T}_{2\Omega})$ by direct calculation, and $\overline{\text{span}}\{e^{-\pi i n\gamma/\Omega}\} = L^2(\mathbb{T}_{2\Omega})$ by the Fejér Theorem (Theorem 3.4.6*a*) and the method of Exercise 3.16.
 ii. For any $F \in L^2(\mathbb{T}_{2\Omega})$ and any N

$$0 \le \|F - S_N(F)\|^2_{L^2(\mathbb{T}_{2\Omega})} = \|F\|^2_{L^2(\mathbb{T}_{2\Omega})} - \sum_{|n| \le N} |f[n]|^2,$$

and so

$$\sum_{|n| \le N} |f[n]|^2 \le \|F\|^2_{L^2(\mathbb{T}_{2\Omega})}. \tag{3.4.22}$$

This is *Bessel's Inequality* (1828). Further, if $N > M$, then

$$\|S_N(F) - S_M(F)\|^2_{L^2(\mathbb{T}_{2\Omega})} = \sum_{M < |n| \le N} |f[n]|^2,$$

and so, by (3.4.22), $\{S_N(F)\}$ is a Cauchy sequence in $L^2(\mathbb{T}_{2\Omega})$. Thus, $\sum f[n] e^{-\pi i n\gamma/\Omega}$ converges to some $K \in L^2(\mathbb{T}_{2\Omega})$ since $L^2(\mathbb{T}_{2\Omega})$ is complete.
 iii. By *i*, if $G \in L^2(\mathbb{T}_{2\Omega})$ and

$$\forall n \in \mathbb{Z}, \qquad \int_{\mathbb{T}_{2\Omega}} G(\gamma) e^{\pi i n\gamma/\Omega} d\gamma = 0,$$

then (3.4.17) allows us to conclude that $\|G\|_{L^2(\mathbb{T}_{2\Omega})} = 0$, and so G is the 0-function.
 iv. Now, for any $F \in L^2(\mathbb{T}_{2\Omega})$ and corresponding K as in *ii*, we have (by (3.4.17) again)

$$\int_{\mathbb{T}_{2\Omega}} (F(\gamma) - K(\gamma)) e^{\pi i n\gamma/\Omega} d\gamma = f[n] - \lim_{N \to \infty} \sum_{|m| \le N} f[m] \int_{\mathbb{T}_{2\Omega}} e^{-\pi i (m-n)\gamma/\Omega} d\gamma$$

$$= 0.$$

Therefore, by *iii*, $F = K$ a.e.; and so (3.4.21) is obtained, giving part *c* as well as part *a*.

v.   Part *b* follows from (3.4.17), (3.4.21), and the calculation

$$\int_{\mathbb{T}_{2\Omega}} F(\gamma)\overline{G(\gamma)}d\gamma = \lim_{N\to\infty}\int_{\mathbb{T}_{2\Omega}} S_N(F)(\gamma)\overline{S_N(G)(\gamma)}d\gamma$$

$$= \lim_{N\to\infty} \sum_{|m|,|n|\leq N} f[m]\overline{g[n]} \int_{\mathbb{T}_{2\Omega}} e^{-\pi i(m-n)\gamma/\Omega}d\gamma$$

$$= \sum f[n]\overline{g[n]}. \quad\blacksquare$$

In light of Bessel's Inequality (3.4.22), we know that if $F \in L^2(\mathbb{T})$ for the pairing $f \longleftrightarrow F$, then $f \in \ell^2(\mathbb{Z})$. Riesz' formulation of the Riesz–Fischer Theorem completes the picture as follows.

3.4.13 Theorem. $\ell^2(\mathbb{Z}) \longrightarrow L^2(\mathbb{T})$

There is a unique linear bijection $\mathcal{F} : \ell^2(\mathbb{Z}) \to L^2(\mathbb{T}_{2\Omega})$ with the properties:

a) $\forall f \in \ell^2(\mathbb{Z}), \quad \|f\|_{\ell^2(\mathbb{Z})} = \|\mathcal{F}f\|_{L^2(\mathbb{T}_{2\Omega})};$

b) $\forall F \in L^2(\mathbb{T}_{2\Omega}), \quad f \equiv \mathcal{F}^{-1}F$ *is the sequence of Fourier coefficients of F.*

PROOF.   In light of Corollary 3.4.7 and Theorem 3.4.12, it is sufficient to prove that for any sequence $\{c_n\} \in \ell^2(\mathbb{Z})$ there is $F \in L^2(\mathbb{T}_{2\Omega})$, uniquely determined a.e., such that $\{c_n\}$ is the sequence of Fourier coefficients of F.

If we define $S_N(\gamma) = \sum_{|n|\leq N} c_n e^{-\pi i n\gamma/\Omega}$, then $\{S_N\} \subseteq L^2(\mathbb{T}_{2\Omega})$ is a Cauchy sequence since

$$\|S_N - S_M\|_{L^2(\mathbb{T}_{2\Omega})} = \sum_{M<|n|\leq N} |c_n|^2$$

when $N > M$. By the completeness of $L^2(\mathbb{T}_{2\Omega})$ there is a unique $F \in L^2(\mathbb{T}_{2\Omega})$ for which

$$\lim_{N\to\infty} \|F - S_N\|_{L^2(\mathbb{T}_{2\Omega})} = 0.$$

Further, for each n and for $N \geq |n|$, we have

$$|f[n] - c_n| = \left|\int_{\mathbb{T}_{2\Omega}} (F(\gamma) - S_N(\gamma))e^{\pi i n\gamma/\Omega}d\gamma\right|$$

$$\leq \|F - S_N\|_{L^2(\mathbb{T}_{2\Omega})}$$

by Hölder's Inequality. Thus, $c_n = f[n]$. $\blacksquare$

3.5. $A(\mathbb{T})$ **and the Wiener Inversion Theorem**

Besides the formidable task of finding an effective intrinsic characterization of $A(\mathbb{T})$, e.g., Examples 1.4.4 and 3.3.4, the space $A(\mathbb{T})$ is worthy of careful attention because of its algebraic properties and the ramifications of those properties, cf., the introductory paragraph of Section 3.4. Fortunately, there is an accessible masterpiece on the subject by Kahane [Kah70]. Our modest goal in this section will be to state these algebraic properties and to prove Wiener's theorem on the inversion of absolutely convergent Fourier series. We shall refer to this result as Wiener's Inversion Theorem.

3.5.1 Definition. Convolution

a. Let $f, g \in \ell^1(\mathbb{Z})$. The *convolution* of f and g, denoted by $f * g$, is

$$f * g[n] = \sum_{k=-\infty}^{\infty} f[n-k]g[k] = \sum_{k=-\infty}^{\infty} f[k]g[n-k].$$

More simply than the cases of $L^1(\mathbb{R})$ and $L^1(\mathbb{T}_{2\Omega})$, we see that $f * g \in \ell^1(\mathbb{Z})$ since

$$\sum_{n=-\infty}^{\infty} |f * g[n]| \le \sum_{k=-\infty}^{\infty}\sum_{n=-\infty}^{\infty} |f[n-k]g[k]| = \sum |f[n]| \sum |g[k]|.$$

Rewriting this expression, we have

$$\forall f, g \in \ell^1(\mathbb{Z}), \qquad \|f * g\|_{\ell^1(\mathbb{Z})} \le \|f\|_{\ell^1(\mathbb{Z})}\|g\|_{\ell^1(\mathbb{Z})}. \tag{3.5.1}$$

b. $\ell^1(\mathbb{Z})$ is a *commutative algebra* taken with the operations of addition and convolution, i.e., $\ell^1(\mathbb{Z})$ is a complex vector space under addition, and convolution is distributive with respect to addition, as well as being associative, and commutative.

Further, $\ell^1(\mathbb{Z})$ has a *unit u* under convolution. u is defined by $u[n] = \delta(0, n)$, so that

$$\forall f \in \ell^1(\mathbb{Z}), \qquad f * u = u * f = f$$

since

$$\forall n \in \mathbb{Z}, \qquad \sum_{k=-\infty}^{\infty} f[n-k]\delta(0, k) = f[n].$$

A straightforward calculation yields the following result.

3.5.2 Proposition.

Let $\Omega > 0$ and consider $A(\mathbb{T}_{2\Omega})$ (Definition 3.1.8a).

    a.   *Let $F, G \in A(\mathbb{T}_{2\Omega})$, and let $f = \{f[n]\}$, $g = \{g[n]\} \in \ell^1(\mathbb{Z})$ be the sequences of Fourier coefficients of F and G, i.e., $\hat{f} = F$ and $\hat{g} = G$. Then $f * g \in \ell^1(\mathbb{Z})$ is the sequence of Fourier coefficients of $FG \in A(\mathbb{T}_{2\Omega})$, i.e.,*

$$\forall n \in \mathbb{Z}, \qquad (f * g)[n] = \int_{\mathbb{T}_{2\Omega}} F(\gamma)G(\gamma)e^{\pi in\gamma/\Omega}\,d\gamma.$$

(Recall that "$\int_{\mathbb{T}_{2\Omega}}$" designates "$(1/2\Omega)\int_\alpha^{\alpha+2\Omega}$" for any fixed $\alpha \in \hat{\mathbb{R}}$.)

    b.   *$A(\mathbb{T}_{2\Omega})$ is a commutative algebra under the operations of addition and (ordinary pointwise) multiplication of functions. The function $U \equiv 1 \in A(\mathbb{T}_{2\Omega})$ is the multiplicative unit of $A(\mathbb{T}_{2\Omega})$, and*

$$\forall F, G \in A(\mathbb{T}_{2\Omega}), \qquad \|FG\|_{A(\mathbb{T}_{2\Omega})} \le \|F\|_{A(\mathbb{T}_{2\Omega})}\|G\|_{A(\mathbb{T}_{2\Omega})}. \qquad (3.5.1)'$$

3.5.3 Example. The $A(\mathbb{T})$ Norm

    a.   Let $F \in A(\mathbb{T}_{2\Omega})$, and let $f = \{f[n]\} \in \ell^1(\mathbb{Z})$ be the sequence of Fourier coefficients of F. Then

$$\|F\|_{A(\mathbb{T}_{2\Omega})} = |f[0]| + {\sum}' \frac{1}{|n|}|nf[n]|. \qquad (3.5.2)$$

Using (3.5.2) and Exercise 3.26 we can conclude that *if $F \in A(\mathbb{T}_{2\Omega})$ and if F' exists and is an element of $L^2(\mathbb{T}_{2\Omega})$, then*

$$\|F\|_{A(\mathbb{T}_{2\Omega})} \le |f[0]| + \frac{\Omega}{\sqrt{3}}\|F'\|_{L^2(\mathbb{T}_{2\Omega})}. \qquad (3.5.3)$$

    b.   Let F_ϵ be the 2Ω-periodic triangle function, $\max(1 - |\gamma|/\epsilon, 0)$, defined in Remark 3.1.2b. As we stated there, its sequence $\{w_{(\epsilon)}[n]\}$ of Fourier coefficients is an element of $\ell^1(\mathbb{Z})$, and so $F_\epsilon \in A(\mathbb{T}_{2\Omega})$. Now define the 2Ω-periodic trapezoid function $V_\epsilon = 2F_{2\epsilon} - F_\epsilon$. Note that

$$\|V_\epsilon\|_{A(\mathbb{T}_{2\Omega})} \le 3. \qquad (3.5.4)$$

In fact, since $w_{(\epsilon)} \ge 0$ on $\mathbb{Z}$, we have

$$\|V_\epsilon\|_{A(\mathbb{T}_{2\Omega})} \le 2\|F_{2\epsilon}\|_{A(\mathbb{T}_{2\Omega})} + \|F_\epsilon\|_{A(\mathbb{T}_{2\Omega})}$$

$$= 2\sum w_{(2\epsilon)}[n] + \sum w_{(\epsilon)}[n] = 2F_{2\epsilon}(0) + F_{2\epsilon}(0) = 3.$$

We shall show that *if* $F \in A(\mathbb{T}_{2\Omega})$ *and* $F(0) = 0$ *(and so* $F(2\Omega n) = 0$ *for all* $n \in \mathbb{Z}$*), then*

$$\lim_{\epsilon \to 0} \|F V_\epsilon\|_{A(\mathbb{T}_{2\Omega})} = 0. \tag{3.5.5}$$

First, $F V_\epsilon \in A(\mathbb{T}_{2\Omega})$ by Proposition 3.5.2. If $F \in C^1(\mathbb{T}_{2\Omega})$, then $F \in A(\mathbb{T}_{2\Omega})$ by (3.5.3). Also, the pointwise a.e. derivative $(F V_\epsilon)'$ is not only supported by $[-2\epsilon, 2\epsilon]$ on $[-\Omega, \Omega]$, but it is uniformly bounded independent of ϵ. In fact,

$$\|(F V_\epsilon)'\|_{L^\infty(\mathbb{T}_{2\Omega})} \leq \|F'\|_{L^\infty(\mathbb{T}_{2\Omega})} + \sup_{\gamma \in [-2\epsilon, -\epsilon] \cup [\epsilon, 2\epsilon]} \frac{1}{\epsilon} |F(\gamma)|,$$

and the second term on the right side is bounded independent of ϵ since $F(\gamma) = F(\gamma) - F(0)$ and the mean value theorem applies. Further, it is clear that

$$\lim_{\epsilon \to 0} \int_{-2\epsilon}^{2\epsilon} F(\gamma) V_\epsilon(\gamma) \, d\gamma = 0.$$

Thus, remembering the support of $(F V_\epsilon)'$, we can apply (3.5.3) again to obtain (3.5.5) for $F \in C^1(\mathbb{T}_{2\Omega})$.

For arbitrary $F \in A(\mathbb{T}_{2\Omega})$, we define the trigonometric polynomials

$$F_N(\gamma) = \sum_{1 \leq |n| \leq N} \check{F}[n] e^{-\pi i n \gamma / \Omega} + a_N[0],$$

where $a_n[0] \equiv -\sum_{1 \leq |n| \leq N} \check{F}[n]$. Thus, $F_N \in C^1(\mathbb{T}_{2\Omega})$ and $F_N(0) = 0$, so the result of the previous paragraph can be used to obtain

$$\lim_{\epsilon \to 0} \|F_N V_\epsilon\|_{A(\mathbb{T}_{2\Omega})} = 0.$$

Therefore, since

$$\|F V_\epsilon\|_{A(\mathbb{T}_{2\Omega})} \leq \|(F - F_N) V_\epsilon\|_{A(\mathbb{T}_{2\Omega})} + \|F_N V_\epsilon\|_{A(\mathbb{T}_{2\Omega})},$$

we can further apply Proposition 3.5.2 and (3.5.4) to compute

$$\overline{\lim_{\epsilon \to 0}} \|F V_\epsilon\|_{A(\mathbb{T}_{2\Omega})} \leq 3 \|F - F_N\|_{A(\mathbb{T}_{2\Omega})}.$$

The left side is independent of N and the right side is

$$3 \left\| \check{F}[0] - \sum_{|n| > N} \check{F}[n] e^{-\pi i n \gamma / \Omega} + \sum_{1 \leq |n| \leq N} \check{F}[n] \right\|_{A(\mathbb{T}_{2\Omega})}$$

$$\leq 3 \left| \sum_{|n| \leq N} \check{F}[n] \right| + 3 \sum_{|n| > N} |\check{F}[n]|.$$

As $N \to \infty$, the first term tends to 0 since $F(0) = 0$, and the second term tends to 0 since $F \in A(\mathbb{T}_{2\Omega})$.

The proof of (3.5.5) is complete.

There are far-reaching generalizations of (3.5.5) related to spectral synthesis and the ideal structure of $L^1(\mathbb{T})$, e.g., [Ben75, Section 1.2].

3.5.4 Proposition.

$A(\mathbb{T}_{2\Omega}) = L^2(\mathbb{T}_{2\Omega}) * L^2(\mathbb{T}_{2\Omega})$.

PROOF. The inclusion $L^2(\mathbb{T}_{2\Omega}) * L^2(\mathbb{T}_{2\Omega}) \subseteq A(\mathbb{T}_{2\Omega})$ is a consequence of Parseval's formula and Hölder's Inequality. In fact,

$$\sum |f[n]g[n]| \leq \left(\sum |f[n]|^2\right)^{1/2} \left(\sum |g[n]|^2\right)^{1/2} < \infty,$$

where $F, G \in L^2(\mathbb{T}_{2\Omega})$ and $\hat{f} = F$, $\hat{g} = G$; and so $fg \in \ell^1(\mathbb{Z})$.

For the inclusion, $A(\mathbb{T}_{2\Omega}) \subseteq L^2(\mathbb{T}_{2\Omega}) * L^2(\mathbb{T}_{2\Omega})$, let $F \in A(\mathbb{T}_{2\Omega})$, where $\hat{f} = F$. For each n, we can write $f[n] = (g[n])^2$ for some $g[n] \in \mathbb{C}$; and we define the sequence $g = \{g[n]\}$. $g \in \ell^2(\mathbb{Z})$ since $f \in \ell^1(\mathbb{Z})$, and, hence, $F = \hat{g} * \hat{g} \in L^2(\mathbb{T}_{2\Omega}) * L^2(\mathbb{T}_{2\Omega})$. ∎

3.5.5 Remark. Factorization

a. The factorization, $A(\mathbb{T}) = L^2(\mathbb{T}) * L^2(\mathbb{T})$, or, equivalently, $\ell^1(\mathbb{Z}) = \ell^2(\mathbb{Z})\ell^2(\mathbb{Z})$, of Proposition 3.5.4 is elementary. A consequence of Proposition 3.5.2 is the inclusion $A(\mathbb{T})A(\mathbb{T}) \subseteq A(\mathbb{T})$. In this context, we observe that $A(\mathbb{T}) = A(\mathbb{T})A(\mathbb{T})$ or, equivalently, $\ell^1(\mathbb{Z}) = \ell^1(\mathbb{Z}) * \ell^1(\mathbb{Z})$, is also valid since $U \equiv 1 \in A(\mathbb{T})$.

Further, $A(\mathbb{Z}) = A(\mathbb{Z})A(\mathbb{Z})$ and $A(\hat{\mathbb{R}}) = A(\hat{\mathbb{R}})A(\hat{\mathbb{R}})$. However, these two results are far from trivial and are due to Salem and Rudin, respectively. Paul Cohen (1959) proved that $A(\Gamma) = A(\Gamma)A(\Gamma)$ for any locally compact abelian group Γ, see [Koo64], [Ptá72] for elegant proofs of the Cohen Factorization Theorem. Using Salem's Theorem, or an argument with convex functions, we also have $A(\mathbb{T}) = L^1_+(\mathbb{T}) * A(\mathbb{T})$, where $L^1_+(\mathbb{T}) = \{F \in L^1(\mathbb{T}) : F \geq 0\}$.

b. Although the proof that $A(\mathbb{T}) = A(\mathbb{T})A(\mathbb{T})$ is elementary, we have the following more intricate relationship: *if $F \in A(\mathbb{T})$ never vanishes, then*

$$\forall H \in A(\mathbb{T}), \ \exists G \in A(\mathbb{T}) \quad \text{such that} \quad H = FG.$$

In particular, *if $F \in A(\mathbb{T})$ never vanishes, then $1/F \in A(\mathbb{T})$*. This last fact is *Wiener's Inversion Theorem*. There are Banach algebra proofs [Ben75, pages 22–23], a "spectral radius" proof [Ben75, pages 23–24], extensive generalizations that are documented and compared in [Ben75], and classical proofs going back to Wiener's original techniques [Wie81, Volume II, pages 519–623, esp., page 532], [Wie33, page 91]. We shall proceed in this last direction in Sections 3.5.6–3.5.9.

c. Before proving Wiener's Inversion Theorem, let us point out that a modified version of it is a natural component in the proof of Wiener's Tauberian Theorem, Theorem 2.9.12 and Remark 2.9.13, cf., the historical remark in [Ben75, pages 142–143] and the proofs in [Ben75, Sections 1.1–1.4]. Also, as indicated earlier, these results are fundamental in spectral synthesis, e.g., [Ben75, Section 2.5]. We shall apply Wiener's Inversion Theorem in Section 3.6.

3.5.6 Definition. Local Membership

Let $I \subseteq A(\mathbb{T}_{2\Omega})$ be an *ideal* in the algebra $A(\mathbb{T}_{2\Omega})$, i.e., I is a subalgebra of $A(\mathbb{T}_{2\Omega})$ and $FG \in I$ whenever $F \in A(\mathbb{T}_{2\Omega})$ and $G \in I$.

A function $F : \mathbb{T}_{2\Omega} \to \mathbb{C}$ belongs to I locally at $\gamma \in \mathbb{T}_{2\Omega}$ if

$$\exists G_\gamma \in I \quad \text{and} \quad \exists N_\gamma = (\alpha, \beta) \subseteq \mathbb{T}_{2\Omega} \quad \text{such that} \quad \gamma \in (\alpha, \beta) \quad \text{and}$$

$$\forall \lambda \in N_\gamma, G_\gamma(\lambda) = F(\lambda).$$

In this case we write $F \in I_{\mathrm{loc}}(\gamma)$.

3.5.7 Theorem. Local Membership Theorem

Let $I \subseteq A(\mathbb{T}_{2\Omega})$ be an ideal, and let $F : \mathbb{T}_{2\Omega} \to \mathbb{C}$ be a function. If $F \in I_{\mathrm{loc}}(\gamma)$ for each $\gamma \in \mathbb{T}_{2\Omega}$, then $F \in I$.

PROOF. For each $\gamma \in \mathbb{T}_{2\Omega}$ choose G_γ and N_γ (as in Definition 3.5.6) for which $G_\gamma = F$ on N_γ. Clearly we can take N_γ centered at γ; and for each γ we choose a closed interval $C_\gamma \subseteq N_\gamma$, also centered at γ and whose length $|C_\gamma|$ equals $\frac{1}{2}|N_\gamma|$. Since $\mathbb{T}_{2\Omega}$ is a compact set, we can find $\gamma_1, \ldots, \gamma_n$ so that

$$\mathbb{T}_{2\Omega} = \bigcup_{j=1}^{n} C_j, \tag{3.5.6}$$

e.g., Definition B.1.

Next, choose $V_j \in A(\mathbb{T}_{2\Omega})$, $j = 1, \ldots, n$, where $V_j = 1$ on C_{γ_j} and $V_j = 0$ off N_{γ_j}, e.g., Example 3.5.3, cf., Exercise 1.50 and the general construction in [Ben75, Proposition 1.1.4]. Since I is an ideal, we have $V_j G_j \in I$ for each $j = 1, \ldots, n$. It is also clear that

$$\forall \gamma \in \mathbb{T}_{2\Omega} \quad \text{and} \quad \forall j = 1, \ldots, n, \qquad V_j(\gamma)G(\gamma) = V_j(\gamma)F(\gamma). \tag{3.5.7}$$

Defining

$$F_0 = F(1 - (1 - V_1)(1 - V_2) \cdots (1 - V_n)), \tag{3.5.8}$$

we see that $F_0 \in I$ because of (3.5.7) and the fact that the "1's" cancel when we compute the right side of (3.5.8). Finally, $F_0 = F$ on $\mathbb{T}_{2\Omega}$. In fact, if $\gamma \in \mathbb{T}_{2\Omega}$, then there is k for which $1 - V_k(\gamma) = 0$ because of (3.5.6). ∎

3.5.8 Example. $2|f[0]| > \|F\|_{A(\mathbb{T}_{2\Omega})}$ **Implies** $1/F \in A(\mathbb{T}_{2\Omega})$

Let $F \in A(\mathbb{T}_{2\Omega}) \backslash \{0\}$, and let $f = \{f[n]\}$ be the sequence of Fourier coefficients of F. Assume

$$2|f[0]| > \|F\|_{A(\mathbb{T}_{2\Omega})}. \tag{3.5.9}$$

We shall show that $1/F \in A(\mathbb{T}_{2\Omega})$. To this end we combine the series expansion, $1/(1+G) = 1 - G + G^2 - G^3 + \cdots$, and (3.5.9) to compute for $G(\gamma) \equiv \sum' \frac{f[n]}{f[0]} e^{-\pi i n \gamma / \Omega}$ that

$$\frac{1}{F(\gamma)} = \frac{1}{f[0]} (1 - G(\gamma) + G(\gamma)^2 - G(\gamma)^3 + \cdots).$$

Thus, by (3.5.9) and Proposition 3.5.2b,

$$\left\| \frac{1}{F} \right\|_{A(\mathbb{T}_{2\Omega})} \leq \frac{1}{|f[0]|} \sum_{n=0}^{\infty} \|G\|_{A(\mathbb{T}_{2\Omega})}^n$$

$$= \frac{1}{|f[0]|} \frac{1}{1 - \|G\|_{A(\mathbb{T}_{2\Omega})}} = \frac{1}{2|f[0]| - \|F\|_{A(\mathbb{T}_{2\Omega})}} < \infty.$$

3.5.9 Theorem. Wiener Inversion Theorem

Let $F \in A(\mathbb{T}_{2\Omega})$.
   a. If $F(\gamma_0) \neq 0$, then there is $G \in A(\mathbb{T}_{2\Omega})$ such that $F = G$ on some open interval N centered at γ_0 and $1/G \in A(\mathbb{T}_{2\Omega})$.
   b. If F never vanishes, then $1/F \in A(\mathbb{T}_{2\Omega})$.

PROOF.   *a.*   Without loss of generality, let $\gamma_0 = 0$, and define

$$\forall \gamma \in \mathbb{T}_{2\Omega}, \qquad G_\epsilon(\gamma) = F(0) + V_\epsilon(\gamma)(F(\gamma) - F(0)), \tag{3.5.10}$$

where V_ϵ was defined in Example 3.5.3. Choose $\eta \equiv |F(0)|/3 > 0$, and apply (3.5.5) to find $\epsilon > 0$ for which $\|V_\epsilon(F - F(0))\|_{A(\mathbb{T}_{2\Omega})} < \eta$. For this ϵ, set $G = G_\epsilon$.
   Since

$$\left| \int_{\mathbb{T}_{2\Omega}} V_\epsilon(\gamma)(F(\gamma) - F(0)) \, d\gamma \right| \leq \|V_\epsilon(F - F(0))\|_{A(\mathbb{T}_{2\Omega})},$$

we have

$$\left| \int_{\mathbb{T}_{2\Omega}} G(\gamma)\, d\gamma \right| = \left| F(0) + \int_{\mathbb{T}_{2\Omega}} V_\epsilon(\gamma)(F(\gamma) - F(0))\, d\gamma \right| \tag{3.5.11}$$

$$\geq |F(0)| - \|V_\epsilon(F - F(0))\|_{A(\mathbb{T}_{2\Omega})} > |F(0)| - \eta.$$

On the other hand, it is immediate from (3.5.10) that

$$\|G\|_{A(\mathbb{T}_{2\Omega})} < |F(0)| + \eta. \tag{3.5.12}$$

Combining (3.5.11) and (3.5.12) with the definition of η, we obtain

$$2 \left| \int_{\mathbb{T}_{2\Omega}} G(\gamma)\, d\gamma \right| > \frac{4}{3}|F(0)| > \|G\|_{A(\mathbb{T}_{2\Omega})}. \tag{3.5.13}$$

From (3.5.10) we see that $G = F$ on $N_0 = (-\epsilon, \epsilon)$, and, because of (3.5.13) and Example 3.5.8, $1/G \in A(\mathbb{T}_{2\Omega})$. This completes the proof of part *a*.

b. For each $\gamma \in \mathbb{T}_{2\Omega}$, we use part *a* to choose $G_\gamma \in A(\mathbb{T}_{2\Omega})$ such that $G_\gamma = F$ on some open interval N_γ centered at γ and $1/G_\gamma \in A(\mathbb{T}_{2\Omega})$.

Thus, $1/F \in A(\mathbb{T}_{2\Omega})_{\text{loc}}(\gamma)$ for each $\gamma \in \mathbb{T}_{2\Omega}$, and so $1/F \in A(\mathbb{T}_{2\Omega})$ by Theorem 3.5.7. ∎

3.5.10 Remark. $A(\mathbb{T})$ and $A(\hat{\mathbb{R}})$

a. Wiener's concept of *local membership* leads one to investigate the relationship between $A(\mathbb{T})$ and $A(\hat{\mathbb{R}})$. For example, it is natural to ask the question: supposing $F \in A(\mathbb{T})$, $F(\pm 1/2) = 0$, and G is defined on $\hat{\mathbb{R}}$ as

$$G(\gamma) = \begin{cases} F(\gamma), & \text{if } |\gamma| \leq \dfrac{1}{2}, \\ 0, & \text{otherwise}, \end{cases} \tag{3.5.14}$$

is $G \in A(\hat{\mathbb{R}})$? Wiener proved the following theorem. *Let $F : \mathbb{T} \to \mathbb{C}$ be a function vanishing on $[\frac{1}{2} - \epsilon, \frac{1}{2} + \epsilon]$, and define $G : \hat{\mathbb{R}} \to \mathbb{C}$ by (3.5.14); then $F \in A(\mathbb{T})$ if and only if $G \in A(\hat{\mathbb{R}})$, and*

$$\exists C_1(\epsilon), C_2(\epsilon) > 0 \quad \text{such that } C_1(\epsilon)\|G\|_{A(\hat{\mathbb{R}})} \leq \|F\|_{A(\mathbb{T})} \leq C_2(\epsilon)\|G\|_{A(\hat{\mathbb{R}})}.$$

The proof is not difficult, e.g., [Wie33, pages 80–82].

b. The following extension of Wiener's result from part *a* was proved by Wik [Wik65]. *Let $F \in L^\infty(\mathbb{T})$ vanish on $[\frac{1}{2} - \epsilon, \frac{1}{2} + \epsilon]$, define $G : \hat{\mathbb{R}} \to \mathbb{C}$ by (3.5.14), and let $w : \mathbb{R} \to \mathbb{R}$ be an even positive function, which is increasing on $(0, \infty)$ and which satisfies the condition*

$$\exists C > 0 \quad \text{such that} \quad \forall t \in \mathbb{R} \backslash \{0\}, \ w(2t) \le Cw(t). \tag{3.5.15}$$

Then

$$\sum |\check{F}[n]| w(n) < \infty$$

if and only if

$$\int \check{G}(t) w(t) \, dt < \infty.$$

In this statement, we use "*w*" to denote a so-called *weight*, not the Fejér function.

c. Using the result of part *b*, Wik [Wik65] proved that *if* $F \in A(\mathbb{T})$, $-\frac{1}{2} < \alpha < \beta < \frac{1}{2}$, $F(\alpha) = F(\beta) = 0$, *and*

$$\sideset{}{'}\sum |\check{F}[n]| \log |n| < \infty,$$

then $F_{\alpha,\beta} \in A(\mathbb{T})$ *where*

$$\forall \gamma \in \left[-\frac{1}{2}, \frac{1}{2} \right), \quad F_{\alpha,\beta}(\gamma) = F(\gamma) \mathbf{1}_{[\alpha,\beta]}(\gamma)$$

and $F_{\alpha,\beta}$ *is defined* 1-*periodically on* $\hat{\mathbb{R}}$.

d. Condition (3.5.15) is a *doubling* condition for weights. Such conditions play a conceptually important role in an interesting and unresolved set of problems categorized as *weighted norm inequality problems*, e.g., [G-CRdeF85]. An example of a weighted norm inequality is

$$\left(\int |\hat{f}(\gamma)|^2 \, d\mu(\gamma) \right)^{1/2} \le C \left(\int |f(t)|^2 \, w(t) \, dt \right)^{1/2}, \tag{3.5.16}$$

where $w > 0$ and μ is a positive measure, cf., Definition 2.6.5 and Theorem 3.7.2, which are applicable raisons d'être for dealing with such inequalities. The problem is to characterize the relationship between w and μ so that (3.5.16) is valid for a large class of functions, e.g., [BH92] for measure weights μ.

e. Wik's Theorem from part *c* can be thought of in terms of local membership *or* weighted norm inequalities. In the context of *local membership*, we can obtain $F_{\alpha,\beta} \in A(\mathbb{T})$ by Theorem 3.5.7 if we can show $F_{\alpha,\beta} \in A(\mathbb{T})_{\text{loc}}(\gamma)$ for $\gamma = \alpha, \beta$, since local membership is obvious for other values of γ. In the context of *weighted norm inequalities*, define $w(n) = \log |n|$ and $Fv = F_{\alpha,\beta}$, i.e., $v = \mathbf{1}_{[\alpha,\beta]}$ on $[-\frac{1}{2}, \frac{1}{2})$; and consider the weighted norm inequality

$$\|Fv\|_{A(\mathbb{T})} \le C \sideset{}{'}\sum |\check{F}[n]| w(n), \tag{3.5.17}$$

for all $F \in A(\mathbb{T})$ for which $F(\alpha) = F(\beta) = 0$. Then Wik's Theorem can be restated by saying that if the right side of (3.5.17) is finite, then $Fv = F_{\alpha,\beta} \in A(\mathbb{T})$.

f. *Geometrical* considerations play a significant role in a class of weighted norm inequalities referred to as *restriction theory* , e.g., [Ash76, pages 107–117 by E. M. Stein], [Ste93]. Also, extensions of the *classical uncertainty principle inequality* (2.8.5) are critical in quantifying the implications of inequalities such as (3.5.16), e.g., [BF94, Chapter 7], [BL94].

3.6. Maximum Entropy and Spectral Estimation

We shall discuss the Maximum Entropy Theorem and prove a spectral estimation theorem. In so doing we shall prove the Fejér–Riesz Theorem and indicate the role of $A(\mathbb{T})$ in such matters.

3.6.1 Remark. The Spectral Estimation and Extension Problem

a. We gave a qualitative statement of the spectral estimation problem in Definition 2.8.6*b*. We shall now aim to quantify that statement for both the stochastic setting of Section 2.8 and the deterministic setting of Section 2.9. As a first step, we say that the *spectral estimation problem* is to estimate the power spectrum in terms of given autocorrelation data on a finite interval.

In the context of Fourier series, we are given $N > 0$ and data $X_N \equiv \{r_n : n = 0, \pm 1, \ldots, \pm N\} \subseteq \mathbb{C}$, and the *extension problem* associated with spectral estimation is to find nonnegative functions $S \in L^1(\mathbb{T})$ for which $\check{S}[n] = r_n$ for $n = 0, \pm 1, \ldots, \pm N$. Because of Herglotz' Theorem on $\mathbb{Z}$ (Theorem 2.7.10), X_N must satisfy some positive definiteness condition, e.g., Definition 3.6.2. With this stipulation on X_N there are generally *many* nonnegative solutions $S \in L^1(\mathbb{T})$ as Krein (1940) first showed in the setting of $\mathbb{R}$. After Krein, contributions to the extension problem were made by Chover, Doss, Dym, Gohberg, R. R. Goldberg, Rudin, et al.; and [Rud63] also analyzed the difficulties in extending Krein's Theorem to higher dimensions. The fact that there can be many solutions to the extension problem leads to Fourier uniqueness problems in the spirit of Example 1.10.6, e.g., [Pric85, pages 149–170].

b. From the point of view of spectral estimation there are various ways of choosing a specific solution S from part *a*, depending on the type of application. One procedure, the *Maximum Entropy Method* (MEM), involves choosing the function $S = S_{\mathrm{MEM}}$, which maximizes a certain logarithmic integral associated with entropy, e.g., Theorem 3.6.3.

Mathematically, we shall see that this choice restricts us to $A(\mathbb{T})$ instead of $L^1(\mathbb{T})$. Physically, since entropy is a measure of disorder in a system, S_{MEM} represents maximum uncertainty with regard to what we do not know about the system, whereas it depends on all the known autocorrelation data X_N. Thus, the choice of S_{MEM} is a mathematical guarantee that the least number of assumptions has been made regarding the information content of the unmeasured data at $|n| >$ N, e.g., [Chi78], [IEEE82] for expert physical rationales and expositions.

John Parker Burg invented MEM in 1967, and van den Bos (1971) [Chi78, pages 92–93] showed that the Maximum Entropy Method of choosing a *spectral estimator S* is equivalent to least squares linear prediction, used in speech processing, and autoregression, used in statistics, e.g., Section 3.7. MEM is also related to the maximum likelihood method, e.g., [Chi78, page 3 and pages 132–133]. There is a deep study of MEM and moment problems by Landau [Lan87], cf., Definition 2.7.8 c. There is also an important new mathematical contribution by Gabardo [Gab93], cf., our extension of MEM to $\mathbb{R}$ in *A quantitative maximum entropy theorem for the real line, Integral Equations and Operator Theory* **10** (1987), 761–779.

3.6.2 Definition. Positive Definite Matrices

a.   An $(N + 1) \times (N + 1)$ matrix $R = (r_{jk})$, where $r_{jk} \in \mathbb{C}$ and $0 \le j, k \le N$, is *Hermitian* if $r_{jk} = \overline{r_{kj}}$. An $(N + 1) \times (N + 1)$ matrix $R = (r_{jk})$ is *positive semidefinite* if

$$\forall\, c = (c_0, c_1, \dots, c_N) \in \mathbb{C}^{N+1}, \quad \langle Rc, c \rangle = \sum_{j,k} r_{jk} c_k \overline{c}_j \ge 0.$$

Positive semidefinite matrices are Hermitian, e.g., Exercise 2.52.

b.   An $(N + 1) \times (N + 1)$ positive semidefinite matrix R is *positive definite*, written $R \gg 0$, if $\langle Rc, c \rangle = 0$ implies $c = 0$. Clearly, if $R \gg 0$, then R is nonsingular and R^{-1} exists. In fact, if $Rc = 0$, then $\langle Rc, c \rangle = 0$, and so $c = 0$ by hypothesis; thus, $R : \mathbb{C}^{N+1} \to \mathbb{C}^{N+1}$ is a linear injection and we have the result.

c.   Let R be an $(N + 1) \times (N + 1)$ matrix with eigenvalues $\{\lambda_0, \dots, \lambda_N\}$. By definition, the *trace* of R is $\sum_{j=0}^{N} r_{jj}$. It can be shown that the trace of R equals $\sum_{j=0}^{N} \lambda_j$ and that the determinant of R is $\prod_{j=0}^{N} \lambda_j$. Also, R is Hermitian if and only if

$$\forall c, d \in \mathbb{C}^{N+1}, \quad \langle Rc, d \rangle = \langle c, Rd \rangle;$$

and the eigenvalues of an Hermitian matrix are real.

Finally, if R is Hermitian, then $R \gg 0$ if and only if each eigenvalue $\lambda_j > 0$, e.g., [Str88], cf., part *a*.

d.   Let $\{r_j : j = 0, \pm 1, \pm 2, \dots, \pm N\} \subseteq \mathbb{C}$ satisfy the condition $r_j = \overline{r}_{-j}$ for each j; and define the $(N + 1) \times (N + 1)$ matrix $R = (r_{jk})$, where $j, k \ge 0$ and $r_{jk} \equiv r_{j-k}$. R is a *Toeplitz matrix*, i.e., it takes constant values on "diagonals of negative slope". R is Hermitian since $r_j = \overline{r}_{-j}$. From the previous discussion, $\sum_{j=0}^{N} \lambda_j = (N + 1)r_0$; and if $R \gg 0$, then the determinant of R is positive, $r_0 > 0$, and $R^{-1} \equiv (c_{jk})$ exists, cf., Exercise 3.52.

e.   *Let* $S \in L^1(\mathbb{T}) \backslash \{0\}$ *be nonnegative, and let* $s = \{s[n]\}$ *be the sequence of Fourier coefficients of S. Then, for each N, the* $(N + 1) \times (N + 1)$ *matrix*

$R = (s[j - k])$, where $0 \leq j, k \leq N$, is positive definite. In fact,

$$\sum_{0 \leq j,k \leq N} s[j - k] c_k \bar{c}_j = \int_{\mathbb{T}} S(\gamma) \left| \sum_{j=0}^{N} c_j e^{-2\pi i j \gamma} \right|^2 d\gamma > 0$$

for all $(c_0, \ldots, c_N) \in \mathbb{C}^{N+1} \setminus \{0\}$, since $\sum_{j=0}^{N} c_j e^{-2\pi i j \gamma} = 0$ for at most finitely many points.

3.6.3 Theorem. The Maximum Entropy Theorem

Let $\{r_j : j = 0, \pm 1, \pm 2, \ldots, \pm N\} \subseteq \mathbb{C}$ satisfy the condition $r_j = \bar{r}_{-j}$ for each j, and assume the $(N + 1) \times (N + 1)$ matrix $R = (r_{jk}) >> 0$, where $j, k \geq 0$ and $r_{jk} \equiv r_{j-k}$. There is a unique function $S \in A(\mathbb{T})$, with Fourier coefficients $\{s[n]\} \in \ell^1(\mathbb{Z})$, satisfying the following properties:

a) $\forall |n| \leq N$, $\quad s[n] = r_n$;

b) $S > 0$ *on* $\mathbb{T}$, *and, hence (by Theorem 3.5.9),* $S^{-1} \in A(\mathbb{T})$;

c) $S = |S_+|^2$ *where* $S_+ \in A(\mathbb{T})$ *has the form*

$$S_+(\gamma) = \sum_{n=0}^{\infty} s_+[n] e^{-2\pi i n \gamma};$$

d) $\forall |n| > N$, $\quad (S^{-1})^{\vee}[n] = 0$ *and*

$$S(\gamma) = 1 \left/ \sum_{n=-N}^{N} \left(\frac{1}{c_{00}} \sum_{m=0}^{N} \bar{c}_{0m} c_{0,m-n} \right) e^{-2\pi i n \gamma} \right.,$$

where $R^{-1} = (c_{mn})$;

e) *for all $F \in A(\mathbb{T})$, for which $F > 0$ on $\mathbb{T}$ and $\check{F}[n] = r_n$ when $|n| \leq N$, we have*

$$\int_{\mathbb{T}} \log F(\gamma) \, d\gamma \leq \int_{\mathbb{T}} \log S(\gamma) \, d\gamma,$$

and equality is obtained if and only if $F = S$.

Our proof of Theorem 3.6.3 in [Pric85, pages 95–97] depends on the fact that *if the matrix R of Theorem 3.6.3 is positive definite and if $(a_0, a_1, \ldots, a_N)^T = R^{-1}(1, 0, 0, \ldots, 0)^T$, then*

$$\forall \gamma \in \mathbb{T}, \qquad \sum_{n=0}^{N} a_j e^{-2\pi i n \gamma} \neq 0,$$

e.g., [GS58], [GL94], cf., [DG79] for an important extension. We shall not prove Theorem 3.6.3 since we shall not prove this fact. Instead, we shall prove Theorem 3.6.6 below, which is essentially Theorem 3.6.3 and which depends on the Fejér–Riesz Theorem.

3.6.4 Theorem. Fejér–Riesz Theorem

Let

$$A(\gamma) = \sum_{n=-N}^{N} a_n e^{-\pi i n \gamma / \Omega} \geq 0 \quad on \quad \mathbb{T}_{2\Omega},$$

and define $A_*(z) = \sum_{n=-N}^{N} a_n z^n$ so that $A_*(e^{-\pi i \gamma/\Omega}) = A(\gamma)$. There is a unique trigonometric polynomial $B(\gamma) = \sum_{n=0}^{N} b_n e^{-\pi i n \gamma/\Omega}$ with the properties that

$$A = |B|^2 \quad on \quad \mathbb{T}_{2\Omega}, \tag{3.6.1}$$

and if $B_*(z) = 0$, then $|z| \leq 1$.

PROOF. Since A is real, the Fourier coefficients $\{a_n\}$ have the property that $\bar{a}_n = a_{-n}$ for $n = -N, \ldots, 0, 1, \ldots, N$. In particular, $a_0 \in \mathbb{R}$. Without loss of generality, assume $a_{-N} \neq 0$.

Now define the polynomial P_* by

$$P_*(z) = z^N A_*(z) = a_{-N} + a_{-N+1}z + \cdots + a_0 z^N$$
$$+ a_1 z^{N+1} + \cdots + a_N z^{2N}, \quad z \in \mathbb{C}.$$

Clearly,

$$\forall z \in \mathbb{C}\backslash\{0\}, \quad z^{2N} \overline{P_*\left(\frac{1}{\bar{z}}\right)} = P_*(z).$$

Let $P_*(z_0) = 0$. If $z_0 \neq 0$, then $1/\bar{z}_0$ is also a zero of P_*. If $0 < |z_0| < 1$, then by differentiation we see that z_0 and $1/\bar{z}_0$ have the same multiplicity. If $|z_0| = 1$, then z_0 has even multiplicity since $A \geq 0$.

Thus,

$$P_*(z) = C \prod_{j=1}^{p}(z - z_j)\left(z - \frac{1}{\bar{z}_j}\right) \prod_{j=1}^{q}(z - u_j)^2, \tag{3.6.2}$$

where $0 < |z_j| < 1$, $|u_j| = 1$, and $2p + 2q = 2N$, i.e., $p + q = N$, cf., Exercise 3.28i.

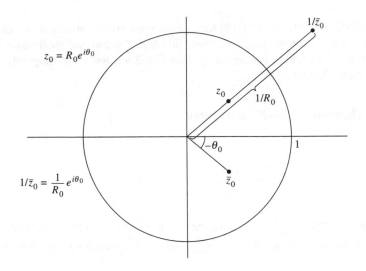

FIGURE 3.2.

Since $A \geq 0$, (3.6.2) allows us to write

$$A(\gamma) = |A_*(z)| = |z^N P_*(z)| = |P_*(z)|$$

$$= |C| \prod_{j=1}^{p} |z - z_j| \left| z - \frac{1}{\overline{z}_j} \right| \prod_{j=1}^{q} |z - u_j|^2 \tag{3.6.3}$$

for $z = e^{-\pi i \gamma / \Omega}$, where $p + q = N$. Because $|z - (1/\overline{z}_j)| = |(z - z_j)/z_j|$ for such z on the unit circle, (3.6.3) becomes

$$A(\gamma) = \left| C^{1/2} \prod_{j=1}^{p} (e^{-\pi i \gamma / \Omega} - z_j) z_j^{-1/2} \prod_{j=1}^{q} (e^{-\pi i \gamma / \Omega} - u_j) \right|^2 .$$

Equation (3.6.1) is obtained by setting

$$B(\gamma) = C^{1/2} \prod_{j=1}^{p} (e^{-\pi i \gamma / \Omega} - z_j) z_j^{-1/2} \prod_{j=1}^{q} (e^{-\pi i \gamma / \Omega} - u_j). \tag{3.6.4}$$

The claim about the zeros of B_* is immediate from (3.6.4). ∎

3.6.5 Remark. Fejér–Riesz Theorem: Potpourri and Titillation

a. The Fejér–Riesz Theorem was proved by Fejér and F. Riesz, and published by Fejér (*J. Reine Angew. Math.* **146** (1915), 53–82, especially pages 55–59), cf.,

[RN55, pages 117–118], [PS76, page 259], [GS58, pages 20–22], [Ach56, pages 152–153].

b. The Fejér–Riesz Theorem is a critical component in the classical proof of the *Spectral Theorem for Unitary Operators* in a Hilbert space, e.g., [RN55, pages 280–284], cf., the historical note on operator-theoretic applications in [Bur79].

c. Herglotz' Theorem (1911) on $\mathbb{Z}$ (Theorem 2.7.10) can be proved as a consequence of the Fejér–Riesz Theorem, e.g., [RN55, pages 115–118], where the context is in terms of the moment problems mentioned in Definition 2.7.8c.

d. Using the Fejér–Riesz Theorem, it is elementary to prove that *if* $F \in C(\mathbb{T}_{2\Omega})$ *is nonnegative, then there is a sequence* $\{B_N\}$ *of trigonometric polynomials on* $\mathbb{T}_{2\Omega}$ *for which*

$$\lim_{N \to \infty} \| F - |B_N|^2 \|_{L^\infty(\mathbb{T}_{2\Omega})} = 0,$$

cf., Exercise 3.48.

e. *Krein Theorem.* Using the Fejér–Riesz Theorem, Krein proved that *if* $f \in PW_\Omega$ *is nonnegative, then there is* $b \in PW_{\Omega/2}$ *for which* $f = |b|^2$ *on* $\mathbb{R}$ *and for which the zeros of the entire function*

$$b(z) = \int_{-\Omega/2}^{\Omega/2} \hat{b}(\gamma) e^{2\pi i z \gamma} \, d\gamma$$

are in the half-plane $\text{Im } z \geq 0$, e.g., [Ach56, page 154]. (PW_Ω is the Paley–Wiener space defined in Remark 1.10.8.) In fact, this result is true for a space larger than PW_Ω, e.g., [Ach56, pages 137–152].

f. If $A(\gamma) = \sum_{n=-N}^{N} a_n e^{-2\pi i n \gamma} \geq 0$ on $\mathbb{T}$ and $a_0 = 1$, then $A(0) \leq 2N + 1$. This is an immediate consequence of the fact that $\{a_n\} >> 0$:

$$0 \leq P(0) = \sum_{n=-N}^{N} a_n \leq \sum_{n=-N}^{N} |a_n| \leq (2N+1)a_0.$$

In his paper referenced in part *a*, Fejér proved that $A(0) \leq N + 1$.

g. The fact that the zeros z_0 of the polynomial B_*, of Theorem 3.6.4, satisfy $|z_0| < 1$ for $A > 0$ on $\mathbb{T}_{2\Omega}$ is useful in filter design. A rational function H, all of whose zeros and poles z_0 satisfy $|z_0| < 1$, is a *minimum phase filter*, e.g., [OS75, pages 345–353], [Dau92, pages 194 ff.].

h. *Daubechies Theorem.* One of the early stunning successes of *wavelet theory* was Daubechies' Theorem (1987): *for any* $r \geq 0$, *there is a constructible function* $\psi \in C_c^r(\mathbb{R})$ *for which* $\{\psi_{m,n} : m, n \in \mathbb{Z}\}$ *is an ONB for* $L^2(\mathbb{R})$, *where*

$$\psi_{m,n}(t) = 2^{m/2} \psi(2^m t - n).$$

Her proof requires the Fejér–Riesz Theorem, e.g., [Dau92, Chapter 6], especially pages 167–174 for the role of Theorem 3.6.4.

i. The functions A_* and B_* of Theorem 3.6.4 are called z-transforms of $\{a_n\}$ and $\{b_n\}$, respectively, e.g., Exercise 3.22.

3.6.6 Theorem. A Spectral Estimation Theorem

Let $\{r_j : j = 0, \pm 1, \pm 2, \ldots, \pm N\} \subseteq \mathbb{C}$ satisfy the condition $r_j = \overline{r}_{-j}$ for each j; and assume the $(N + 1) \times (N + 1)$ matrix $R = (r_{jk}) >> 0$, where $j, k \geq 0$ and $r_{jk} \equiv r_{j-k}$. Let $\Omega > 0$. There is a positive function $S \in A(\mathbb{T}_{2\Omega})$, with Fourier coefficients $\{s[n]\} \in \ell^1(\mathbb{Z})$, satisfying the following properties:

$$\forall \, |n| \leq N, \qquad s[n] = r_n, \tag{3.6.5}$$

and

$$\forall \gamma \in \mathbb{T}_{2\Omega}, \qquad S(\gamma) = \frac{1}{A}(\gamma) = 1 \left/ \sum_{n=-N}^{N} a_n e^{-\pi i n \gamma / \Omega}, \right. \tag{3.6.6}$$

where A designates the sum in the denominator on the right side of (3.6.6).

PROOF. *a.* In order to prove this result we shall proceed in the following devious way.

Given the hypotheses on R we shall *momentarily* assume both (3.6.5) and (3.6.6) for some nonnegative function $S \in L^1(\mathbb{T}_{2\Omega})$. Using these hypotheses *and* conclusions, we shall show in parts *b–e* how to obtain the coefficients $\{a_n : |n| \leq N\}$ of (3.6.6) from the given data $\{r_n : |n| \leq N\}$.

In order to give an honest proof of the theorem, we work backward. In particular, we take the hypotheses on R (without the conclusions of the theorem!) and solve the system of equations in part *e* to obtain $\{a_n\}$. Then we define S in terms of these a_n by means of (3.6.6). The calculations in parts *b–e* allow us to conclude that S is nonnegative, $S \in A(\mathbb{T}_{2\Omega})$, and $\check{S}[n] = r_n$ for all $|n| \leq N$.

b. We shall prove that S is not only in $L^1(\mathbb{T}_{2\Omega})$, but $S \in A(\mathbb{T}_{2\Omega})$. In fact, A is a trigonometric polynomial, and so it is a continuous function on $\mathbb{T}_{2\Omega}$ with at most finitely many zeros. However, S is not integrable over any small interval centered at such a zero, e.g., Exercise 3.27; and thus S and A are really positive on $\mathbb{T}_{2\Omega}$. Thus, $S = 1/A \in A(\mathbb{T}_{2\Omega})$ by Wiener's Theorem (Theorem 3.5.9).

c. Since $A > 0$ on $\mathbb{T}_{2\Omega}$, we can apply the Fejér–Riesz Theorem (Theorem 3.6.4). Thus, there is a trigonometric polynomial $B(\gamma) = \sum_{n=0}^{N} b_n e^{-\pi i n \gamma / \Omega}$ with the properties that $A = |B|^2$ on $\mathbb{T}_{2\Omega}$, $b_0 \neq 0$, and for which

$$B_*(z) = \sum_{n=0}^{N} b_n z^n = 0 \quad \text{implies} \quad |z| < 1,$$

where $B_*(e^{-\pi i \gamma / \Omega}) \equiv B(\gamma)$. In fact, the proof of Theorem 3.6.4 allows us to write

$$B_*(z) = C^{1/2} \prod_{j=1}^{N} (z - z_j) z_j^{-1/2},$$

where $\{z_j\}$ is the set of zeros of B_* and each z_j satisfies $0 < |z_j| < 1$. Therefore,

$$\overline{B_*\left(\frac{1}{\overline{z}}\right)} = \overline{C}^{1/2} \prod_{j=1}^{N} \left(\frac{1}{z} - \overline{z}_j\right) \overline{z}_j^{-1/2} = \sum_{n=0}^{N} \overline{b}_n z^{-n},$$

and its zeros $z = 1/\overline{z}_j$ are outside the unit circle. Hence,

$$\frac{1}{\overline{B_*(1/\overline{z})}} = \sum_{n=0}^{\infty} c_n z^{-n}, \qquad |z| \leq 1. \tag{3.6.7}$$

Further, if $z = e^{-\pi i \gamma / \Omega}$, then

$$\overline{B_*\left(\frac{1}{\overline{z}}\right)} = \overline{B(\gamma)}. \tag{3.6.8}$$

d. Combining the factorization $A = |B|^2$ with (3.6.6), we obtain

$$SB = \frac{1}{\overline{B}} \in A(\mathbb{T}_{2\Omega}). \tag{3.6.9}$$

Because of (3.6.7) and (3.6.8), the Fourier coefficients $(SB)\check{}[n]$ vanish if $n > 0$. Since $(SB)\check{}[n] = \check{S} * \check{B}[n]$, we have from (3.6.9) that

$$\forall n > 0, \qquad \sum_{k=0}^{N} s[n-k] b_k = 0. \tag{3.6.10}$$

e. Because of property (3.6.5), the N cases of (3.6.10) for $n = 1, \ldots, N$ give rise to the N equations

$$r_1 b_0 + r_0 b_1 + r_{-1} b_2 + \cdots + r_{1-N} b_N = 0,$$
$$r_2 b_0 + r_1 b_1 + r_0 b_2 + \cdots + r_{2-N} b_N = 0,$$
$$\vdots \tag{3.6.11}$$
$$r_N b_0 + r_{N-1} b_1 + r_{n-2} b_2 + \cdots + r_0 b_N = 0.$$

We rewrite (3.6.11) as the matrix equation

$$Rb^T = -b_0 (r_1, \ldots, r_N)^T,$$

where $b = (b_1, b_2, \ldots, b_N)$. R is invertible since $R >> 0$, e.g., Definition 3.6.2*c,d*. Thus, we compute b, cf., Example 3.7.9; and then we compute $\{a_n : -N \leq n \leq N\}$ in terms of b since $A = |B|^2$. ∎

3.6.7 Example. Spectral Estimation and Maximum Entropy

a.   Equation (3.6.6) gives spectral (frequency) information about a digital signal in the case that relatively little data X is known about the signal. In fact, a small value of $A(\gamma) > 0$ in (3.6.6) allows one to guess that this value of γ is a frequency component of the signal, which generates given autocorrelation data X. Of course, we very rarely get something for nothing, and so this method of spectral estimation can only be used effectively when, as indicated in Remark 3.6.1*b*, certain physical parameters make sense.

This method of spectral estimation is a form of the MEM, and it should be compared with Fourier transform methods, e.g., Proposition 2.8.8 and Example 2.9.7, which really give accurate spectral information but which usually require large data sets X.

b.   Let $\{r_n : |n| \leq N\} \subseteq \mathbb{C}$ satisfy the hypotheses of either the MEM Theorem (Theorem 3.6.3) or Theorem 3.6.6. The relationship between these theorems was alluded to in part *a*, and it is quantified by the following suggestive calculation. The calculation itself was made early on in the development of MEM, e.g., [Chi78, page 55], [IEEE82, page 944]. It is a rationale (not a proof) for supposing that S has the form (3.6.6) in the case that $\int_{\mathbb{T}} \log S(\gamma) \, d\gamma$ is the largest (or smallest) value of $\{\int_{\mathbb{T}} \log F(\gamma) \, d\gamma\}$, when F ranges over all positive functions F in $A(\mathbb{T})$ for which $\check{F}[n] = r_n$, $|n| \leq N$.

Let $|n| > N$ be fixed, and consider the continuous (complex) variable $r \equiv r_n$. Assuming $\int_{\mathbb{T}} \log S(\gamma) \, d\gamma$ is an optimum we have

$$0 = \frac{\partial}{\partial r} \int_{\mathbb{T}} \log S(\gamma) \, d\gamma = \int_{\mathbb{T}} \frac{1}{S(\gamma)} \frac{\partial S(\gamma)}{\partial r} \, d\gamma,$$

where γ is fixed in the expression $\partial S(\gamma)/\partial r$, cf., the proof of Theorem 3.7.7. Writing $S(\gamma) = \sum r_j e^{-2\pi i j \gamma}$ we have $\partial S(\gamma)/\partial r = e^{-2\pi i n \gamma}$. Thus, $(1/S)^{\check{}}[n] = 0$ for all $|n| > N$, and so S has the form (3.6.6).

3.7. Prediction and Spectral Estimation

An *extension* of the factorization given by the Fejér–Riesz Theorem (Theorem 3.6.4) is the *Szegö Factorization Theorem*: *let $A \in L^1(\mathbb{T})$ be nonnegative; then $\log A \in L^1(\mathbb{T})$ if and only if $A = |B|^2$ for some $B \in H^2(\mathbb{T})$*, where

$$H^2(\mathbb{T}) = \{F \in L^2(\mathbb{T}) : \forall n < 0, \quad \check{F}[n] = 0\}.$$

Proofs can be found in [GS58, Section 1.14], [Hof62, pages 48–54]. We shall not prove Szegö's Factorization Theorem, but we mention it because Szegö's original proof (*Math. Ann.* **84** (1921), 232–244) depended on the following result, which he proved in 1920 (*Math. Zeit.* **6** (1920), 167–202). (The proof of the Szegö Factorization Theorem in [GS58, Chapter 1.14] is the joint work of F. Riesz and Szegö, which actually appeared before Szegö's original proof, cf., [MW57, pages 115 ff].)

3.7.1 Theorem. Szegö Alternative

Let $W \in L^1(\mathbb{T})$ be nonnegative, and define the "geometric mean" of W as

$$g(W) = \begin{cases} \exp \int_{\mathbb{T}} \log W(\gamma) \, d\gamma, & \text{if} \quad \log W \in L^1(\mathbb{T}), \\ 0, & \text{if} \quad \log W \notin L^1(\mathbb{T}). \end{cases}$$

Then

$$\inf \int_{\mathbb{T}} |1 - P(\gamma)|^2 W(\gamma) \, d\gamma = g(W),$$

where the infimum is taken over all trigonometric polynomials P on $\mathbb{T}$ of the form

$$P(\gamma) = \sum_{n=1}^{N} a_n e^{-2\pi i n \gamma},$$

and where $N \geq 1$ and $a_1, \ldots, a_N$ vary.

The Szegö Alternative has been generalized in several directions, e.g., [Ach56, pages 256 ff], [DM76], [Hel64, pages 19–24], [Hof62, pages 48–50]. It can also be reformulated as follows.

3.7.2 Theorem. Kolmogorov Theorem (1940)

Let $W \in L^1(\mathbb{T})$ be nonnegative, and define the space

$$L^2_W(\mathbb{T}) = \left\{ F : \|F\|_{L^2_W(\mathbb{T})} = \left(\int_{\mathbb{T}} |F(\gamma)|^2 W(\gamma) \, d\gamma \right)^{1/2} < \infty \right\}.$$

Then

$$\overline{\text{span}}\{e^{-2\pi i n \gamma} : n \leq 0\} = L^2_W(\mathbb{T}) \tag{3.7.1}$$

if and only if $\log W \notin L^1(\mathbb{T})$, where $\overline{\text{span}}X$ is defined as the closure in $L^2_W(\mathbb{T})$, taken with the $\| \cdots \|_{L^2_W(\mathbb{T})}$-norm, of the linear span of elements from X.

In the spirit of Section 2.8, we state the following definition.

3.7.3 Definition. Stationary Sequences and Power Spectra

a. Let $f \in \ell^2(\mathbb{Z})$. The ℓ^2-*autocorrelation* of f is the sequence $p : \mathbb{Z} \to \mathbb{C}$ defined as

$$\forall n \in \mathbb{Z}, \qquad p[n] = \sum_n f[n+m]\overline{f[m]},$$

cf., Example 2.7.9 for L^2-autocorrelation. By the Parseval formula, $\hat{p} = |F|^2 \in L^1(\mathbb{T})$. $S \equiv |F|^2$ is the *power spectrum* of f.

b. A sequence $x = \{x[n] : n \in \mathbb{Z}\}$ in a complex Hilbert space H is *stationary* if the *inner product*

$$r[n] = \langle x[n+k], x[k] \rangle, \qquad n \in \mathbb{Z},$$

is independent of k, e.g., Definition B.4. The sequence r is the *autocorrelation* of x, and it is elementary to check that r is a positive definite function on $\mathbb{Z}$. By Herglotz' Theorem, there is $S \in M_+(\mathbb{T})$ for which $\hat{r} = S$. S is the *power spectrum* of x, e.g., Definition 3.10.3.

3.7.4 Remark. Prediction and Kolmogorov's Theorem

a. Theorem 3.7.2 is valid in more general contexts, including replacing the weight W by any $\mu \in M_+(\mathbb{T})$, e.g., [Kol41], cf., [Ach56, pages 261–263].

b. Let H be a Hilbert space, and let $x \in H$ be a stationary sequence with *power spectrum* S. Define $H(x)$ as $\overline{\text{span}}\{x[n]\} \subseteq H$. Kolmogorov noted that *there is a unique linear mapping,*

$$Z : L_S^2(\mathbb{T}) \longrightarrow H(x),$$

defined on the exponentials as $Z(e^{2\pi in\gamma}) = x[n]$, *which is an isometric isomorphism, i.e., Z is a bijection and* $\|F\|_{L_S^2(\mathbb{T})} = \|Z(F)\|$ *for all* $F \in L_S^2(\mathbb{T})$, *where* $\|\cdots\|$ *is the norm on* H.

c. Using Theorem 3.7.2 and the result of part *b*, Kolmogorov solved the problem of "predicting the future from the whole past" [Kol41], cf., [DM76], [Ben92a]. We shall not go into the details of defining *deterministic sequences*, which are required for a clear statement of the prediction problem that Kolmogorov solved. However, one has the intuition of *prediction* from (3.7.1) in the sense that the *past information* $\{e^{2\pi in\gamma} : n \le 0\}$ is sufficient to *approximate* or *predict* any $F \in L_W^2(\mathbb{T})$, including exponentials $F(\gamma) = e^{2\pi in\gamma}$ for *times* $n > 0$.

We shall now attempt to quantify Remark 3.7.4c for the practical matter of addressing prediction problems that arise in analyzing bioelectric traces, speech data, economic and weather trends, and a host of time series from a variety of subjects. We begin with the following example.

3.7.5 Example. Prediction Estimates

  a. Let $\Omega > 0$; and let $\hat{f} = F$, where $F \in L^2(\mathbb{T}_{2\Omega})$ has Fourier coefficients $f = \{f[n]\} \in \ell^2(\mathbb{Z})$. Suppose that for a given value of n, $f[n]$ is not explicitly known, whereas $H_n \equiv \{f[n-k] : k \geq 1\}$ *is* known. When is it possible to predict, i.e., approximate or evaluate, $f[n]$ in terms of H_n? One way of addressing this question is to write

$$\epsilon_N[n] = f[n] - \sum_{k=1}^{N} a_k f[n-k] \qquad (3.7.2)$$

so that

$$\epsilon_N[n] = \int_{\mathbb{T}_{2\Omega}} F(\gamma) e^{\pi i n \gamma / \Omega} \, d\gamma - \sum_{k=1}^{N} a_k \int_{\mathbb{T}_{2\Omega}} F(\gamma) e^{\pi i (n-k)\gamma / \Omega} \, d\gamma,$$

from which we have the estimate

$$|\epsilon_N[n]| \leq \int_{\mathbb{T}_{2\Omega}} |F(\gamma)| \left| 1 - \sum_{k=1}^{N} a_k e^{-\pi i k \gamma / \Omega} \right| d\gamma$$

$$\leq \|F\|_{L^2(\mathbb{T}_{2\Omega})} \left(\int_{\mathbb{T}_{2\Omega}} \left| 1 - \sum_{k=1}^{N} a_k e^{-\pi i k \gamma / \Omega} \right|^2 d\gamma \right)^{1/2} . \qquad (3.7.3)$$

In the transition from (3.7.2) to the right side of (3.7.3) we have squandered our information H_n. Further, $1 \notin \overline{\text{span}}\{e^{-\pi i k \gamma / \Omega} : k \geq 1\}$ in $L^2(\mathbb{T}_{2\Omega})$ by Theorem 3.4.12. This fact is corroborated by the Szegö Alternative since the weight $W \equiv 1$ on the right side of (3.7.3) has the property that $\log W \in L^1(\mathbb{T}_{2\Omega})$.

  b. We now adjust the calculation of part *a* by implementing the Parseval formula to compute

$$\|\epsilon_N\|_{\ell^2(\mathbb{Z})} = \left(\int_{\mathbb{T}_{2\Omega}} |F(\gamma)|^2 \left| 1 - \sum_{k=1}^{N} a_k e^{-\pi i k \gamma / \Omega} \right|^2 d\gamma \right)^{1/2} .$$

Of course, n is no longer fixed as it is in (3.7.2). On the other hand, if $\log W \notin L^1(\mathbb{T}_{2\Omega})$, where $W \equiv |F|^2$, then the Szegö Alternative (Theorem 3.7.1) or Kolmogorov's Theorem (Theorem 3.7.2) can be used to glean prediction theoretic information in the following way, cf., Exercise 3.54. For each fixed n, $|\epsilon_N[n]| \leq \|\epsilon_N\|_{\ell^2(\mathbb{Z})}$, and so $\inf |\epsilon_N[n]| = \inf \|\epsilon_N\|_{\ell^2(\mathbb{Z})} = 0$ in the case $\log |F| \notin L^1(\mathbb{T}_{2\Omega})$, where the infima are taken over all $N \geq 1$ and $a_1, \ldots, a_N \in \mathbb{C}$.

  c. Our next adjustment of the calculation in part *a* deals with *analogue* signals $f \in PW_\Omega$ instead of *discrete* signals in $\ell^2(\mathbb{Z})$. We write

$$\epsilon_N(t) = f(t) - \sum_{k=1}^{N} a_k f\left(t - k\frac{\alpha}{\Omega}\right)$$

for a fixed $\alpha > 0$ and for a fixed $t \in \mathbb{R}$. Then we compute

$$\epsilon_N(t) = \int \hat{f}(\gamma)e^{2\pi it\gamma}\,d\gamma - \sum_{k=1}^{N} a_k \int \hat{f}(\gamma)e^{2\pi i(t-k\alpha/\Omega)\gamma}\,d\gamma,$$

cf., [Ben92b, Proposition 9]. Thus,

$$|\epsilon_N(t)| \leq \left(\frac{\Omega}{\alpha}\right)^{1/2} \|f\|_{L^2(\mathbb{R})} \left(\int_{-\alpha}^{\alpha}\left|1 - \sum_{k=1}^{N} a_k e^{-2\pi ik\gamma}\right|^2 d\gamma\right)^{1/2}. \qquad (3.7.4)$$

As in (3.7.3), we have squandered information about t on the right side of (3.7.4). However, it can be shown that *if $\alpha < 1/2$, then*

$$\overline{\text{span}}\{e^{-2\pi ik\gamma} : k \geq 1\} = L^2[-\alpha, \alpha]. \qquad (3.7.5)$$

Thus,

$$\inf |\epsilon_N(t)| = 0,$$

where the infimum is taken over all $N \geq 1$ and $a_1, \ldots, a_N \in \mathbb{C}$.

The density result (3.7.5) is due to Carleman, e.g., [You80, pages 114–116], cf., Remark 3.7.11.

d. Suppose we are given a fixed discrete "time" n, resp., a fixed continuous "time" t. In part b, resp., part c, we have shown in theory how to predict the value $f[n]$, resp., $f(t)$, in terms of its known values at previous times. This prediction requires f to satisfy certain conditions; and the prediction itself is made in terms of a given error bound ϵ. In fact, in both cases it can be proved that coefficients a_j, $j = 1, \ldots, N$, exist with the property that $|\epsilon_N[n]| < \epsilon$, resp., $|\epsilon_N(t)| < \epsilon$.

We wish to design an applicable tool based on the idea of Example 3.7.5. We begin with the definition of *linear prediction*.

3.7.6 Definition. Linear Prediction Models

a. Let $M, N \geq 1$. A *pole-zero linear prediction model* of a sequence $f \in \ell^2(\mathbb{Z})$ is an equation of the form

$$f[n] = \sum_{k=1}^{N} a_k f[n-k] + \sum_{j=0}^{M} b_j u[n-j] \qquad (3.7.6)$$

for each n, where $u \in \ell^2(\mathbb{Z})$, $\{a_k\}, \{b_k\} \subseteq \mathbb{R}$, and $b_0 \neq 0$.

$f[n]$ can be thought of as the output of a *linear translation-invariant system* with some unknown input u, including its past and present values, as well as input consisting of past values of f, viz., $f[n-1], \ldots, f[n-N]$. In this point of view, the goal is to estimate the *system* parameters $\{a_k\}, \{b_k\} \subseteq \mathbb{R}$ so that (3.7.6) is a meaningful predictor of f.

b. If $b_1 = b_2 = \cdots = b_M = 0$, then (3.7.6) is an *all-pole model*, or, equivalently, an *autoregressive* (AR) *model*. If $a_1 = a_2 = \cdots = a_N = 0$, then (3.7.6) is an *all-zero model*, or, equivalently, a *moving average* (MA) *model*. The full pole-zero model (3.7.6) is also called an *autoregressive moving average* (ARMA) *model*.

c. In the following, as in Example 3.7.5, we shall deal with the all-pole model. This type of linear prediction goes back to Yule's work (1927) on sun spot analysis, cf., the beginning of Section 2.9 for another remark on sun spots. The mathematics underlying the effectiveness of linear prediction is Gauss' technique of linear least squares estimation from 1795, cf., Proposition 1.10.9. Major contributions were made by Kolmogorov and Wiener, independently, in the early 1940s, e.g., [Wie49] including Appendix C by Norman Levinson. Linear prediction is a staple in time series analysis, e.g., [BSS88], [JN84], [Mak75], [Pri81]. We mention, for example, the introduction of linear prediction into speech analysis and data compression by Fant, Jury, Atal and Schroeder, and Itakura and Saito in the 1960s, e.g., [MG82].

3.7.7 Theorem. Least Squares Method for All-Pole Model

Let $N \geq 1$, assume $f \in \ell^2(\mathbb{Z}) \backslash \{0\}$ is real-valued with ℓ^2-autocorrelation r, and consider the sequence of equations

$$\forall n \in \mathbb{Z}, \qquad f[n] = \sum_{k=1}^{N} a_k f[n-k] + u[n]. \qquad (3.7.7)$$

There are unique coefficients $a_1^*, \ldots, a_N^* \in \mathbb{R}$ and a sequence $\epsilon_N \in \ell^2(\mathbb{Z})$ defined by

$$\epsilon_N[n] = f[n] - \sum_{k=1}^{N} a_k^* f[n-k]$$

for each n such that

$$\forall a = (a_1, \ldots, a_N) \in \mathbb{R}^N,$$

$$0 \leq r[0] - \sum_{k=1}^{N} a_k^* r[k] = \|\epsilon_N\|_{\ell^2(\mathbb{Z})}^2 \leq \|u\|_{\ell^2(\mathbb{Z})}^2, \qquad (3.7.8)$$

where u is defined by (3.7.7). (Thus, u depends on $a \in \mathbb{R}^N$ as opposed to the interpretation of Definition 3.7.6).

PROOF.   **a.**   For a given $n \in \mathbb{Z}$ and $a = (a_1, \ldots, a_N) \in \mathbb{R}^N$, we define the approximation $f_a[n]$ of $f[n]$ as

$$f_a[n] = \sum_{k=1}^{N} a_k f[n - k].$$

Clearly, $(f - f_a) \in \ell^2(\mathbb{Z})$, and we set

$$E_N(a) = \sum_n \left(f[n] - \sum_{k=1}^{N} a_k f[n - k] \right)^2 < \infty.$$

b.   For each k, we have the formal calculation

$$\frac{\partial E_N(a)}{\partial a_k} = -2 \sum_n \left(f[n] - \sum_{j=1}^{N} a_j f[n - j] \right) f[n - k]. \qquad (3.7.9)$$

By Hölder's Inequality, the right side converges uniformly on any bounded interval of the a_k-axis. Thus, the right side of (3.7.9) is in fact $\partial E_N(a)/\partial a_k$.

c.   A *necessary* condition in order that E_N have a local minimum (or maximum) at $a^* \in \mathbb{R}^N$ is that

$$\forall k = 1, \ldots, N, \qquad \frac{\partial E_N(a^*)}{\partial a_k} = 0. \qquad (3.7.10)$$

A *sufficient* condition in order that E_N (which satisfies (3.7.10)) have a local minimum at $a^* \in \mathbb{R}^N$ is that $D_{N-k}(a^*) > 0$ for each $k = 0, 1, \ldots, N$, where $D_0(a) = 1$ and $D_{N-k}(a)$ is the determinant obtained from the $N \times N$ matrix,

$$\left(\frac{\partial^2 E_N(a)}{\partial a_m \partial a_n} \right), \qquad m, n = 1, \ldots, N,$$

by deleting the last k rows and columns and taking the determinant of the resulting $(N - k) \times (N - k)$ matrix, e.g., [Apo57].

d.   In our case, by (3.7.9), we have

$$\forall a \in \mathbb{R}^N,$$

$$\frac{\partial^2 E_N(a)}{\partial a_m \partial a_n} = 2 \sum_j f[j - m] f[j - n] = 2r[m - n] = 2r[n - m]. \qquad (3.7.11)$$

Since $f \in \ell^2(\mathbb{Z})$, we compute $\hat{r} = |F|^2 \in L^1(\mathbb{T})$ by the Parseval formula; and the $N \times N$ matrix $R = (r[m - n])$, $m, n = 1, \ldots, N$, is positive definite, as we showed in Definition 3.6.2e. An elementary characterization of positive definite matrices is that $D_0(a), D_1(a), \ldots, D_N(a)$ be positive [Per52, Theorem 9-26].

Thus, our goal is to find values $a^* \in \mathbb{R}^N$ for which (3.7.10) is valid; and, by the previous paragraph and the sufficient conditions for minima in part c, these values will in fact be minimizers of E_N.

e. To obtain a candidate a^* for a minimizer, we rewrite (3.7.9) and (3.7.10) as the system of N linear equations in N unknowns $a_1, \ldots, a_N$:

$$\forall 1 \leq k \leq N,$$

$$\sum_{j=1}^{N} a_j \sum_n f[n-j]f[n-k] = \sum_n f[n]f[n-k], \tag{3.7.12}$$

or, equivalently,

$$\forall 1 \leq k \leq N, \qquad \sum_{j=1}^{N} a_j r[k-j] = r[k]. \tag{3.7.13}$$

Since $D_N(a) > 0$ by (3.7.11) and the discussion in part d, the system (3.7.13) has a unique solution $a^* = (a_1^*, \ldots, a_N^*) \in \mathbb{R}^N$.

Thus, the inequality in (3.7.8) is obtained.

f. Expanding the square in the definition of E_N and substituting the minimizer a^* into (3.7.12), we obtain

$$0 \leq E_N(a^*) = \sum_n f[n]^2 - 2\sum_{k=1}^{N} a_k^* \sum_n f[n]f[n-k]$$

$$+ \sum_n \left(\sum_{k=1}^{N} a_k^* f[n-k]\right)^2$$

$$= \sum_n f[n]^2 - 2\sum_{k=1}^{N} a_k^* \sum_n f[n]f[n-k]$$

$$+ \sum_{k=1}^{N} a_k^* \sum_n f[n]f[n-k]$$

$$= \sum_n f[n]^2 - \sum_{k=1}^{N} a_k^* \sum_n f[n]f[n-k]$$

$$= r[0] - \sum_{k=1}^{N} a_k^* r[k].$$

This completes (3.7.8). ∎

3.7.8 Example. Spectral Estimation and the All-Pole Model

a.   A typical and important issue in many problems and fields is to find the *spectral peaks* or fundamental frequencies in a given signal $f \in \ell^2(\mathbb{Z})$. In theory, the graph of the Fourier series $F = \hat{f}$ will provide this spectral information. In reality, there are potential problems. For example, trigonometric polynomial approximations of $\hat{f}$, or approximations such as Proposition 2.8.8, may be inadequate because they are either too good or too bad! In the former case, a very good approximation of $\hat{f}$ may have so much spectral information from "noise" embedded in f that desirable information about pure tones (in f) is obscured when observing $\hat{f}$. In the latter case, the approximation may not be developed enough to specify relevant frequencies in $\hat{f}$.

b.   In cases, such as those hypothesized in part *a*, where the Fourier transform cannot be directly used to observe some fundamental frequencies in a signal, there are other methods that sometimes provide spectral information. The all-pole model is one of these methods. The prediction estimates in Example 3.7.5 show that the *prediction error* ϵ_N in Theorem 3.7.7 tends to 0 as $N \to \infty$ in many cases. The all-pole model in (3.7.7) of Theorem 3.7.7 shows that such a model and its corresponding prediction error can be used to specify spectral peaks by the following process and rationale.

By taking the Fourier transforms of the sequences f and u, (3.7.7) becomes

$$F(\gamma)\left(1 - \sum_{k=1}^{N} a_k e^{-2\pi i k \gamma}\right) = U(\gamma), \tag{3.7.14}$$

where F and U are Fourier series with Fourier coefficients $\{f[n]\}$ and $\{u[n]\}$, respectively. With the minimization effected by Theorem 3.7.7, (3.7.14) becomes

$$F(\gamma) = \frac{\hat{\epsilon}_N(\gamma)}{1 - \sum_{k=1}^{N} a_k^* e^{-2\pi i k \gamma}}.$$

Assume $0 < A \le |\hat{\epsilon}_N(\gamma)| \le B$, which is reasonable for many applications, e.g., [Chi78]. Suppose F is continuous and $|F(\gamma_0)|$ is large in comparison to $|F(\gamma)|$ for values of γ near γ_0, i.e., suppose F has a spectral peak at γ_0. Define the polynomial

$$P_*(z) = 1 - \sum_{k=1}^{N} a_k^* z^k, \qquad z \in \mathbb{C}.$$

We write $P(\gamma) = P_*(e^{-2\pi i \gamma})$. Then $|P(\gamma_0)|$ is small. This simple observation is the basis of the *all-pole model method of spectral estimation*.

To describe this method, we consider the following procedure for a given $N \ge 1$ and a given sequence f. First, choose a threshold $\epsilon > 0$ and consider the annular region $A_\epsilon = \{z : 1-\epsilon \le |z| \le 1+\epsilon\}$. The choice of ϵ can be adjusted according to the amount of spectral information desired. Next, compute $\{a_k^* : k = 1, \ldots, N\}$, cf., Example 3.7.9. Compute the zeros z_0 of P_*, e.g., in MATLAB use the *roots* command. If $z_0 \in A_\epsilon$, compute γ_0 by taking the projection $e^{-2\pi i \gamma_0}$ of z_0 to the unit circle.

$F(\gamma_0)$ is a candidate for a spectral peak.

c. As a caveat for our presentation in part *b*, we note that we have been precise about certain matters, e.g., Theorem 3.7.7, but quite cavalier about others. For example, the structure of u is important for the type of application at hand; and the proper behavior of $\hat{u}$ is important for the success of the all-pole model method of spectral estimation in that application, e.g., [JN84, Section 2.4], [Pri81].

We should also mention that we are not using the all-pole method to predict values of f so much as to determine its spectral behavior. In fact, to compute a^* in Theorem 3.7.7 we assume knowledge of each $f[n]$.

3.7.9 Example. Levinson Recursion Algorithm

a. Norman Levinson (1947) was the first to use the *structure* of a Toeplitz matrix (Definition 3.6.2*d*) to solve the system of linear equations (3.7.13) recursively, e.g., [Wie49, Appendix B]. This system not only plays a role in linear prediction (Theorem 3.7.7) but was also essential in the proof of Theorem 3.6.6, e.g., (3.6.11) which is associated with MEM. (Recall the equivalence of these methods noted in Remark 3.6.1*b*.) In statistics, (3.6.11) and (3.7.13) are called the Yule–Walker equations. In making use of the Toeplitz structure, Levinson's algorithm has led to numerically realistic computations of the prediction coefficients $a_1^*, \ldots, a_N^*$. For example, the classical Gauss elimination method requires $N^3 + KN^2$ multiplications or divisions, whereas even Levinson's original method only required $N^2 + KN$ such operations.

b. If $\{c_k : k \geq 0\} \subseteq \mathbb{C}$ and $\exp(\sum_{k=0}^{\infty} c_k z^k) = \sum_{k=0}^{\infty} b_k z^k$, then one example of the Lebedev–Milin inequalities is the inequality

$$|b_n|^2 \leq \exp\left(\sum_{k=1}^{n} k|c_k|^2 - \sum_{k=1}^{n} \frac{1}{k}\right), \qquad b_0 = 1.$$

These inequalities have applications in univalent function theory, as well as in number theory and spectral synthesis, e.g., our construction of *idelic pseudomeasures* in *Zeta functions for idelic pseudomeasures, Ann. Scuola Norm. Sup.* **6** (1979), 367–377.

It turns out that

$$b_{k+1} = \sum_{j=0}^{k}\left(1 - \frac{j}{k+1}\right) c_{k+1-j} b_j \tag{3.7.15}$$

[Pou84], and that this recursion formula can be implemented numerically to deal with linear prediction problems. For example, in dealing with the Fejér–Riesz

Theorem (Theorem 3.6.4), $A = |B|^2$, we let $\{c_k\}$ be the sequence of Fourier coefficients of $\log A$, where $A \geq 0$ is the given nonnegative polynomial. Then, by way of a standard argument in complex analysis, we obtain $b_0 = e^{c_0/2}$ and (3.7.15) for $k = 0, \ldots, N-1$, where $b_0, \ldots, b_N$ are the Fourier coefficients of B, e.g., [Pou84].

3.7.10 Remark. Ramifications of Szegö Factorization on $\mathbb{R}$

a. We first stated and discussed the Paley–Wiener Logarithmic Integral Theorem in Example 1.6.5c. It asserts that *if* $\varphi \in L^2(\hat{\mathbb{R}}) \backslash \{0\}$ *is nonnegative, then there is* $f \in L^2(\mathbb{R})$, *for which* supp $f \subseteq [0, \infty)$ *and* $|\hat{f}| = \varphi$ *a.e., if and only if*

$$\int \frac{|\log \varphi(\gamma)|}{1 + \gamma^2} \, d\gamma < \infty.$$

Although it is elementary to prove the Szegö Factorization Theorem on $\mathbb{R}$ from this theorem of Paley and Wiener, e.g., [Pric85, pages 156–157], they were, in fact, motivated in their research by a result of Carleman on quasi-analytic functions [PW33], [PW34, Theorem XII], cf., [Koo88], [Rud66, Chapter 19] for the theory of quasi-analytic functions.

b. By an approximate identity argument, the Paley–Wiener Logarithmic Integral Theorem can be used to prove the following result. *If* $\mu \in M(\mathbb{R})$, *then* supp $\mu \subseteq [T, \infty)$ *if and only if*

$$\int \frac{|\log |\hat{\mu}(\gamma)||}{1 + \gamma^2} \, d\gamma < \infty.$$

c. The result for compactly supported measures analogous to that of part *b* is due to Beurling and Malliavin: *let* W *be a continuous function for which* $|W| > 1$ *and* $\log |W|$ *is uniformly continuous; then the condition*

$$\int \frac{\log |W(\gamma)|}{1 + \gamma^2} \, d\gamma < \infty$$

is necessary and sufficient that for all $\epsilon > 0$ *there exists* $\mu \in M(\mathbb{R})$ *such that* supp $\mu \subseteq [-\epsilon, \epsilon]$ *and* $\hat{\mu} K \in L^\infty(\hat{\mathbb{R}})$, see [BM62], [Mal79].

3.7.11 Remark. Closure Theorems for Sets of Exponentials

In Example 3.7.5 we saw a role for the closure theorems of Szegö, Kolmogorov, and Carleman. There are other landmark contributions by Paley and Wiener [PW34], Levinson [Lev40], and Beurling and Malliavin [BM67], as well as deep results by others, cf., the superb expositions of [Red77], [You80], [Koo88]. We shall close this section with some perspective on such theorems.

a.   Equation (3.7.5) was one of the first substantial results in an area which culminated in the work of Beurling and Malliavin [BM67]. Beurling and Malliavin solved the following *closure problem* for a given discrete subset $D \subseteq \mathbb{R}$: find the upper bound $\Omega \geq 0$ of the set of $\alpha \geq 0$ for which

$$\overline{\text{span}}\{e^{-2\pi i t \gamma} : t \in D\} = L^2[-\alpha, \alpha]. \tag{3.7.16}$$

Their solution includes writing Ω in terms of a density condition on D.

b.   Density results such as (3.7.16) are a *weak* form of *sampling theorems*— weak but *not* necessarily elementary. In fact, if (3.7.16) is valid and $f \in PW_\alpha$, then there is a sequence of trigonometric polynomials P_n, where

$$P_n(\gamma) = \sum_{t \in D_n} a_{t,n} e^{-2\pi i t \gamma} \quad \text{and} \quad D_n \subseteq D,$$

for which

$$\lim_{n \to \infty} \|\hat{f} - P_n\|_{L^2[-\alpha,\alpha]} = 0.$$

Distributionally, each $P_n = \hat{p}_n$ where $p_n = \sum_{t \in D_n} a_{t,n} \delta_t$. Thus,

$$\|\hat{f} - P_n\|_{L^2[-\alpha,\alpha]} = \|\hat{f} - P_n \mathbf{1}_{[-\alpha,\alpha]}\|_{L^2(\hat{\mathbb{R}})}$$
$$= \|f - p_n * d_{2\pi\alpha}\|_{L^2(\mathbb{R})},$$

and so

$$\lim_{n \to \infty} \sum_{t \in D_n} a_{t,n} \tau_t d_{2\pi\alpha} = f \quad \text{in } L^2(\mathbb{R}),$$

cf., the sampling theorems in [Ben92b], [BF94], [BSS88], [Hig85].

c.   Let $\mu \in M_+(\mathbb{R})$ and define

$$L^1_\mu(\mathbb{R}) = \left\{ f : \|f\|_{L^1_\mu(\mathbb{R})} = \int |f(t)| \, d\mu(t) < \infty \right\},$$

cf., Remarks 2.7.4*a* and 3.5.10*d*.

By Theorem 2.7.6, $\mu = w + \mu_s$ where $w \in L^1(\mathbb{R})$ is nonnegative and μ_s is the sum of the discrete and continuous singular parts of μ. Krein proved the following L^1-version of Kolmogorov's Theorem (Theorem 3.7.2) on $\mathbb{R}$:

$$\overline{\text{span}}\{e^{-2\pi i t \gamma} : \gamma \leq 0\} = L^1_\mu(\mathbb{R})$$

if and only if

$$\int \frac{|\log w(t)|}{1 + t^2} \, dt = \infty.$$

d. Suppose $w \geq 1$ on $\mathbb{R}$. In particular, $w \notin L^1(\mathbb{R})$, whereas $w \in L^1(\mathbb{R})$ in part *c*. Assume w is even, $1 = w(0) \leq w(t)$, and $w(u + t) \leq w(u)w(t)$ for $u, t \in \mathbb{R}$. $L_w^1(\mathbb{R}) \subseteq L^1(\mathbb{R})$ is an algebra under convolution, and Beurling (1938) posed the *spectral analysis question*: is every proper closed ideal $I \subseteq L_w^1(\mathbb{R})$ contained in a regular (i.e., $L_w^1(\mathbb{R})/I$ has a unit under convolution) maximal ideal? For an equivalent analytic means of posing this question, consider the following property of $I \equiv L_w^1(\mathbb{R})$:

$$\forall \gamma \in \hat{\mathbb{R}}, \ \exists f \in I \ \text{ such that } \ \hat{f}(\gamma) \neq 0. \tag{3.7.17}$$

Then the spectral analysis question is equivalent to finding conditions on w so that, whenever a closed ideal $I \subseteq L_w^1(\mathbb{R})$ satisfies (3.7.17), we can conclude that $I = L_w^1(\mathbb{R})$.

We have posed the spectral analysis question in this section since there is an equivalent dual formulation in terms of sets of exponentials in $L_{1/w}^\infty(\mathbb{R})$, e.g., [Ben75].

e. Beurling (1938) proved that if

$$\int \frac{\log w(t)}{1 + t^2} \, dt < \infty, \tag{3.7.18}$$

then the spectral analysis question has an affirmative answer [Beu89].

Let $w = 1$ on $\mathbb{R}$. Then (3.7.18) is satisfied and Beurling's Theorem reduces to the Wiener Tauberian Theorem: $f \in L^1(\mathbb{R})$ has a nonvanishing Fourier transform if and only if the closed principal ideal $I \subseteq L^1(\mathbb{R})$ generated by f is all of $L^1(\mathbb{R})$, cf., Remark 2.9.13 and the formulation of Wiener's Tauberian Theorem in Theorem 2.9.12.

3.8. Discrete Fourier Transform

In Section 1.1 we defined the Fourier transform $\hat{f}$ of $f : \mathbb{R} \to \mathbb{C}$; in this case $\hat{f}$ is defined on $\hat{\mathbb{R}}$. In Section 3.1 we defined the Fourier transform $\hat{f}$ of $f : \mathbb{Z} \to \mathbb{C}$; in this case, and for a given $\Omega > 0$, $\hat{f}$ is defined on $\mathbb{T}_{2\Omega} \equiv \hat{\mathbb{R}}/(2\Omega\mathbb{Z})$, i.e., the Fourier series $\hat{f}$ is a 2Ω-periodic function on $\hat{\mathbb{R}}$ with Fourier coefficients $f = \{f[n]\}$. The next step is the following definition.

3.8.1 Definition. Discrete Fourier Transform

a. Let N be a positive integer, and let $\mathbb{Z}_N$ be *the set of integers* $0, 1, \ldots, N-1$ *under addition modulo* N. This means that if $m, n \in \mathbb{Z}_N$ and the ordinary sum $m + n \leq N - 1$, then the addition modulo N of m and n has the value $m + n$.

However, if $m, n \in \mathbb{Z}_N$ and the ordinary sum $m + n > N - 1$, then the addition modulo N of m and n has the value $m + n - N$. For example, the addition table for $\mathbb{Z}_6$ is given in Figure 3.3.

When dealing with $\mathbb{Z}_N$ we shall *denote* the addition modulo N of $m, n \in \mathbb{Z}_N$ by $m + n \in \mathbb{Z}_N$.

To *define* functions f on $\mathbb{Z}_N$, we assign values $f[n]$ for $n = 0, 1, \ldots, N-1$, and then we extend f as an N-periodic function on $\mathbb{Z}$. Thus, $f[m + nN] = f[m]$ for any $m \in \{0, 1, \ldots, N - 1\}$ and for all $n \in \mathbb{Z}$. In this case, we write $f : \mathbb{Z}_N \to \mathbb{C}$, cf., part *e*.

b. The *Fourier transform* of $f : \mathbb{Z}_N \to \mathbb{C}$ is the function $F : \mathbb{Z}_N \to \mathbb{C}$ defined as

$$\forall n \in \{0, 1, \ldots, N - 1\}, \qquad F[n] = \sum_{m=0}^{N-1} f[m] e^{-2\pi i m n / N}. \qquad (3.8.1)$$

Because of the setting $\mathbb{Z}_N$, F is called the *Discrete Fourier Transform* (DFT) of f. In this context we shall write

$$f \longleftrightarrow F, \qquad \hat{f} = F, \qquad f = \check{F},$$

just as we did in Definitions 1.1.2 and 3.1.1 for the cases of Fourier transforms (on $\mathbb{R}$) and Fourier series.

c. Let $f : \mathbb{Z}_N \to \mathbb{C}$ and define the (standard) notation

$$W_N = e^{-2\pi i / N},$$

not to be confused with the notation for the Fejér function in Example 3.4.5. The DFT F of f is defined as

$$\forall n \in \mathbb{Z}_N, \qquad F[n] = \sum_{m \in \mathbb{Z}_N} f[m] W_N^{mn}. \qquad (3.8.2)$$

Clearly, (3.8.1) and (3.8.2) are equivalent.

+	0	1	2	3	4	5
0	0	1	2	3	4	5
1	1	2	3	4	5	0
2	2	3	4	5	0	1
3	3	4	5	0	1	2
4	4	5	0	1	2	3
5	5	0	1	2	3	4

FIGURE 3.3.

We could have defined (3.8.1) for each $n \in \mathbb{Z}$ by our definition in part a; in fact,

$$\forall n \in \mathbb{Z}, \quad F[n] = \sum_{m=0}^{N-1} f[m]e^{-2\pi imn/N}$$

$$= \sum_{m=0}^{N-1} f[m]e^{-2\pi im(n+N)/N} = F[n+N].$$

In any case, $\mathbb{Z}_N$ is the natural domain for F.

  d. $\mathbb{Z}_N$ is a commutative group under the operation of addition modulo N.

A *character* γ of $\mathbb{Z}_N$ is a homomorphism from $\mathbb{Z}_N$ into the multiplicative group $\{z \in \mathbb{C} : |z| = 1\}$, i.e.,

$$\forall m \in \mathbb{Z}_N, \qquad |\gamma(m)| = 1$$

and

$$\forall m, n \in \mathbb{Z}_N, \qquad \gamma(m+n) = \gamma(m)\gamma(n).$$

The set of characters of $\mathbb{Z}_N$ is denoted by $\hat{\mathbb{Z}}$, and $\hat{\mathbb{Z}}_N$ becomes a commutative group by defining the addition $\gamma_1 + \gamma_2$ of characters γ_1, γ_2 by means of the formula

$$\forall m \in \mathbb{Z}_N, \qquad (\gamma_1 + \gamma_2)(m) = \gamma_1(m)\gamma_2(m).$$

In this setting it can be *proved* that $\hat{\mathbb{Z}} = \mathbb{Z}_N$, e.g., [Rud62], cf., Remark 3.1.3*e*.

Algebraic considerations are fundamental in many aspects of harmonic analysis; but, in this book, I am coming closest to cheating the reader by their omission in my treatment of the DFT and FFT.

  e.   Let $L(\mathbb{Z}_N)$ denote the vector space of complex sequences on $\mathbb{Z}_N$, and define the DFT mapping

$$\mathcal{F}_N : L(\mathbb{Z}_N) \longrightarrow L(\hat{\mathbb{Z}}_N)$$

$$f \longmapsto \hat{f}.$$

Since $\mathbb{Z}_N$ is a finite set, $L(\mathbb{Z}_N)$ can be considered any one of the L^p-spaces on $\mathbb{Z}_N$. In fact, $L(\mathbb{Z}_N)$ is the N-dimensional space of all functions on $\mathbb{Z}_N$.

When one thinks of $L(\mathbb{Z}_N)$ as $L^2(\mathbb{Z}_N)$, we define the *inner product*

$$\forall f, g \in L(\mathbb{Z}_N), \qquad \langle f, g \rangle = \frac{1}{N}\sum_{m=0}^{N-1} f[m]\overline{g[m]}.$$

  f.   To make the analogy between the DFT F of f on $\mathbb{Z}_N$ (N even) and Fourier series on $\mathbb{T}$ with Fourier coefficients on $\mathbb{Z}$, we could consider the finite sets

$$\left\{0, \frac{1}{N}, \frac{2}{N}, \ldots, \frac{N-1}{N}\right\} \quad \text{and} \quad \left\{-\frac{N}{2}, -\frac{N}{2}+1, \ldots, \frac{N}{2}-1\right\},$$

corresponding to approximations of $\mathbb{T}$ and $\mathbb{Z}$, respectively.

The inversion theorem for the DFT is elementary.

3.8.2 Theorem. Inversion Formula for the DFT

Let $N > 1$ and let $f : \mathbb{Z}_N \longrightarrow \mathbb{C}$ have DFT F. Then

$$\forall m = 0, 1, \ldots, N-1, \qquad f[m] = \frac{1}{N} \sum_{n=0}^{N-1} F[n]e^{2\pi imn/N}. \qquad (3.8.3)$$

PROOF. Note that $W_N^N = 1$, and that if $N > 1$, then $W_N \neq 1$. Thus, since

$$1 + r + r^2 + \cdots + r^{N-1} = \frac{1-r^N}{1-r}, \qquad r \neq 1, \qquad (3.8.4)$$

we see that

$$1 + W_N + W_N^2 + \cdots + W_N^{N-1} = 0. \qquad (3.8.5)$$

For fixed m, the right side of (3.8.3) is

$$\frac{1}{N} \sum_{n=0}^{N-1} F[n]e^{2\pi imn/N} = \frac{1}{N} \sum_{n=0}^{N-1} F[n]W_N^{-mn}$$

$$= \frac{1}{N} \sum_{n=0}^{N-1} \left(\sum_{j=0}^{N-1} f[j]W_N^{jn}\right) W_N^{-mn} \qquad (3.8.6)$$

$$= \frac{1}{N} \sum_{j=0}^{N-1} f[j] \left(\sum_{n=0}^{N-1} W_N^{(j-m)n}\right).$$

If $j = m$, then $\sum_{n=0}^{N-1} W_N^{(j-m)n} = N$. If $j \neq m$ and $r \equiv W_N^{j-m}$, then $r \neq 1$ and (3.8.4) gives

$$\sum_{n=0}^{N-1} W_N^{(j-m)n} = \frac{1}{1-r}\left(1 - e^{2\pi i(m-j)}\right) = 0. \qquad (3.8.7)$$

Substituting this information into the right side of (3.8.6) gives $f[m]$. ∎

The simplicity of (3.8.5) or (3.8.7) in Theorem 3.8.2 evaporates if one considers the *Gauss sum*,

$$\mathcal{G}_N^{\pm} = \sum_{n=0}^{N-1} e^{\pm 2\pi i n^2/N}, \tag{3.8.8}$$

e.g., Example 3.8.6, Theorem 3.8.7, Theorem 3.8.9, and Theorem 3.8.10.

3.8.3 Corollary.

$\mathcal{F}_N : L(\mathbb{Z}_N) \longrightarrow L(\mathbb{Z}_N)$ *is a linear bijection.*

3.8.4 Theorem. ONB and Parseval Formula

a. *ONB.* $\{\frac{1}{\sqrt{N}} W_N^n : n = 0, \ldots, N-1\}$ *is an ONB for* $L(\mathbb{Z}_N)$ *taken with the inner product defined in Definition 3.8.1e.*

b. *Parseval formula. Let* $f, g \in L(\mathbb{Z}_N)$; *and consider the pairings* $f \longleftrightarrow F$, $g \longleftrightarrow G$. *Then*

$$\sum_{m=0}^{N-1} f[m]\overline{g[m]} = \frac{1}{N} \sum_{n=0}^{N-1} F[n]\overline{G[n]}, \tag{3.8.9}$$

and, in particular,

$$\left(\sum_{m=0}^{N-1} |f[m]|^2\right)^{1/2} = \left(\frac{1}{N}\sum_{n=0}^{N-1} |F[n]|^2\right)^{1/2}.$$

PROOF. ***a.*** For a fixed $k, n \in \mathbb{Z}_N$, we consider W_N^k and W_N^n as functions on $\mathbb{Z}_N$ defined by $W_N^k[m] \equiv W_N^{mk}$. Then

$$\left\langle \frac{1}{\sqrt{N}} W_N^k, \frac{1}{\sqrt{N}} W_N^n \right\rangle = \frac{1}{N}\sum_{m=0}^{N-1} W_N^{mk} W_N^{-mn} = \frac{1}{N}\sum_{m=0}^{N-1} \left(e^{-2\pi i(k-n)/N}\right)^m. \tag{3.8.10}$$

The right side of (3.8.10) is 1 if $k = n$. If $k \neq n$, then $r \equiv e^{-2\pi i(k-n)/N} \neq 1$ since $(k-n)/N \notin \mathbb{Z}$. Thus the right side of (3.8.10) is

$$\frac{1-r^N}{1-r} = 0$$

since $r^N = 1$. Therefore, $\{\frac{1}{\sqrt{N}} W_N^n : n \in \mathbb{Z}_N\}$ is orthonormal.

Linear independence follows from orthonormality. In fact, if $k \in \mathbb{Z}_N$ is fixed and $\sum_{n=0}^{N-1} a_n W_N^n = 0$, then

$$\forall m \in \mathbb{Z}, \qquad \frac{1}{N}\sum_{n=0}^{N-1} a_n W_N^{mn} W_N^{-mk} = 0,$$

and so

$$\sum_{n=0}^{N-1} a_n \left(\frac{1}{N} \sum_{m=0}^{N-1} W_N^{mn} W_N^{-mk} \right) = 0.$$

By the orthonormality the left side is a_k.

The result follows since $L(\mathbb{Z}_N)$ is N-dimensional.

b. It is sufficient to prove (3.8.9); and (3.8.9) follows from Theorem 3.8.2, part *a*, and the fact that

$$\sum_{m=0}^{N-1} f[m]\overline{g[m]} = \frac{1}{N} \sum_{n=0}^{N-1} F[n] \sum_{k=0}^{N-1} \overline{G[k]} \left(\frac{1}{N} \sum_{m=0}^{N-1} W_N^{mn} W_N^{-mk} \right). \qquad \blacksquare$$

3.8.5 Example. The DFT Matrix

a. The DFT $(N \times N)$-matrix $\mathcal{D}_N$ is defined as $(\frac{1}{\sqrt{N}} W_N^{mn})$, $m, n = 0, \ldots, N-1$, i.e.,

$$\mathcal{D}_N = \frac{1}{\sqrt{N}} \begin{pmatrix} 1 & 1 & 1 & \cdots & 1 \\ 1 & e^{-2\pi i/N} & e^{-2\pi i2/N} & \cdots & e^{-2\pi i(N-1)/N} \\ 1 & e^{-2\pi i2/N} & e^{-2\pi i4/N} & \cdots & e^{-2\pi i2(N-1)/N} \\ \vdots & & & & \\ 1 & e^{-2\pi i(N-1)/N} & e^{-2\pi i2(N-1)/N} & \cdots & e^{-2\pi i(N-1)(N-1)/N} \end{pmatrix}.$$

Thus,

$$\mathcal{D}_2 = \frac{1}{\sqrt{2}} \begin{pmatrix} 1 & 1 \\ 1 & -1 \end{pmatrix}.$$

If $f \in L(\mathbb{Z}_N)$ is considered as a column vector, then the DFT $F \in L(\mathbb{Z}_N)$ of f is the column vector

$$F = \sqrt{N} \, \mathcal{D}_N f.$$

By Theorem 3.8.2 we have

$$\mathcal{D}_N^{-1} = \frac{1}{\sqrt{N}} \overline{\mathcal{D}}_N,$$

where $\overline{\mathcal{D}}_N$ denotes complex conjugation of the entries of $\mathcal{D}_N$.

b. Note that the *trace* (Definition 3.6.2) of $\mathcal{D}_N$ is

$$\text{trace}(\mathcal{D}_N) = \frac{1}{\sqrt{N}} \sum_{n=0}^{N-1} e^{-2\pi i n^2/N} = \frac{1}{\sqrt{N}} \mathcal{G}_N^-,$$

which we shall evaluate in Theorems 3.8.9 and 3.8.10.

c.   Let $U_N^n : \mathbb{Z}_N \longrightarrow \mathbb{C}$, for fixed $n \in \mathbb{Z}_N$, be defined by $U_N^n[m] = \delta(m, n)$ for $m \in \mathbb{Z}_N$. Clearly, $\{\sqrt{N}\, U_N^n : n = 0, \ldots, N - 1\}$ is an ONB for $L(\mathbb{Z}_N)$, with the inner product defined in Definition 3.8.1e.

It is easy to check that

$$\mathcal{D}_N U_N^n = \frac{1}{\sqrt{N}}\, W_N^n, \tag{3.8.11}$$

where U_N^n and W_N^n are considered as column vectors and the left side is matrix multiplication. Similarly, a direct calculation shows that

$$\mathcal{D}_N W_N^{-n} = \sqrt{N}\, U_N^n. \tag{3.8.12}$$

For example,

$$\mathcal{D}_5 W_5^{-2}$$

$$= \frac{1}{\sqrt{5}} \begin{pmatrix} 1 & + & e^{2\pi i 2/5} & + & e^{2\pi i 4/5} & + & e^{2\pi i 6/5} & + & e^{2\pi i 8/5} \\ 1 & + & e^{2\pi i/5} & + & e^{2\pi i 2/5} & + & e^{2\pi i 3/5} & + & e^{2\pi i 4/5} \\ 1 & + & 1 & + & 1 & + & 1 & + & 1 \\ 1 & + & e^{-2\pi i/5} & + & e^{-2\pi i 2/5} & + & e^{-2\pi i 3/5} & + & e^{-2\pi i 4/5} \\ 1 & + & e^{-2\pi i 2/5} & + & e^{-2\pi i 4/5} & + & e^{-2\pi i 6/5} & + & e^{2\pi i 8/5} \end{pmatrix}. \tag{3.8.13}$$

The column matrix on the right side of (3.8.13) is $\sqrt{5}\, U_5^2 \in L(\mathbb{Z}_5)$. In fact, $\sum_{m=0}^{N-1} e^{-2\pi i m n/N} = 0$ for a fixed $n \in \mathbb{Z}_N$ by (3.8.4).

Combining (3.8.11) and (3.8.12) we see that

$$\forall n \in \mathbb{Z}_N, \qquad \mathcal{D}_N^4 U_N^n = U_N^n.$$

Since $\{\sqrt{N}\, U_N^n\}$ is an ONB for $L(\mathbb{Z}_N)$, we conclude that

$$\mathcal{D}_N^4 = I, \tag{3.8.14}$$

where I is the identity matrix, cf., Example 1.10.12 where we did the analogous calculation for the Fourier transform on $\mathbb{R}$.

d.   Equation (3.8.14) leads naturally to investigating the *eigenvalue problem* for the DFT. In fact, because of (3.8.14), *the eigenvalues of $\mathcal{D}_N$ are $\pm 1, \pm i$.* The more difficult aspect of the eigenvalue problem is the *multiplicity problem*, viz., finding the eigenvectors of $\mathcal{D}_N$ and the dimension of the space of eigenvectors for each of the eigenvalues.

In any case the complete eigenvalue problem was essentially solved by Gauss, fundamental related calculations were exposited in E. Landau's classic book, *Vorlesungen über Zahlentheorie* (Volume 1, 1927, pages 164–165), and the explicit solution was recorded in [Goo62, page 261]. Recent comprehensive contributions are due to McClellan and Parks [MP72] and Auslander and Tolimieri, e.g., [AT79]. Using a technique due to Schur, e.g., [BSh66, pages 349–353], it is shown in [AT79, page 856] that *the solution of the multiplicity problem is equivalent to evaluating* trace$(\mathcal{D}_N)$ *for all* N.

In the following material, from Example 3.8.6 to Theorem 3.8.10, we shall deal with the evaluation of the Gauss sum $\mathcal{G}_N^{\pm}$ ($\mathcal{G}_N^{\pm}$ was defined in (3.8.8)).

3.8.6 Example. $\mathcal{G}_N^{\pm} = 0$ **if 4 Divides** $N - 2$

a. If $N - 2 = 4k$, $k \geq 0$, then $N = 2(2k + 1) \equiv 2M$, where $M \geq 1$ is odd. Then

$$\text{trace}(\mathcal{D}_N) = \sum_{m=0}^{M-1} e^{-2\pi i m^2/(2M)} + \sum_{m=0}^{M-1} e^{-2\pi i (m+M)^2/(2M)}. \tag{3.8.15}$$

Expanding $(m + M)^2$ and using the fact that $e^{-2\pi i M/2} = -1$ for M odd, we see that

$$\sum_{m=0}^{M-1} e^{-2\pi i (m+M)^2/(2M)} = -\sum_{m=0}^{M-1} e^{-2\pi i m^2/(2M)} e^{-2\pi i m};$$

and, hence, the right side of (3.8.15) is 0.

b. Note that $\mathcal{G}_3^+ = i\sqrt{3}$ and $\mathcal{G}_3^- = -i\sqrt{3}$. In fact,

$$\sum_{m=0}^{2} e^{2\pi i m^2/3} = 1 + e^{2\pi i/3} + e^{2\pi i 4/3} = 1 + 2e^{2\pi i/3}$$

$$= 1 + 2\cos\frac{2\pi}{3} + 2i\sin\frac{2\pi}{3} = i\sqrt{3}$$

and

$$\sum_{m=0}^{2} e^{-2\pi i m^2/3} = 1 + 2\cos\frac{2\pi}{3} - 2i\sin\frac{2\pi}{3} = -i\sqrt{3}.$$

3.8.7 Theorem. $\left| \sum_{m=0}^{N-1} e^{2\pi i m^2/N} \right| = \sqrt{N}$, N **Odd**

Let $N \geq 1$ *be an odd integer. Then*

$$\left| \sum_{m=0}^{N-1} e^{\pm 2\pi i m^2/N} \right| = \sqrt{N}. \tag{3.8.16}$$

PROOF. Let $g[m] = e^{-2\pi i m^2/N}$, $m = 0, 1, \ldots, N - 1$; and let $G = \hat{g}$. We shall prove (3.8.16) with the minus sign in the exponent. The plus sign case is a consequence of the same argument with the signs in the DFT properly adjusted. For $n = 0, \ldots, N - 1$, we compute

$$G[2n] = \sum_{m=0}^{N-1} e^{-2\pi i m^2/N} e^{-2\pi i m(2n)/N}$$

$$= e^{2\pi i n^2/N} \sum_{m=0}^{N-1} e^{-2\pi i (m+n)^2/N} \qquad (3.8.17)$$

$$= e^{2\pi i n^2/N} \sum_{m=0}^{N-1} e^{-2\pi i m^2/N} = e^{2\pi i n^2/N} G[0].$$

The second step follows since $(m + n)^2 - n^2 = m^2 + 2mn$; and the third step follows since $W_N^{(m+N)^2} = W_N^{m^2}$ and by noting that the Gauss sum for either "$m+n$" or "m" has a domain of N consecutive integers. By (3.8.17) we have

$$\sum_{n=0}^{N-1} |G[2n]|^2 = N|G[0]|^2. \qquad (3.8.18)$$

Since N is odd,

$$\sum_{n=0}^{N-1} |G[n]|^2 = \sum_{n=0}^{N-1} |G[2n]|^2. \qquad (3.8.19)$$

For example, let $N = 5$ so that $G[2n]$, $n = 0, 1, 2$, consists of $G[0]$, $G[2]$, $G[4]$; and $G[2 \cdot 3] = G[1]$ and $G[2 \cdot 4] = G[3]$ since 5 divides $6 - 1$ and 5 divides $8 - 3$, respectively. The same phenomenon occurs for any odd $N \geq 3$ by properties of $\mathbb{Z}_N$.

Combining (3.8.18) and (3.8.19) with the Parseval formula we obtain

$$N|G[0]|^2 = N \sum_{m=0}^{N-1} |g[m]|^2 = N^2.$$

(3.8.16) is obtained.   ∎

It is more difficult to evaluate $\mathcal{G}_N^{\pm}$ than $|\mathcal{G}_N^{\pm}|$. We shall first compute $\mathcal{G}_N^{+}$ for N odd in Theorem 3.8.9. Gauss gave the first proof of Theorem 3.8.9 in 1805 after working "with all efforts" for four years, e.g., [BE81, pages 109–110]. There have been many other derivations of Theorem 3.8.9, e.g., [BE81]. We shall give Daniel Shanks' proof [Sha58].

Shanks originally devised ingenious finite term identities to prove deep infinite term identities of Euler and Gauss. The original idea for his proofs goes back to his Ph.D. thesis at the University of Maryland in 1954. The identity by which he obtained Gauss' infinite term identity is Lemma 3.8.8, and in [Sha58] he used it to obtain Gauss' Theorem (Theorem 3.8.9).

3.8.8 Lemma. Shanks Finite Identity

If $x > 0$, $P_0(x) \equiv 1$, and

$$\forall N \in \mathbb{N}, \qquad P_N(x) = \prod_{n=1}^{N} \left(\frac{1 - x^{2n}}{1 - x^{2n-1}} \right),$$

then

$$\sum_{n=1}^{2N} x^{n(n-1)/2} = \sum_{n=0}^{N-1} \frac{P_N(x)}{P_n(x)} x^{n(2N+1)}. \tag{3.8.20}$$

The proof of Lemma 3.8.8 begins with the identity

$$(1 - x^{2N}) x^{n(2N+1)} = (1 - x^{2N-1}) x^{n(2N-1)}$$
$$+ (1 - x^{2n+1}) x^{(n+1)(2N-1)} - (1 - x^{2n}) x^{n(2N-1)},$$

which we multiply by $P_{N-1}(x)/P_n(x)(1 - x^{2N-1})$.

3.8.9 Theorem. Gauss Computation of $\mathcal{G}_N^+$, N Odd

Let $N \geq 1$ be an integer. Then

$$\frac{N - 1}{4} \in \mathbb{N} \cup \{0\} \quad \textit{implies} \quad \mathcal{G}_N^+ = \sqrt{N}$$

and

$$\frac{N - 3}{4} \in \mathbb{N} \cup \{0\} \quad \textit{implies} \quad \mathcal{G}_N^+ = i\sqrt{N}.$$

(These two cases include all odd integers N.)

PROOF. We shall consider the case $\frac{N-1}{4} \in \mathbb{N} \cup \{0\}$. The other case is similar. Let $x = v^2$ and $v = e^{i\theta}$. Clearly,

$$P_M(x) = v^M \prod_{n=1}^{M} \left(\frac{v^{2n} - v^{-2n}}{v^{2n-1} - v^{1-2n}} \right). \tag{3.8.21}$$

Further, if $Q_0 \equiv 1$ and

$$Q_M = \prod_{n=1}^{M} \left(\frac{\sin 2n\theta}{\sin(2n - 1)\theta} \right),$$

then

$$\sum_{n=1}^{2M} v^{n(n-1)} = \sum_{n=0}^{M-1} \frac{Q_M}{Q_n} v^{M+n(4M+1)} \qquad (3.8.22)$$

and

$$\sum_{n=1}^{2M+1} v^{n(n-1)} = \sum_{n=0}^{M} \frac{Q_M}{Q_n} v^{M+n(4M+1)} \qquad (3.8.23)$$

by Lemma 3.8.8. For example, in the case of (3.8.22), the left side is precisely the left side of (3.8.20); and so

$$\sum_{n=1}^{2M} v^{n(n-1)} = \sum_{n=0}^{M-1} \frac{P_M(x)}{P_n(x)} x^{n(2M+1)}, \qquad (3.8.24)$$

from which (3.8.22) is obtained by substituting (3.8.21) into the right side of (3.8.24).

Letting $N = 2K + 1$ and $\theta = \frac{2\pi}{N}$, we have

$$v^{2K} = \exp i \left(2K \left(\frac{2\pi}{N} \right) \right) = \exp i \left((2K+1) \left(\frac{2\pi}{N} \right) - \left(\frac{2\pi}{N} \right) \right) = v^{-1}.$$

Thus,

$$\mathcal{G}_N^+ = \sum_{n=0}^{N-1} e^{2\pi i n^2/N} = \sum_{n=0}^{N-1} e^{i\theta n^2} = \sum_{n=0}^{N-1} v^{n^2}$$

$$= \sum_{n=-K}^{K} v^{(K+n)^2} = v^{K^2} \sum_{n=-K}^{K} v^{2nK+n^2} \qquad (3.8.25)$$

$$= v^{K^2} \sum_{n=-K}^{K} v^{n(n-1)} = v^{K^2} \left[\sum_{n=1}^{K} v^{n(n-1)} + \sum_{n=1}^{K+1} v^{n(n-1)} \right].$$

Finally, let $N = 4M + 1$ and $K = 2M$. Combining (3.8.25) with (3.8.22) and (3.8.23), we obtain

$$\mathcal{G}_N^+ = v^{K^2} \left[\sum_{n=1}^{K} v^{n(n-1)} + \sum_{n=1}^{K+1} v^{n(n-1)} \right]$$

$$= v^{4M^2} \left[\sum_{n=0}^{M-1} \frac{Q_M}{Q_n} v^{M+n(4M+1)} + \sum_{n=0}^{M} \frac{Q_M}{Q_n} v^{M+n(4M+1)} \right]$$

$$= \sum_{n=0}^{M-1} \frac{Q_M}{Q_n} v^{(M+n)(4M+1)} + \sum_{n=0}^{M} \frac{Q_M}{Q_n} v^{(M+n)(4M+1)}.$$

Note that $v^N = 1$ since $v^{2K} = v^{-1}$. Hence,

$$\mathcal{G}_N^+ = 1 + 2 \sum_{n=0}^{M-1} \frac{Q_M}{Q_n}. \tag{3.8.26}$$

Also, $Q_0 = 1$ and $Q_N > 0$ for $n = 1, \ldots, M$ since $\theta = 2\pi/N$. In fact, for such n, $0 < 2n\theta \le 4M\pi/N < \pi$. Thus, $\mathcal{G}_N^+ > 0$ if $\frac{N-1}{4} \in \mathbb{N} \cup \{0\}$. The result follows by (3.8.16). (Equation (3.8.26) gives a cryptic way to write $\sqrt{N}$ as a sum of products of quotients of sines!) ∎

In 1835, Dirichlet used Fourier series to evaluate $\mathcal{G}_N^\pm$ for all N. In this paper, Dirichlet also gave Gauss' proof of the *Law of Quadratic Reciprocity* (Remark 3.8.11*a*) once he (Gauss) had Theorem 3.8.9, e.g., Dirichlet's *Werke*, Chelsea Publishing Company, New York, pages 257–270. The following is Dirichlet's computation of $\mathcal{G}_N^\pm$. It not only shows Dirichlet's brilliance, but it also illustrates the power of elementary harmonic analysis.

3.8.10 Theorem. Dirichlet Computation of $\mathcal{G}_N^\pm$

$$\mathcal{G}_N^\pm = \begin{cases} (1+i)\sqrt{N}, \ resp., \ (1-i)\sqrt{N}, & if \ \dfrac{N}{4} \in \mathbb{N}, \\[2mm] \sqrt{N}, & if \ \dfrac{(N-1)}{4} \in \mathbb{N} \cup \{0\}, \\[2mm] 0, & if \ \dfrac{(N-2)}{4} \in \mathbb{N} \cup \{0\}, \\[2mm] i\sqrt{N}, \ resp., \ -i\sqrt{N}, & if \ \dfrac{(N-3)}{4} \in \mathbb{N} \cup \{0\}, \end{cases} \tag{3.8.27}$$

where the two values for two cases on the right side of (3.8.27) indicate $\mathcal{G}_N^+$ and $\mathcal{G}_N^-$, respectively.

PROOF. We shall evaluate $\mathcal{G}_N^-$. The calculation of $\mathcal{G}_N^+$ is similar.

  a. Let $G(\gamma) = e^{-2\pi i \gamma^2/N}$ on $\hat{\mathbb{R}}$. Then

$$\sum_{k=0}^{N-1} (\tau_{-k} G(0) + \tau_{-k} G(1)) = (1 + e^{-2\pi i/N})$$

$$+ (e^{-2\pi i/N} + e^{-2\pi i 2^2/N}) + (e^{-2\pi i 2^2/N} + e^{-2\pi i 3^2/N})$$

$$+ \cdots + (e^{-2\pi i (N-1)^2/N} + e^{-2\pi i N^2/N}) = 2\mathcal{G}_N^-.$$

Let

$$F(\gamma) = \tau_{-0} G(\gamma) + \tau_{-1} G(\gamma) + \cdots + \tau_{-(N-1)} G(\gamma)$$

for $\gamma \in \hat{\mathbb{R}}$. Then

$$F(\gamma + 1) = F(\gamma) + e^{-2\pi i \gamma^2/N} (e^{-4\pi i \gamma} - 1)$$

and so $F(1) = F(0)$. At this point, we shall only consider F defined on $[0, 1)$ and extend F as a 1-periodic function on $\hat{\mathbb{R}}$.

By the smoothness of G and the fact that $F \in C(\mathbb{T})$, we can invoke Dirichlet's Theorem (Theorem 3.1.6) and Remark 3.1.7*b* to write

$$\mathcal{G}_N^- = \frac{F(1) + F(0)}{2} = F(0) = \sum f[m],$$

where the sequence $f = \{f[m]\}$ of Fourier coefficients of F is evaluated by

$$f[m] = \int_0^1 \sum_{k=0}^{N-1} \tau_{-k} G(\gamma) e^{2\pi i m \gamma} \, d\gamma$$

$$= \sum_{k=0}^{N-1} \int_k^{k+1} G(\lambda) e^{2\pi i m \lambda} \, d\lambda = \int_0^N e^{-2\pi i \lambda^2/N} e^{2\pi i m \lambda} \, d\lambda$$

for each $m \in \mathbb{Z}$. Completing the square, we obtain

$$f[m] = e^{\pi i m^2 N/2} \int_0^N e^{-2\pi i (\gamma - mN/2)^2/N} \, d\gamma,$$

and, hence,

$$\mathcal{G}_N^- = \sum e^{\pi i m^2 N/2} \int_0^N e^{-2\pi i (\gamma - mN/2)^2/N} \, d\gamma. \tag{3.8.28}$$

If m is even, then $e^{\pi i m^2 N/2} = 1$; and if m is odd, then $e^{\pi i m^2 N/2} = e^{\pi i N/2} = i^N$. Separating the sum on the right side of (3.8.28) into even ($m = 2k$) and odd ($m = 2k + 1$) parts, and making the corresponding changes of variable, (3.8.28) becomes

$$\mathcal{G}_N^- = I_e + I_o,$$

where

$$I_e = \sum \int_{-kN}^{(1-k)N} e^{-2\pi i \lambda^2/N} \, d\lambda = \int e^{-2\pi i \lambda^2/N} \, d\lambda$$

and

$$I_o = i^N \sum \int_{(-k-(1/2))N}^{(1-k-(1/2))N} e^{-2\pi i \lambda^2/N} \, d\lambda = i^N \int e^{-2\pi i \lambda^2/N} \, d\lambda.$$

Using Theorem 2.10.1, we see that

$$\int e^{-2\pi i \lambda^2 / N} \, d\lambda = \frac{\sqrt{N}}{2}(1 - i).$$

Combining this information we have the formula,

$$\mathcal{G}_N^- = \frac{\sqrt{N}}{2}(1 - i)(1 + i^N). \tag{3.8.29}$$

b. Equation (3.8.27) follows by letting $N = 4k, 4k + 1, 4k + 2, 4k + 3$, respectively, cf., Example 3.8.6*a* for the case $N = 4k + 2$ and Example 3.8.6*b* for the case $N = 4k + 3$. ∎

3.8.11 Remark. Gauss Sums: Potpourri and Titillation

a. *Law of Quadratic Reciprocity.* The *Law of Quadratic Reciprocity* asserts that if p and q are distinct odd primes, then

$$\left(\frac{p}{q}\right)\left(\frac{q}{p}\right) = e^{2\pi i(p-1)(q-1)/8}, \tag{3.8.30}$$

where $\left(\frac{p}{q}\right)$ is the *Legendre Symbol*. It is defined as follows. If r and m are relatively prime, then r is the *quadratic residue* of m if

$$\exists n \in \mathbb{N} \quad \text{such that} \quad \frac{n^2 - r}{m} \in \mathbb{N} \cup \{0\},$$

and it is a *quadratic nonresidue* if there is no such n. Then $\left(\frac{q}{p}\right) = 1$ if p is a quadratic residue of q, and $\left(\frac{p}{q}\right) = -1$ if it is a quadratic nonresidue of q.

Equation (3.8.30) is important in Diophantine polynomial equations, e.g., [BE81], [HW65], as well as more advanced (and just as difficult!) number-theoretic topics related to the Riemann ζ-function and more general Dirichlet L-functions, e.g., [BSh66], [Cha68], [Cha70].

Equation (3.8.30) was first stated by Euler (1783). Legendre had an incomplete proof since he used a property, only later proved by Dirichlet, about primes in arithmetic progressions, e.g., Section 3.2.3. Independently, Gauss discovered (3.8.30), which he called the "Theorema aureum", in March 1795 before his 18th birthday on April 30, 1777; and he provided a rigorous proof by April 1796. He went on to give seven other proofs. His third proof is in [HW65, Chapter VI]. His fourth proof, published in 1809, used Theorem 3.8.9. This means of obtaining the Law of Quadratic Reciprocity by means of Gauss sums has been refined by magisterial lineage: Schaar (1848), Kronecker (1880), Hecke (1919), C. L. Siegel (1966), e.g., [Cha68, pages 34–42].

b.   Littlewood Flatness Problem. Let $\mathcal{U}_N$ denote the class of *unimodular trigonometric polynomials* $U(\gamma) = \sum_{n=0}^{N} u_n e^{-2\pi i n \gamma}$, i.e., $|u_n| = 1$ for $n = 0, 1, \ldots, N$. In light of a question asked in [Lit66] and expanded upon in [Lit68, Problem 19], we state the following *Littlewood Flatness Problem*: determine whether or not there are unimodular polynomials $U_N \in \mathcal{U}_N$ and positive constants $\epsilon_N > 0$ tending to zero, as $N \to \infty$, such that

$$\forall \gamma \in \mathbb{T},$$
$$(1 - \epsilon_N)\sqrt{N + 1} \leq |U_N(\gamma)| \leq (1 + \epsilon_N)\sqrt{N + 1} \tag{3.8.31}$$

for all large N. It turns out that Gauss sums and their variants play a natural role in dealing with (3.8.31).

The inequalities (3.8.31) assert that there are trigonometric polynomials $\frac{1}{\sqrt{N+1}} U_N$, whose modulus is almost identically 1 on $\mathbb{T}$ and whose coefficients have moduli that are all identically 1. Also, using the *Parseval formula*, they imply that

$$\forall N, \exists U_N \in \mathcal{U}_N \quad \text{such that} \quad \lim_{N \to \infty} \frac{\|U_N\|_{L^\infty(\mathbb{T})}}{\|U_N\|_{L^2(\mathbb{T})}} = 1. \tag{3.8.32}$$

There have been herculean attempts to prove (3.8.31), sometimes in concert with subtle failures. Finally, in 1980, Kahane [Kah80] proved (3.8.31) by showing that such polynomials U_N *exist*, cf., [QS96]. His proof has a fundamental probabilistic component, and it remains to *construct* the U_N. There *are* constructions in the more general case where it is only assumed that the moduli of the coefficients are bounded by 1, e.g., [BNe74], [Benk92].

c.   Antenna Theory. The ratio $\|F\|_{L^\infty(\mathbb{T})}/\|F\|_{L^2(\mathbb{T})}$ is called the *crest factor* of F. This relationship between L^∞ ("maximum and minimum values") and L^2 ("energy") norms plays a role in a number of applications. Further, under certain constraints, trigonometric polynomials provide a natural model for arrays of energy transmitters and receptors in fields such as acoustics, electromagnetism, and seismology, e.g., [Sche60].

In particular, the space $\mathcal{U}_N$ combined with (3.8.32) can be used in *antenna array signal processing*, e.g., [Stei76], [Hay85]. The space $\mathcal{U}_N$ gives way to other models for other aspects of antenna theory. For example, transmitters and receptors can be placed at points $x_1, \ldots, x_N \in \mathbb{R}^3$; and space factors for the corresponding array outputs can be modeled by trigonometric polynomials $P(\gamma) = \sum_{n=1}^{N} \exp 2\pi i x_n \cdot \gamma$. Since each of the coefficients is 1, it becomes relevant to check how close P is to the δ measure, and to analyze the impact of the sidelobes of P outside of a neighborhood of the origin, e.g., [BHe93], cf., equation (UP) in Remark 1.1.4 and Example 2.4.8b, as well as the case of absolute values for the examples considered in part *b* and Theorem 2.10.3a.

d. *Lagrange's Theorem and Surgery.* The quadratic forms $\sum r_{jk} c_k \overline{c_j}$ with which we have dealt have had complex entries; and we have established central results in *harmonic analysis* in the case these forms are nonnegative. It turns out that quadratic forms with integer or rational number entries have been central to the development of *number theory* and *algebra*. The illustrious Dedekind, Frobenuis, Minkowski, E. Noether, E. Artin, C. L. Siegel, Eichler, and Hasse are all major contributors.

Lagrange (1770) proved that *every positive integer is the sum of four squares of integers*. More generally, if $\{r_{jk} : j, k = 1, \ldots, N\} \subseteq \mathbb{N}$, it is natural to ask which integers $n \in \mathbb{N}$ are of the form $n = \sum r_{jk} c_k c_j$ for some sequence $c = \{c_j : 1, \ldots, N\} \subseteq \mathbb{N}$, and how many such "solutions" c there are, e.g., [HW65, Chapter 20]. The problem in this generality is unsolved. Many of the special cases that are known, as well as Minkowski's classification of quadratic forms over $\mathbb{Q}$, involve Gauss sums, e.g., [Scha85, Chapter 5]. These Gauss sums are of the form $\sum e^{2\pi i q(x)}$ where q is a quadratic form associated with some algebraic structure. For example, if b is a bilinear function $b : V \times V \to K$, where V is a vector space over a field K, then $q(x) = b(x, x)$ defines a quadratic form.

Surgery invariants can also be determined by evaluating Gauss sums of this type. The purpose of surgery in *differential topology* is to characterize simply connected manifolds in higher dimensions, and a basic Gauss sum was computed in this setting by J. W. Morgan and D. P. Sullivan (*Ann. of Math.*, 1974). The first use of Gauss sums in surgery problems was made by Edgar Brown.

3.9. Fast Fourier Transform

Let $N > 1$ and let $f : \mathbb{Z}_N \to \mathbb{C}$ have DFT F. As is apparent from Example 3.8.5*a*, $\{F[n] : n \in \mathbb{Z}_N\}$ can be computed with $(N - 1)^2$ operations, where an *operation* is defined to mean a complex multiplication followed by a complex addition. The *Fast Fourier Transform* (FFT) is an algorithm to compute a DFT by $N \log_2 N$ operations in the case $N = 2^r$, e.g., Example 3.9.2. The fundamental paper on the FFT is due to Cooley and Tukey [CT65], and it involves an idea that they refer to as a *two-step algorithm*.

It turns out that two-step Fourier analysis algorithms have been used in various applications since early in the nineteenth century. The first published results are due to Francesco Carlini (1828) in his research on hourly barometric variations. This historical fact, as well as the fact that Gauss had a general form of the FFT as early as 1805, are found in the fascinating article by M. T. Heideman, D. H. Johnson, and C. S. Burrus [HJB84] on the history of the FFT, cf., [DVe90], [IEEE69], [OS75, Chapter 6], [RR72], [Walk91] for other developments. Gauss' results were published posthumously in 1866.

3.9.1 Theorem. Two-Step FFT Algorithm

Let $N_1, N_2 > 1$ and let $f : \mathbb{Z}_N \to \mathbb{C}$ have DFT F, where $N = N_1 N_2$.

a. Each $n = 0, 1, \ldots, N-1$ has the unique representation $n = n_2 N_1 + n_1$, for some $n_1 = 0, 1, \ldots, N_1 - 1$ and some $n_2 = 0, 1, \ldots, N_2 - 1$; and each $m = 0, 1, \ldots, N-1$ has the unique representation $m = m_1 N_2 + m_2$, for some $m_1 = 0, 1, \ldots, N_1 - 1$ and some $m_2 = 0, 1, \ldots, N_2 - 1$.

b. In the format of part a, F can be written as

$$F[n_2 N_1 + n_1] = \sum_{m_2=0}^{N_2-1} \left(\sum_{m_1=0}^{N_1-1} f[m_1 N_2 + m_2] W_{N_1}^{m_1 n_1} W_N^{m_2 n_1} \right) W_{N_2}^{m_2 n_2} \qquad (3.9.1)$$

for each $n = n_2 N_1 + n_1$.

c. The total number of operations required to compute $\{F[n] : n = n_2 N_1 + n_1,\ n_1 \in \mathbb{Z}_{N_1},\ \text{and}\ n_2 \in \mathbb{Z}_{N_2}\}$ is

$$N(N_1 + N_2).$$

PROOF. **a.** Part a is immediate when, for the case of n, we think of the N_2 equispaced elements $0, N_1, 2N_1, \ldots, (N_2 - 1)N_1$ of $\mathbb{Z}_N$. In fact, for each fixed $n_2 N_1$ we obtain all the elements of $\mathbb{Z}_N$ that are greater than or equal to $n_2 N_1$ and less than $(n_2 + 1)N_1$ by adding $n_1 = 0, 1, \ldots, N_1 - 1$ to $n_2 N_1$. (If $n_2 = N_2 - 1$, then $(n_2 + 1)N_1 = 0$.)

b. For a given n, choose n_1, n_2 for which $n = n_2 N_1 + n_1$. Then

$$F[n_2 N_1 + n_1] = \sum_{m_1, m_2} f[m_1 N_2 + m_2] W_N^{m_1 N_2 n} W_N^{m_2 n}. \qquad (3.9.2)$$

We calculate

$$e^{-2\pi i m_1 N_2 n/N} = e^{-2\pi i m_1 n/N_1} = e^{-2\pi i m_1 (n_2 N_1 + n_1)/N_1}$$

and

$$e^{-2\pi i m_2 (n_2 N_1 + n_1)/N} = e^{-2\pi i m_2 n_2/N_2} e^{-2\pi i m_2 n_1/N},$$

and so the right side of (3.9.2) is

$$\sum_{m_1, m_2} f[m_1 N_2 + m_2] W_{N_1}^{m_1 n_1} W_N^{m_2 n_1} W_{N_2}^{m_2 n_2}.$$

Writing out the domains of m_1 and m_2, this is precisely (3.9.1), which we shall also write as

$$F[n] = \sum_{m_2=0}^{N_2-1} F[m_2, n_1] W_{N_2}^{m_2 n_2},$$

where

$$F[m_2, n_1] = \sum_{m_1=0}^{N_1-1} f[m_1 N_2 + m_2] W_{N_1}^{m_1 n_1} W_N^{m_2 n_1}.$$

c.   For fixed n_1, n_2, and m_2, we can compute

$$\sum_{m_1=0}^{N_1-1} f[m_1 N_2 + m_2] W_{N_1}^{m_1 n_1} \qquad\qquad (3.9.3)$$

by $N_1 - 1$ operations, where the first of these operations yields

$$f[0 \cdot N_2 + m_2] + f[1 \cdot N_2 + m_2] W_{N_1}^{1 \cdot n_1},$$

and where the $(N_1 - 1)$st operation yields the complete sum (3.9.3). Multiplying this sum by $W_N^{m_2 n_1}$ we obtain $F[m_2, n_1]$ in $(N_1 - 1) + 1 = N_1$ operations.

Now, using the data $\{F[m_2, n_1]\}$ that required N_1 operations to compute, we can compute each

$$F[n] = \sum_{m_2=0}^{N_2-1} F[m_2, n_1] W_{N_2}^{m_2 n_2}$$

by an *additional* N_2 operations. Thus, for each pair (n_1, n_2), $F[n] = F[n_2 N_1 + n_1]$ can be computed by means of (3.9.1) in $N_1 + N_2$ operations. Since there are $N = N_1 N_2$ pairs (n_1, n_2), all the data $\{F[n] : n = 0, \ldots, N-1\}$ can be computed by means of (3.9.1) in $N(N_1 + N_2)$ operations. ∎

3.9.2 Example. Order $N \log_2 N$ Algorithm

a.   Let $N = 2N_1$. Then each $n = 0, 1, \ldots, N-1$ can be written as $n = n_2 N_1 + n_1$ for $n_1 = 0, 1, \ldots, N_1 - 1$ and $n_2 = 0, 1$. Hence, if $F : \mathbb{Z}_N \to \mathbb{C}$, defined as

$$F[n] = \sum_{m=0}^{N-1} f[m] W_N^{mn},$$

is the DFT of $f : \mathbb{Z}_N \to \mathbb{C}$, then

$$F[n] = F[n_2 N_1 + n_1] = \sum_{m_1, m_2} f[2m_1 + m_2] W_N^{(2m_1 + m_2)n},$$

where $m_1 = 0, 1, \ldots, N_1 - 1$ and $m_2 = 0, 1$. We compute

$$F[n_2 N_1 + n_1] = \sum_{m_2=0}^{1} \sum_{m_1=0}^{N_1-1} f[2m_1 + m_2] W_{2N_1}^{(2m_1+m_2)n}$$

$$= \sum_{m_2=0}^{1} \sum_{m_1=0}^{N_1-1} f[2m_1 + m_2] W_{N_1}^{m_1 n_1} W_N^{m_2 n_1} e^{-\pi i m_2 n_2},$$

since

$$e^{-2\pi i 2 m_1 n/(2N_1)} = e^{-2\pi i m_1 (n_2 N_1 + n_1)/N_1} = W_{N_1}^{m_1 n_1}$$

and

$$e^{-2\pi i m_2 n/(2N_1)} = e^{-2\pi i m_2 (n_2 N_1 + n_1)/(2N_1)}$$

$$= e^{-\pi i m_2 n_2} e^{-2\pi i m_2 n_1/(2N_1)} = e^{-\pi i m_2 n_2} W_N^{m_2 n_1}.$$

Therefore,

$$F[n_2 N_1 + n_1] = \sum_{m_1=0}^{N_1-1} f[2m_1] W_{N_1}^{m_1 n_1} + e^{-\pi i n_2} W_N^{n_1} \sum_{m_1=1}^{N_1-1} f[2m_1 + 1] W_{N_1}^{m_1 n_1}.$$

Consequently, because

$$e^{-\pi i n_2} W_N^{n_1} = e^{-2\pi i n_2 N_1/(2N_1)} W_N^{n_1} = W_N^n,$$

we obtain the *FFT algorithm*,

$$F[n_2 N_1 + n_1] = \sum_{m_1=0}^{N_1-1} f[2m_1] W_{N_1}^{m_1 n_1} + W_N^n \sum_{m_1=1}^{N_1-1} f[2m_1 + 1] W_{N_1}^{m_1 n_1} \qquad (3.9.4)$$

for each $n = n_2 N_1 + n_1$.

b. Let $\#(K)$ be the number of multiplications required to compute the DFT of $f : \mathbb{Z}_K \to \mathbb{C}$. Clearly, $\#(K) \le (K-1)^2$.

In dealing with the right side of (3.9.4), we see that $2\#(N_1)$ multiplications are required to compute

$$s_e(n_1) = \sum_{m_1=0}^{N_1-1} f[2m_1] W_{N_1}^{m_1 n_1} \quad \text{and} \quad s_o(n_1) = \sum_{m_1=0}^{N_1-1} f[2m_1 + 1] W_{N_1}^{m_1 n_1}$$

for all $n_1 \in \mathbb{Z}_{N_1}$.

Now note that $W_N^n = \pm W_N^{n_1}$ depending on whether $n_2 = 0, 1$ in the representation $n = n_2 N_1 + n_1$. Thus, N_1 multiplications are required to compute $\{W_N^{n_1} s_o(n_1) : n_1 \in \mathbb{Z}_{N_1}\}$ for given data $\{W_N^{n_1}, s_o(n_1) : n_1 \in \mathbb{Z}_{N_1}\}$. Consequently, (3.9.4) allows us to write

$$\#(2N_1) = 2\#(N_1) + N_1. \qquad (3.9.5)$$

c. Because of (3.9.5) and the fact that $\#(2) = 1$ (since the DFT of $\{f[0], f[1]\}$ is $\{f[0] + f[1], f[0] + f[1]W_2^1\}$), we have

$$\#(2) = 1$$

$$\#(4) = 2 + 2 = 4 = \frac{1}{2} 4 \log_2 4$$

$$\#(8) = 8 + 4 = 12 = \frac{1}{2} 8 \log_2 8$$

$$\vdots$$

In fact, *if $N = 2^r$, then the number of multiplications required to compute the* DFT *on $\mathbb{Z}_N$ is*

$$\frac{N}{2} \log_2 N. \tag{3.9.6}$$

To prove (3.9.6) we proceed by induction. We just checked the result for $N = 2, 4, 8$. Now assume (3.9.6) is true for $N = N_1 = 2^p$, i.e., make the induction hypothesis that $\#(N_1) = (N_1/2) \log_2 N_1$. Letting $N = 2N_1$, (3.9.5) and the induction hypothesis imply

$$\#(N) = 2\#(N_1) + N_1 = N_1 \log_2 N_1 + N_1$$

$$= \frac{N}{2}(\log_2 N_1 + 1) = \frac{N}{2} \log_2 N,$$

and the proof is complete.

   d.   The *FFT algorithm*, which achieves (3.9.6), is *defined* by means of (3.9.4).

Equation (3.9.4), or the more general (3.9.1), can be viewed in terms of replacing the DFT $N \times N$-matrix $\mathcal{D}_N$, having no 0-entries, in terms of a collection of matrices with many 0-entries, cf., Example 3.9.3 and Carleson's remark at the end of Section 3.2.8.

In this spirit, C. Rader (1968) introduced an important idea, based on the fact that $\mathbb{Z}_p$ is a field, which ultimately led to Winograd's celebrated algorithms (1978) for computing the DFT on $\mathbb{Z}_p$, e.g., [RR72], [DV90].

3.9.3   Example.   Sparse FFT Matrices

We shall quantify the remark in Example 3.9.2d about matrices with 0 entries.

The right side of (3.9.4) splits the domain of $f : \mathbb{Z}_N \to \mathbb{C}$, $N = 2N_1$, into its even and odd parts. This can be accomplished by a matrix operation. For example, if $N = 8$ and

$$\mathcal{C}_8 = \begin{pmatrix} 1 & 0 & 0 & 0 & 0 & 0 & 0 & 0 \\ 0 & 0 & 1 & 0 & 0 & 0 & 0 & 0 \\ 0 & 0 & 0 & 0 & 1 & 0 & 0 & 0 \\ 0 & 0 & 0 & 0 & 0 & 0 & 1 & 0 \\ 0 & 1 & 0 & 0 & 0 & 0 & 0 & 0 \\ 0 & 0 & 0 & 1 & 0 & 0 & 0 & 0 \\ 0 & 0 & 0 & 0 & 0 & 1 & 0 & 0 \\ 0 & 0 & 0 & 0 & 0 & 0 & 0 & 1 \end{pmatrix},$$

then

$$\mathcal{C}_8 f^T = (f[0], f[2], f[4], f[6], f[1], f[3], f[5], f[7])^T. \tag{3.9.7}$$

The rule for constructing $\mathcal{C}_N$ is obvious.

The next operation in (3.9.4) tells us that $\sqrt{N}\mathcal{D}_N$ is related to two copies of $\sqrt{N_1}\mathcal{D}_{N_1}$ embedded into an $N \times N$ array. These two copies address the even and odd parts of the domain of f, resp. Thus, we introduce the $N \times N$ matrix

$$
\mathcal{B}_N = \left(\begin{array}{c|c} \sqrt{N_1}\mathcal{D}_{N_1} & \\ \hline & \sqrt{N_1}\mathcal{D}_{N_1} \end{array} \right),
$$

where the first and third quadrants are each $N_1 \times N_1$ 0-matrices. For example, if $N = 8$, then

$$
\mathcal{B}_8 \mathcal{C}_8 f^T = (s_e(0), s_e(1), s_e(2), s_e(3), s_o(0), s_o(1), s_o(2), s_o(3))^T,
$$

where the "even" and "odd" sums s_e and s_o are the DFTs defined in Example 3.9.2b for the case $N_1 = 4$.

Finally, we have to introduce a matrix that incorporates the factor W_N^n of $s_o(n_1)$ (when $n = n_2 N_1 + n_1$) and adds it to $s_e(n_1)$. Recall from Example 3.9.2b that $W_N^n = W_N^{n_1}$ if $n_2 = 0$ and $W_N^n = -W_N^{n_1}$ if $n_2 = 1$. Therefore, if $N = 8$ we set

$$
\mathcal{A}_8 = \left(\begin{array}{cccc|cccc}
1 & 0 & 0 & 0 & 1 & 0 & 0 & 0 \\
0 & 1 & 0 & 0 & 0 & W_8^1 & 0 & 0 \\
0 & 0 & 1 & 0 & 0 & 0 & W_8^2 & 0 \\
0 & 0 & 0 & 1 & 0 & 0 & 0 & W_8^3 \\
\hline
1 & 0 & 0 & 0 & -1 & 0 & 0 & 0 \\
0 & 1 & 0 & 0 & 0 & -W_8^1 & 0 & 0 \\
0 & 0 & 1 & 0 & 0 & 0 & -W_8^2 & 0 \\
0 & 0 & 0 & 1 & 0 & 0 & 0 & -W_8^3
\end{array} \right)
$$

and obtain

$$
\mathcal{A}_8 \mathcal{B}_8 \mathcal{C}_8 f^T = F^T.
$$

The rule for constructing $\mathcal{A}_N$ is obvious, and so

$$
\mathcal{A}_N \mathcal{B}_N \mathcal{C}_N = \sqrt{N} \mathcal{D}_N
$$

(as easy as ...).

3.9.4 Remark. Butterflies and Bit Reversal

a. Let $N = 2^r$ and let $f : \mathbb{Z}_N \to \mathbb{C}$ have DFT F. Besides the matrix formulation of the FFT algorithm (3.9.4) in Example 3.9.3, we can also formulate it in terms of diagrams whose basic components are called *butterflies*.

To define a butterfly, we begin by *rewriting* (3.9.4) as

$$F[k] = F_0[k] + W_N^k F_1[k], \qquad k = 0, 1, \ldots, \frac{1}{2}N - 1, \qquad (3.9.8)$$

and

$$F\left[k + \frac{1}{2}N\right] = F_0[k] - W_N^k F_1[k], \qquad k = 0, 1, \ldots, \frac{1}{2}N - 1, \quad (3.9.9)$$

where

$$F_0[k] = \sum_{m=0}^{(N/2)-1} f[2m] W_{N/2}^{mk} \quad \text{and} \quad F_1[k] = \sum_{m=0}^{(N/2)-1} f[2m+1] W_{N/2}^{mk}, \quad (3.9.10)$$

and where we have again used the fact that $W_N^{n_2(N/2)+k} = \pm W_N^k$ depending of whether $n_2 = 0, 1$.

The DFT $F[k]$, $k = 0, 1, \ldots, N-1$, written in terms of the calculations (3.9.8)–(3.9.10), can be visualized as

$$
\begin{array}{ccc}
F_0[k] & \longrightarrow & F_0[k] + W_N^k F_1[k] \\
 & \diagdown\diagup & \\
F_1[k] & \longrightarrow & F_0[k] - W_N^k F_1[k],
\end{array}
\qquad (3.9.11)
$$

where $k = 0, 1, \ldots, \frac{1}{2}N - 1$; and this diagram is called a *butterfly*.

b. The butterfly (3.9.11) can be viewed as a construction of the DFT $F : \mathbb{Z}_N \to \mathbb{C}$ in terms of the two DFTs, F_0 and F_1, on $\mathbb{Z}_{N/2}$.

In the same way, F_0 and F_1 can each be constructed in terms of pairs of DFTs on $\mathbb{Z}_{N/4}$. For example,

$$F_0[k] = F_{00}[k] + W_{N/2}^k F_{01}[k]$$

and

$$F_0\left[k + \frac{1}{4}N\right] = F_{00}[k] - W_{N/2}^k F_{01}[k],$$

for $k = 0, 1, \ldots, \frac{1}{4}N - 1$, where F_{00} is the DFT of

$$\{f[0], f[4], f[8], \ldots, f[N-4]\}$$

and F_{01} is the DFT of $\{f[2], f[6], f[10], \ldots, f[N-2]\}$.

Since $N = 2^r$, this procedure can be reduced to the consideration of 2^{r-1} DFTs on $\mathbb{Z}_2$. Each of these is the DFT of a pair $\{f[j], f[k]\}$, and the pairs are mutually disjoint. Clearly, the stepwise evolution of F from these 2^{r-1} DFTs on $\mathbb{Z}_2$ can be pictured and understood in terms of butterflies, e.g., [OS75, Chapter 6], [Walk91].

Computationally, it is convenient to compute the 2-point DFTs first, then the 4-point DFTs, etc.

c. Suppose $f : \mathbb{Z}_N \to \mathbb{C}$ is given and the computation of its DFT F is desired in the natural ordering $(F[0], F[1], \ldots, F[N-1])$, i.e., for a given f, a computational device will compute an ordered N-tuple $(F_0, \ldots, F_{N-1})$, and we want to be sure that $F_k = F[k]$ for each k.

From (3.9.10) it is clear that if we begin with the DFTs of the pairs $\{f[0], f[1]\}$, $\{f[2], f[3]\}$, etc., we shall *not* obtain F in its natural ordering. It turns out that f *can* be ordered in such a way so that the DFTs of consecutive pairs $\{f[j], f[k]\}$ in this ordering yield the desired natural ordering of F. The procedure is called *bit reversal*. For example, if $N = 8$, then the DFTs of $\{f[0], f[4]\}$, $\{f[2], f[6]\}$, $\{f[1], f[5]\}$, $\{f[3], f[7]\}$ will yield the ordered N-tuple $(F[0], \ldots, F[N-1])$ when the halving procedure of part *b* is implemented.

Bit reversal is defined as follows for $N = 2^r$. At level $r = 1$ the bit reversal ordering of the set $\{0, 1\}$ is the ordered 2^1-tuple $(0, 1)$. At level $r = 2$ the bit reversal ordering of the set $\{0, 1, 2, 3\}$ is the ordered 2^2-tuple $(0, 2, 1, 3)$. Inductively, at level m suppose the set $\{0, 1, \ldots, 2^m - 1\}$ has as its bit reversal ordering the ordered $M = 2^m$-tuple,

$$(b_0, \ldots, b_{M-1}).$$

Then, by definition, at level $m + 1$, the *bit reversal* ordering of the set $\{0, 1, \ldots, 2M - 1\}$ is

$$(2b_0, 2b_1, \ldots, 2b_{M-1}, 2b_0 + 1, 2b_1 + 1, \ldots, 2b_{M-1} + 1).$$

For example, the bit reversal orderings at levels 3 and 4 are

$$(0, 4, 2, 6, 1, 5, 3, 7)$$

and

$$(0, 8, 4, 12, 2, 10, 6, 14, 1, 9, 5, 13, 3, 11, 7, 15),$$

respectively. The term "bit reversal" is appropriate since the coefficients of the binary expansions (part *d*) of integers are reversed at the critical step in the above process. This is the reason for the subscripts 0 and 1 in part *b*.

d. The *binary expansion* of $n \in \{0, 1, \ldots, 2^r\}$ is

$$n = \sum_{j=1}^{r} \epsilon_j 2^{j-1},$$

where $\epsilon_j \in \{0, 1\}$. For example, if $\epsilon_0 = \epsilon_1 = \cdots = \epsilon_r = 0$, then $n = 0$; and if $\epsilon_0 = \epsilon_1 = \cdots = \epsilon_{r-1} = 1$, then $n = \sum_{j=1}^{r} 2^{j-1} = (2^r - 1)/(2 - 1) = 2^r - 1$. Thus, each such n is well defined by an r-array $(\epsilon_1, \ldots, \epsilon_r)$ of 0's and 1's.

e. Suppose $\{X_n^r\}$ is a "tree of spaces", where r designates the level and where, for each fixed $r \geq 0$, there are $N = 2^r$ elements X_n^r, indexed by n. Using the binary expansion of part d we write

$$X_n^r = X_{(\epsilon_1,\dots,\epsilon_r)}^r;$$

and, using the bit reversal ordering, the tree $\{X_n^r\}$ has the form

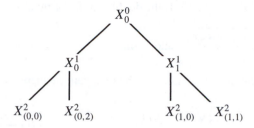

etc. At level $r - 1$ the space $X_{(\epsilon_1,\dots,\epsilon_{m-1},0)}^{r-1}$ is the (single) parent of

$$X_{(\epsilon_1,\dots,\epsilon_{m-1},0)}^r \quad \text{and} \quad X_{(\epsilon_1,\dots,\epsilon_{m-1},1)}^r.$$

Tree models, including some where bit reversal ordering is essential, abound in mathematics and engineering. There are applications, for example, in image compression, channel crosstalk reduction, Walsh functions and the waveletpackets of Coifman, Meyer, and Wickerhauser, subband coding and the theory of nonlinear waveletpackets, frequency localization, multirate systems and bit allocation, and C. Fefferman's proof of Carleson's Theorem proving the Lusin Conjecture, e.g., [BA83], [BF94, Chapter 10], [BSa94], [Dau92], [Fef73], [Mey90].

    f. With regard to the algorithmic butterflies of part *a*, recall the locally lacunary butterfly from Example 1.4.4 in Chapter 1. Even Chapter 2 has a lepidopteral connection: Laurent Schwartz is not only a world class mathematician but has seven butterflies named after him!

    The extension of Carleson's Theorem (Section 3.2.8) to two dimensions is true for partial sums taken over squares $\{(m, n) : -N \le m, n \le N\}$. C. Fefferman's proof uses butterfly-shaped subsets of $\mathbb{Z} \times \mathbb{Z}$ in a fundamental way [Fef71], cf., the proofs by Sjölin and Tevzadze at about the same time.

3.9.5 Example. Computation of W_N, $N = 2^r$

Let $C_r = \cos(2\pi/2^r)$ and $S_r = \sin(2\pi/2^r)$. If $r = 1$, then $C_1 = -1$ and $S_1 = 0$. Using the half-angle formulas

$$\cos\frac{x}{2} = \sqrt{\frac{1 + \cos x}{2}} \quad \text{and} \quad \sin\frac{x}{2} = \sqrt{\frac{1 - \cos x}{2}},$$

we can compute C_r and S_r for large values of r by the recurrence equations

$$C_{r+1} = \sqrt{\frac{1+C_r}{2}} \quad \text{and} \quad S_{r+1} = \sqrt{\frac{1-C_r}{2}}. \tag{3.9.12}$$

The problem with (3.9.12) from a computational point of view is the possibility that $\lim_{r\to\infty}(1-C_r) = 0$ might lead to computational instability. On the other hand, since $\sin^2 x + \cos^2 x = 1$, the double angle formulas can be written as

$$\cos x = \sqrt{\frac{1+\cos 2x}{2}} \quad \text{and} \quad \sin x = \frac{\sin 2x}{2\cos x} = \frac{\sin 2x}{2\sqrt{(1+\cos 2x)/2}};$$

and so we can compute C_r and S_r by the recurrence equations

$$C_{r+1} = \frac{1}{2}\sqrt{2(1+C_r)} \quad \text{and} \quad S_{r+1} = \frac{S_r}{\sqrt{2(1+C_r)}}. \tag{3.9.13}$$

Not only does it suffice to compute just one square root for both equations in (3.9.13), but the aforementioned computational instability is avoided. The associated error analysis and comparisons are due to S. R. Tate (*Trans. IEEE-SP* **43** (1995), 1709–1711).

3.10. Periodization and Sampling

This section provides *introductory* material on *regular*, i.e., *equispaced* or *uniform*, *sampling* of signals. We shall establish the role of *periodization* and the *Poisson Summation Formula* (PSF) in such sampling, cf., [BZ96] for a more extensive treatment.

The subject of sampling, whether as method, point of view, or theory, weaves its fundamental ideas through a panorama of engineering, mathematical, and scientific disciplines. Results have been discovered in one or another discipline independently of similar results in other disciplines. There are spectacular expositions and research-tutorials by Higgins [Hig85] and Butzer, Splettstösser, and Stens [BSS88].

3.10.1 Definition. Periodization on $\mathbb{Z}$

a. Let $N > 1$ and let $f \in \ell^1(\mathbb{Z})$. The *N-periodization* of f is the function $\overset{\circ}{f}_N \in L(\mathbb{Z}_N)$ defined as

$$\forall n \in \mathbb{Z}_N, \qquad \overset{\circ}{f}_N[n] = \sum_k f[n-kN]. \tag{3.10.1}$$

The summation is over $\mathbb{Z}$ and $\overset{\circ}{f}_N$ is well defined since $f \in \ell^1(\mathbb{Z})$.

Conversely, if $g \in L(\mathbb{Z}_N)$, then there are infinitely many sequences $f \in \ell^1(\mathbb{Z})$ for which $\overset{\circ}{f}_N = g$. For example, we can define

$$f[n] = \begin{cases} \dfrac{1}{2} g[n], & \text{if } n = 0, 1, \ldots, 2N - 1, \\ 0, & \text{otherwise,} \end{cases}$$

noting that $g[0] = g[N], g[1] = g[N+1], \ldots, g[N-1] = g[2N-1]$.

b. If $N > 1$ and $f \in \ell^1(\mathbb{Z})$, then the DFT of $\overset{\circ}{f}_N$ is $\overset{\circ}{F}_N$, where

$$\forall n = 0, 1, \ldots, N - 1, \qquad \overset{\circ}{F}_N [n] = \sum_{m=0}^{N-1} \overset{\circ}{f}_N [m] e^{-2\pi i mn/N}.$$

Letting $\hat{f} = F \in A(\mathbb{T})$, we compute

$$\begin{aligned}
\overset{\circ}{F}_N [n] &= \sum_{k} \sum_{m=0}^{N-1} f[m - kN] e^{-2\pi i mn/N} \\
&= \sum_{k} \sum_{j=kN}^{(k+1)N-1} f[j] e^{-2\pi i (j+kN)n/N} \\
&= \sum_{k} \sum_{j=kN}^{(k+1)N-1} f[j] e^{-2\pi i jn/N} = \sum_{j} f[j] e^{-2\pi i jn/N} = F\left(\frac{n}{N}\right),
\end{aligned}$$

i.e.,

$$\forall n = 0, 1, \ldots, N - 1, \qquad \overset{\circ}{F}_N [n] = F\left(\frac{n}{N}\right). \tag{3.10.2}$$

c. Clearly, if $f \in L(\mathbb{Z}_N)$, then we can think of f as an N-periodic element of $\ell^\infty(\mathbb{Z})$.

Notationally, if $f \in \ell^1(\mathbb{Z})$, $N > 1$, and $\overset{\circ}{f}_N$ is the N-periodization of f, then we shall write $f_N : \mathbb{Z} \longrightarrow \mathbb{C}$ to designate $\overset{\circ}{f}_N \in L^1(\mathbb{Z}_N)$ considered as an element of $\ell^\infty(\mathbb{Z})$.

d. With the notation of part *c* we have

$$f_N \in \ell^\infty(\mathbb{Z}) \setminus \bigcup_{p=1}^{\infty} \ell^p(\mathbb{Z}).$$

3.10.2 Example. Sampled Signals

a. We defined the unit $u \in \ell^1(\mathbb{Z})$ in Definition 3.5.1 as $u[0] = 1$ and $u[n] = 0$ if $n \neq 0$. Let $N > 1$ and consider the N-periodization $\overset{\circ}{u}_N \in L(\mathbb{Z}_N)$ of u, i.e.,

$$\forall n \in \mathbb{Z}_N, \qquad \overset{\circ}{u}_N [n] = \sum_{k} u[n - kN].$$

As a function on $\mathbb{Z}$, u_N is 1 on $\{jN : j \in \mathbb{Z}\}$ and 0 otherwise. The function u_5 is depicted in Figure 3.4.

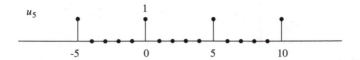

FIGURE 3.4.

b. Let $f \in \ell^1(\mathbb{Z})$ and let $N > 1$. The *sampled signal* $f_s : \mathbb{Z} \longrightarrow \mathbb{C}$, corresponding to f and N, is defined by

$$\forall n \in \mathbb{Z}, \qquad f_s[n] = \sum_k f[kN]u[n - kN].$$

Clearly, $f_s \in \ell^1(\mathbb{Z})$ (Definition 3.5.1), and it takes the value $f[jN]$ for $n = jN$ and is 0 otherwise.

3.10.3 Definition. Fourier Series of Measures

a. Let $\mu \in M(\mathbb{T})$, e.g., Definition 3.1.9. The Fourier series of μ is the series

$$S(\mu)(\gamma) = \sum \check{\mu}[n]e^{-2\pi i n \gamma}, \tag{3.10.3}$$

where

$$\forall n \in \mathbb{Z}, \qquad \check{\mu}[n] = \int_{\mathbb{T}} e^{2\pi i n \gamma} \, d\mu(\gamma). \tag{3.10.4}$$

Thus, the sequence $\check{\mu}$ is an element of $\ell^\infty(\mathbb{Z})$, and the numbers $\check{\mu}[n]$ are the *Fourier coefficients* of μ.

b. Just as we defined $M(\mathbb{T})$ in Definition 3.1.9, we could also have defined the space $D'(\mathbb{T})$ of distributions on $\mathbb{T}$. As such, the theory of Fourier series of distributions can be developed, e.g., [Edw67], [Sch61], [Sch66]. It is a remarkable and elementary theorem of that theory that *sequences* $\{c_n\}$, *for which*

$$\exists M, m \quad such \ that \ \forall n \in \mathbb{Z} \backslash \{0\}, \quad |c_n| \leq M|n|^m,$$

are precisely the sequences of Fourier coefficients of $D'(\mathbb{T})$.

In particular, $M(\mathbb{T}) \subseteq D'(\mathbb{T})$. The simplicity of characterizing the Fourier coefficients of $D'(\mathbb{T})$ does *not* carryover to elementary characterizations of many of its subspaces, such as $M(\mathbb{T})$ and $L^1(\mathbb{T})$.

c. The space of *hyperdistributions*, which includes $D'(\mathbb{T})$, is characterized by the fact that the Fourier coefficients $\{c_n\}$ of hyperdistributions satisfy the condition,

$$\forall \epsilon > 0, \quad \exists N \quad \text{such that} \quad \forall |n| > N, \quad |c_n| \le e^{\epsilon |n|}.$$

Hyperdistributions are related to Fantappié's theory of *analytic functionals*—a product of the early 1940s, before Laurent Schwartz' work. Hyperdistributions were essentially used by Beurling in his characterization of sets having positive *capacity* (1947) [Beu89, Volume 2, pages 125–145]. Their role in providing a reformulation of Riemann's U-sets (in terms of pseudomeasures T on $\mathbb{T}$ for which $\hat{T} \in c_0(\mathbb{Z})$) is found in [Ben71, Chapter 3], [KS63, pages 53-57].

3.10.4 Example. Calculation for $\hat{u}_N$

a. If $f \equiv 1 \in \ell^\infty(\mathbb{Z})$ and $N > 1$, then f_s, corresponding to f and N, is u_N. From (3.10.4), we see that $\check{\delta}_0 = f$, where $\delta_0 \in M(\mathbb{T})$ is the Dirac δ-measure on $\mathbb{T}$ defined in Definition 3.1.9d, i.e.,

$$S(\delta_0)(\gamma) = \sum e^{-2\pi i n \gamma}.$$

b. Let $N > 1$, and let $\overset{\circ}{U}_N$ be the DFT of $\overset{\circ}{u}_N$. By Definition 3.10.1,

$$\forall n \in \mathbb{Z}_N, \quad \overset{\circ}{U}_N[n] = 1. \tag{3.10.5}$$

c. Formally, the Fourier transform of the sequence u_N is the series

$$\sum u_N[n] e^{-2\pi i n \gamma} \tag{3.10.6}$$

defined on $\mathbb{T}$, e.g., Remark 3.10.13. Consider the partial sums

$$\mu_K(\gamma) = \sum_{|n| \le K} u_N[n] e^{-2\pi i n \gamma},$$

noting that, for each $\gamma \in \mathbb{T}$ and each K,

$$\mu_K(\gamma) = \sum_{|j| \le K/N} e^{-2\pi i j N \gamma} = \sum_{|j| \le K/N} e^{2\pi i j N \gamma}.$$

Clearly,

$$\forall F \in C(\mathbb{T}), \quad \mu_K(F) = \sum_{|j| \le K/N} \check{f}[jN]. \tag{3.10.7}$$

The proof of the following result uses the inversion formula for the DFT (Theorem 3.8.2). Equations (3.10.8) and (3.10.9) of Theorem 3.10.5 are forms of the PSF for $\mathbb{Z}$ and $\mathbb{T}$.

3.10.5 Theorem. Poisson Summation Formula for $\mathbb{Z}$ and $\mathbb{T}$

a. *Let $F \in A(\mathbb{T})$, and let $f = \{f[n]\} \in \ell^1(\mathbb{Z})$ be the sequence of Fourier coefficients of F. If $N > 1$, then*

$$N \sum f[nN] = \sum_{n=0}^{N-1} F\left(\frac{n}{N}\right). \tag{3.10.8}$$

b. *Let $N > 1$ and consider the measure*

$$\frac{1}{N} \sum_{n=0}^{N-1} \delta_{n/N} \in M(\mathbb{T}),$$

where $\delta_\gamma \in M(\mathbb{T})$ was defined in Definition 3.1.9d. Then

$$\left(\frac{1}{N} \sum_{n=0}^{N-1} \delta_{n/N}\right)^{\vee} = \sum \tau_{nN} u = u_N \in \ell^{\infty}(\mathbb{Z}). \tag{3.10.9}$$

PROOF. **a.** As in Definition 3.10.1, the DFT $\overset{\circ}{F}_N$ of $\overset{\circ}{f}_N$ is

$$\overset{\circ}{F}_N [n] = F\left(\frac{n}{N}\right). \tag{3.10.10}$$

From the inversion formula for the DFT (Theorem 3.8.2),

$$\overset{\circ}{f}_N [0] = \frac{1}{N} \sum_{n=0}^{N-1} \overset{\circ}{F}_N [n]. \tag{3.10.11}$$

We obtain (3.10.8) by combining (3.10.1), (3.10.10), and (3.10.11).

b. Let $F \in A(\mathbb{T})$, and let $f \in \ell^1(\mathbb{Z})$ be its sequence of Fourier coefficients. Combining (3.10.7) and (3.10.8), we see that $\lim_{K \to \infty} \mu_K(F)$ exists for all $F \in A(\mathbb{T})$, and it equals

$$\lim_{K \to \infty} \left(\sum_{|n| \leq K} u_N[n] e^{-2\pi i n \gamma}\right) (F(\gamma))$$

$$= \frac{1}{N} \sum_{k=0}^{N-1} F\left(\frac{k}{N}\right) = \left(\frac{1}{N} \sum_{k=0}^{N-1} \delta_{k/N}\right)(F). \tag{3.10.12}$$

If $F(\gamma) \equiv e^{2\pi i j \gamma}$, then the left side is $u_N[j]$, and (3.10.9) is obtained. ∎

3.10.6 Theorem. Discrete Classical Sampling Theorem

Let $F \in A(\mathbb{T})$ and let $f = \{f[n]\} \in \ell^1(\mathbb{Z})$ be the sequence of Fourier coefficients of F; and let $S \in A(\mathbb{T})$ and let $s \in \ell^1(\mathbb{Z})$ be the sequence of Fourier coeffi-

cients of S. If N > 1 and $0 < \Omega < \frac{1}{2}$, assume

$$2N\Omega \le 1, \tag{3.10.13}$$

supp $F \subseteq [-\Omega, \Omega]$, supp $S \subseteq [-\frac{1}{2N}, \frac{1}{2N}]$, and $S = 1$ on $[-\Omega, \Omega]$. *(The support hypotheses are written for F and S considered as being defined on $[-\frac{1}{2}, \frac{1}{2})$.) Then*

$$\forall m \in \mathbb{Z}, \qquad f[m] = N \sum_j f[jN]s[m - jN]. \tag{3.10.14}$$

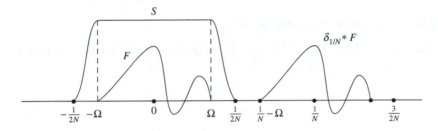

FIGURE 3.5.

PROOF. The PSF (3.10.9) allows us to assert that

$$\forall m \in \mathbb{Z}, \qquad f[m] = \left[N \left(\sum \tau_{kN} u \right) f \right] * s[m] \tag{3.10.15}$$

if and only if

$$\forall \gamma \in \mathbb{T}, \qquad F(\gamma) = \left[\left(\sum_{k=0}^{N-1} \delta_{k/N} \right) * F \right] S(\gamma). \tag{3.10.16}$$

Note that

$$\forall n \in \mathbb{Z}, \qquad f_s[n] = (f u_N)[n];$$

and, by definition,

$$\forall n \in \mathbb{Z}, \qquad f_s * s[n] = \sum_k f_s[k]s[n - k] = \sum_k f[kN]s[n - kN].$$

(The subscript "s" is generic notation for a sampled signal (Example 3.10.2b), and the sequence $s = \{s[n]\}$ is a given function in the statement of this theorem, cf., Definition 3.10.11a.) Thus, the right side of (3.10.15) is the right side of (3.10.14), and so (3.10.15) is (3.10.14).

Consequently, by the equivalence of (3.10.15) and (3.10.16) it is sufficient to prove (3.10.16). To this end, the condition (3.10.13) and the support hypotheses on F and S allow us to conclude that

$$\forall k = 1, \ldots, N - 1, \qquad (-\Omega, \Omega) \cap \operatorname{supp}(\delta_{k/N} * F) = \varnothing$$

and

$$\left[\left(\sum_{k=0}^{N-1} \delta_{k/N} \right) * F \right] S = F S,$$

where $\delta_{k/N} * F$ is considered as being defined on $[-\frac{1}{2}, \frac{1}{2})$. Finally, $F S = F$ since $S = 1$ on $[-\Omega, \Omega]$. Figure 3.5 is a depiction of this argument. ∎

3.10.7 Definition. Periodization on $\mathbb{R}$

a.   Let $T > 0$ and let $f \in L^1(\mathbb{R})$. The *T-periodization* of f is the T-periodic function $\overset{\circ}{f}_T \in L^1(\mathbb{T})$ defined as

$$\overset{\circ}{f}_T (t) = \sum f(t - nT) \qquad \text{a.e.} \tag{3.10.17}$$

Clearly, $\overset{\circ}{f}_T \in L^1(\mathbb{T}_T)$ since

$$\int_0^T | \overset{\circ}{f}_T (t)| \, dt \leq \sum \int_0^T |f(t - nT)| \, dt = \| f \|_{L^1(\mathbb{R})}.$$

b.   If $T > 0$ and $f \in L^1(\mathbb{R})$, then the sequence of Fourier coefficients of $\overset{\circ}{f}_T$ is denoted by $\overset{\circ}{F}_T$, and we obtain

$$\forall n \in \mathbb{Z}, \qquad \overset{\circ}{F}_T [n] = \frac{1}{T} \hat{f} \left(-\frac{n}{T} \right), \tag{3.10.18}$$

where $\hat{f} : \hat{\mathbb{R}} \longrightarrow \mathbb{C}$ is the Fourier transform of f. Equation (3.10.18) is a consequence of the calculation

$$\overset{\circ}{F}_T [n] = \int_{\mathbb{T}_T} \overset{\circ}{f}_T (t) e^{2\pi i n t / T} \, dt$$

$$= \sum_m \frac{1}{T} \int_0^T f(t - mT) e^{2\pi i n t / T} \, dt = \frac{1}{T} \hat{f} \left(-\frac{n}{T} \right).$$

c.   Before proving the PSF for $\mathbb{R}$ and $\hat{\mathbb{R}}$, we should emphasize that the idea of periodization leads *immediately* to a formal proof. In fact, for a given $T > 0$ and function f on $\mathbb{R}$, we form $\overset{\circ}{f}_T$; and then we use (3.10.18), the definition of $\overset{\circ}{f}_T$, and the reasonable expectation that $S(\overset{\circ}{f}_T)(t) = \overset{\circ}{f}_T (t)$ to obtain

$$T \sum f(t + nT) = \sum \hat{f} \left(\frac{n}{T} \right) e^{2\pi i n t / T},$$

cf., [BZ96] for the intricacies involved in proving the PSF.

Equations (3.10.19) and (3.10.20) of Theorem 3.10.8 are forms of the PSF for $\mathbb{R}$ and $\hat{\mathbb{R}}$.

3.10.8 Theorem. Poisson Summation Formula for $\mathbb{R}$ and $\hat{\mathbb{R}}$

a. Let $T > 0$ and let $f \in L^1(\mathbb{R})$. If $S(\overset{\circ}{f}_T)(0)$ and $\overset{\circ}{f}_T(0)$ exist and are equal, then

$$T \sum f(nT) = \sum \hat{f}\left(\frac{n}{T}\right), \tag{3.10.19}$$

cf., (3.10.8). ($S(\overset{\circ}{f}_T)$ is the Fourier series of $\overset{\circ}{f}_T$.)
b. Let $T > 0$. Then $\sum \delta_{nT} \in S'(\mathbb{R}) \cap M(\mathbb{R})$ and

$$\left(\frac{1}{T} \sum \delta_{n/T}\right)^{\vee} = \sum \delta_{nT} \tag{3.10.20}$$

distributionally, cf., (3.10.9).

PROOF. *a.* By (3.10.18), we have

$$S(\overset{\circ}{f}_T)(0) = \sum \overset{\circ}{F}_T[n] = \frac{1}{T} \sum \hat{f}\left(-\frac{n}{T}\right);$$

and, by definition,

$$\overset{\circ}{f}_T(0) = \sum f(-nT).$$

Equation (3.10.19) is obtained by combining these two facts.
b. If $f \in S(\mathbb{R})$, then by means of part *a* we compute

$$\left\langle \left(\sum \delta_{nT}\right)^{\hat{}}, \hat{f} \right\rangle = \left\langle \sum \delta_{nT}, f \right\rangle$$

$$= \sum \overline{f(nT)} = \frac{1}{T} \sum \overline{\hat{f}\left(\frac{n}{T}\right)} = \frac{1}{T} \sum \delta_{n/T}(\overline{\hat{f}})$$

$$= \left\langle \frac{1}{T} \sum \delta_{n/T}, \hat{f} \right\rangle.$$

Since this equality is true for all $\hat{f} \in S(\hat{\mathbb{R}})$, we obtain (3.10.20). ∎

3.10.9 Example. Conditions for the Validity of PSF

a. The hypotheses for Theorem 3.10.8 are not only valid for $f \in S(\mathbb{R})$, but they are also valid for $f \in C^2(\mathbb{R}) \cap L^1(\mathbb{R})$ having relatively mild decay properties at $\pm\infty$.
b. On the other hand, there are continuous functions $f \in L^1(\mathbb{R})$ for which $\hat{f} \in L^1(\hat{\mathbb{R}})$, $f(n) = 0$ for all $n \in \mathbb{Z}$, $\hat{f}(0) = 1$, and $\hat{f}(n) = 0$ for all $n \in \mathbb{Z}\backslash\{0\}$, e.g., [Kat76, pages 130–131].

c. However, if $f \in L^1(\mathbb{R}) \cap BV(\mathbb{R})$ and $f(n) = \frac{1}{2}(f(n+) + f(n-))$ for all $n \in \mathbb{Z}$, then PSF is valid, e.g., [Zyg59, Volume I, Section II.13], [BZ96].

3.10.10 Theorem. Classical Sampling Theorem

Let $T, \Omega > 0$ *satisfy the condition that* $0 < 2T\Omega \le 1$, *and let* $s \in PW_{1/(2T)}$ *satisfy the condition that* $\hat{s} \equiv S = 1$ *on* $[-\Omega, \Omega]$ *and* $S \in L^\infty(\hat{\mathbb{R}})$. *Then*

$$\forall f \in PW_\Omega, \qquad f = T \sum f(nT)\tau_{nT}s, \tag{3.10.21}$$

where the convergence in (3.10.21) is in $L^2(\mathbb{R})$ *norm and uniformly in* $\mathbb{R}$. *A possible sampling function* s *is*

$$s(t) = \frac{\sin 2\pi\Omega t}{\pi t}.$$

PROOF 1.   Using the PSF (3.10.20) we see that

$$f = T\left[\left(\sum \delta_{nT}\right) f\right] * s \tag{3.10.22}$$

if and only if

$$\hat{f} = \left(\sum \tau_{n/T}\hat{f}\right) S. \tag{3.10.23}$$

Note that (3.10.21) is the same as (3.10.22) since

$$(\delta_{nT}f)(h) = \delta_{nT}(fh) = f(nT)h(nT) = f(nT)\delta_{nT}(h).$$

On the other hand, (3.10.23) is valid by the definition of s. Thus, (3.10.21) is true.

We are omitting the analytical details quantifying the formal steps and proving the convergence, e.g., Exercise 3.58. To compensate for this omission we provide the following proof of (3.10.21).

PROOF 2.   By means of the support hypotheses and the Parseval–Plancherel Theorem, we have

$$\left\| f - T \sum_{|n|\le N} f(nT)\tau_{nT}s \right\|_{L^2(\mathbb{R})}$$

$$= \left\| \hat{f}(\gamma) - T \sum_{|n|\le N} f(nT)e^{-2\pi in T\gamma} S(\gamma) \right\|_{L^2(\hat{\mathbb{R}})} \tag{3.10.24}$$

$$= \left\| \hat{f}(\gamma) - T \sum_{|n|\le N} f(nT)e^{-2\pi in T\gamma} S(\gamma) \right\|_{L^2[-1/2T,1/2T]}.$$

Let $G \in L^2(\mathbb{T}_{1/T})$ be defined by

$$G(\gamma) = \begin{cases} \hat{f}(\gamma), & \text{if } |\gamma| < \Omega, \\ 0, & \text{if } \Omega < |\gamma| < \dfrac{1}{2T}. \end{cases}$$

The Fourier series of G is

$$\sum \check{G}[n] e^{-2\pi i n T \gamma},$$

where the Fourier coefficients take the values

$$\forall n \in \mathbb{Z}, \qquad \check{G}[n] = Tf(nT),$$

e.g., Exercise 1.51.

The right side of (3.10.24) is

$$\|\hat{f} - S_N(G)S\|_{L^2[-1/2T, 1/2T]}$$

$$\leq \|\hat{f} - G\|_{L^2[-1/2T, 1/2T]} + \|G - S_N(G)S\|_{L^2[-1/2T, 1/2T]} \quad (3.10.25)$$

$$= \|S(G - S_N(G))\|_{L^2[-1/2T, 1/2T]},$$

and the right side tends to 0 as $N \longrightarrow \infty$ since $S \in L^\infty(\hat{\mathbb{R}})$ and by Theorem 3.4.12c. ($S_N(G)$ designates the Nth partial sum of the Fourier series of G.)

To prove the uniform convergence on $\mathbb{R}$, we compute (using Exercise 1.51 and Hölder's Inequality)

$$\left| f(t) - T \sum_{|n| \leq N} f(nT) \tau_{nT} s(t) \right|$$

$$= \left| \int_{-1/2T}^{1/2T} \left(\hat{f}(\gamma) - T \sum_{|n| \leq N} f(nT) e^{-2\pi i n T \gamma} S(\gamma) \right) e^{2\pi i t \gamma} \, d\gamma \right|$$

$$\leq \frac{1}{\sqrt{T}} \left\| \hat{f}(\gamma) - T \sum_{|n| \leq N} f(nT) e^{-2\pi i n T \gamma} S(\gamma) \right\|_{L^2[-1/2T, 1/2T]}.$$

This last term is the right side of (3.10.24) (times $1/\sqrt{T}$), and so we obtain our result by again combining (3.10.24) and (3.10.25). $\blacksquare$

The second proof of the Classical Sampling Theorem does not use the PSF but does depend on periodization in the use of the periodic function G.

3.10.11 Definition. Sampling and the Nyquist Rate

a. In the sampling formula (3.10.21), the *sampling period* is T, the *sampling sequence* is $\{nT : n \in \mathbb{Z}\}$, and the sequence of *sampled values* is $\{f(nT) : n \in \mathbb{Z}\}$. The function s is the *sampling function*.

b. The *sampling rate* is the number of samples taken per second. Thus, if f is sampled (once) every T seconds, then there are $\frac{1}{T}$ samples per second, so the sampling rate is $\frac{1}{T}$ samples per second. In this case, we say that "the data f is sampled at $\frac{1}{T}$ Hertz (Hz)". For example, the electrocorticogram data, whose spectrogram is shown in Figure 2.4, was sampled at 200 Hz.

c. In high-frequency information, with "close and rapid" fluctuations in the signal, we have to sample often in order to capture all of the activity. Thus, in the case of very high frequencies ("infinite frequencies"), i.e., when the signal f is effectively not bandlimited, we can not characterize f with regularly spaced samples.

Let 2Ω be a given frequency bandwidth. Because of Theorem 3.10.10, if $f \in PW_\Omega$ and $2T\Omega \le 1$ and if f is sampled every T seconds, then f can be perfectly reconstructed in terms of these sampled values. In this case, with $2T\Omega \le 1$, the *minimum* sampling rate for which we have reconstruction by (3.10.21) *for all* $f \in PW_\Omega$ is 2Ω samples per second, e.g., Exercise 3.59. This minimum sampling rate, 2Ω, is the *Nyquist rate* [Nyq28].

3.10.12 Remark. Aliasing

a. The effect of undersampling signals $f \in PW_\Omega$ or of developing formulas for $f \in L^2(\mathbb{R})\backslash PW_\Omega$, vis á vis the goal of exact reconstruction in terms of sampled values, is called *aliasing*.

b. Old motion pictures of fast moving events produce "jumpy" video, and this is a classical example of aliasing. In fact, each frame of film is a sampled value, but the sampling rate is not sufficiently high to produce exact reconstruction of the event. A more modern example of aliasing is that of the *stroboscopic effect*, e.g., [OW83, pages 529–531].

c. Clearly, $\sum \hat{f}(\gamma - \frac{n}{T})$ is $\frac{1}{T}$-periodic. If $f \in PW_\Omega$ and $2T\Omega \le 1$, then

$$\forall n \in \mathbb{Z}, \qquad \operatorname{supp} \tau_{n/T}\hat{f} \cap \operatorname{supp} \tau_{(n+1)/T}\hat{f} \tag{3.10.26}$$

is at most a boundary point, and the key step, (3.10.23), of the proof of Theorem 3.10.10 can be made. If $2T\Omega > 1$, then the intersection (3.10.26) will generally be a larger set than a boundary point, where the high frequencies of $e^{2\pi itn/T} f(t)$ will intersect with the low frequencies of $e^{2\pi it(n+1)/T} f(t)$. The ensuing phenomenon is *aliasing* as described in part *a*. A more basic, but essentially equivalent, explanation in terms of Fourier series is in [Ben92b, page 451]. In either case, the term "aliasing", due to Tukey, catches the flavor of high and low frequencies "assuming the alias of each other", cf., [BTu59], [Pap66], [Pap77].

3.10.13 Remark. Idempotent Measures

a. Helson (1953, 1954) proved that *a sequence $\{c_n : n \in \mathbb{Z}\}$ of 0's and 1's is the sequence of Fourier coefficients of some $\mu \in M(\mathbb{T})$ if and only if $\{c_n\}$ differs* *from a periodic sequence in at most finitely many places*, e.g., [Hel83, pages 158–163], [Rud62, Chapter 3]. The statement for the setting $\mathbb{T}^n$ was proved by Rudin (1959), and the result for arbitrary compact abelian groups is due to Paul Cohen (1960).

b. We mention Helson's Theorem since the proof of Theorem 3.10.5 establishes the fact that the sequence of Fourier coefficients of

$$\frac{1}{N} \sum_{k=0}^{N-1} \delta_{k/N} \in M(\mathbb{T})$$

is a periodic sequence of 0's and 1's.

c. A measure $\mu \in M(\mathbb{T})$ is *idempotent* if $(\check{\mu})^2 = \check{\mu}$; and this can occur if and only if $\check{\mu}[n]$ is 0 or 1 for each n, cf., Example 1.10.7. The theorems of Helson, Rudin, and Cohen characterize idempotent measures.

d. We defined the Dirichlet function D_n in Example 3.4.5; and, in Example 3.4.9*b*, we proved that

$$\|D_n\|_{L^1(\mathbb{T})} = \frac{4}{\pi^2} \log n + c_n, \qquad n \geq 2, \tag{3.10.27}$$

where $\{c_n\}$ is a bounded sequence. Clearly, D_n is an idempotent measure.

If $n_1, \ldots, n_k \in \mathbb{Z}$ and

$$F(\gamma) = \sum_{j=1}^{k} e^{-2\pi i n_j \gamma},$$

then F is also an idempotent measure. An interesting problem emanating from (3.10.27) is to estimate

$$M(k) = \inf\{\|F\|_{L^1(\mathbb{T})} : n_1, \ldots, n_k \in \mathbb{Z}\}.$$

The *Littlewood Conjecture* (1948) asserts that

$$\exists C > 0 \quad \text{such that} \quad \forall k \in \mathbb{N}, \quad M(k) \geq C \log k.$$

The Littlewood Conjecture was proved by McGehee, Pigno, and Brent Smith [MPS81].

Exercises

Exercises 3.1–3.30 are appropriate for Course I.

3.1. Compute the Fourier series of the following functions defined 2Ω-periodically on $\hat{\mathbb{R}}$. To compute a Fourier series means that the Fourier coefficients must be computed.

a) $F(\gamma) = \begin{cases} -1, & \gamma \in [-\Omega, 0), \\ 1, & \gamma \in [0, \Omega). \end{cases}$

b) $F(\gamma) = |\gamma|, \ \gamma \in [-\Omega, \Omega)$.

c) $F(\gamma) = \begin{cases} \gamma, \ \gamma \in [0, \Omega), \\ 0, \ \gamma \in [\Omega, 2\Omega). \end{cases}$

d) $F(\gamma) = \gamma, \ \gamma \in [-\Omega, \Omega)$.

e) $F(\gamma) = \gamma, \ \gamma \in [0, 2\Omega)$.

f) $F(\gamma) = \gamma^2, \ \gamma \in [-\Omega, \Omega)$.

g) $F(\gamma) = \gamma^2, \ \gamma \in [-2\Omega, 0)$.

h) $F(\gamma) = \begin{cases} c \neq 0, \ \gamma \in (0, \Omega), \\ 0, \ \gamma \in [\Omega, 2\Omega]. \end{cases}$

i) $F(\gamma) = \gamma \cos\left(\dfrac{\pi \gamma}{\Omega}\right), \ \gamma \in [-\Omega, \Omega)$.

j) $F(\gamma) = \gamma \sin\left(\dfrac{\pi \gamma}{\Omega}\right), \ \gamma \in [-\Omega, \Omega)$.

k) $F(\gamma) = \left|\sin\left(\dfrac{\pi \gamma}{\Omega}\right)\right|, \ \gamma \in [0, 2\Omega)$.

l) $F(\gamma) = \begin{cases} \cos\left(\dfrac{\pi \gamma}{\Omega}\right), \ \gamma \in (0, \Omega), \\ 0, \ \gamma \in [\Omega, 2\Omega]. \end{cases}$

m) $F(\gamma) = \begin{cases} \sin\left(\dfrac{\pi \gamma}{\Omega}\right), \ \gamma \in (0, \Omega), \\ 0, \ \gamma \in [\Omega, 2\Omega]. \end{cases}$

3.2. Using MATLAB, graph the Nth partial sums, for $N = 1, 2, 4, 8$, of the Fourier series of the functions defined in parts a, b, c, f, j, k of Exercise 3.1.

3.3. Designate F in Exercise 3.1a by F_a, and similarly for parts $b, c, \ldots, m$.

   a) Show that, formally,

$$S(F_b)' = S(F_a) \quad \text{and} \quad S(F_f)' = 2S(F_d),$$

   where $S(F)'$ is the term-by-term differentiation of $S(F)$. (This calculation is legitimate when we consider $S(F)'$ as the distributional derivative of the function F defined 2Ω-periodically on $\mathbb{R}$, cf., Definition 3.1.9.)

   b) Evaluate the Fourier coefficients of $F_m - \frac{1}{2}F_k$.

3.4. Using (3.1.7), prove that the function G defined in part b of the proof of Theorem 3.1.6 is bounded in some interval centered at the origin.

3.5. Compute the Fourier series of

$$F(\gamma) = \frac{\pi - \gamma}{2}, \qquad \gamma \in [-\pi, \pi),$$

defined 2π-periodically on $\hat{\mathbb{R}}$, cf., Example 3.3.6a and Exercise 3.29.

3.6. Prove that the inclusion, $L^2(\mathbb{T}_{2\Omega}) \subseteq L^1(\mathbb{T}_{2\Omega})$, is proper.

3.7. a) Compute the Fourier series of the following functions defined 2π-periodically.

   i) $F(\gamma) = \sin \gamma, \gamma \in [0, 2\pi)$.

   ii) $F(\gamma) = \sin \gamma, \gamma \in [0, \pi)$, and extended evenly to $(-\pi, \pi)$.

b) Compute the Fourier series of $F(\gamma) = \sin \gamma$, $\gamma \in [0, \pi)$, considered as a π-periodic function. Compare this result with part *a*.

3.8. Compute

$$\frac{1}{2\Omega} \int_{-\Omega}^{\Omega} \left| \sum_{n=-100}^{100} e^{\pi i n \gamma / \Omega} \right|^2 d\gamma.$$

3.9. Prove (3.4.17), which states the *continuity of the inner product*.

3.10. Prove the divergence of the series

$$\sum_{|n| \geq 2} \frac{1}{|n| \log |n|}.$$

This fact played a role in Example 3.3.4.

3.11. a) Prove that

$$\sum_{|n| \leq N} e^{inx} = \frac{\sin(N + 1/2)x}{\sin(x/2)}$$

for $x \notin 2\pi \mathbb{Z}$. For $x = -\pi \gamma / \Omega$, this establishes (3.4.5).

b) Prove that

$$\sum_{|n| \leq N} \left(1 - \frac{|n|}{N + 1} \right) e^{inx} = \frac{1}{N + 1} \left(\frac{\sin(N + 1)(x/2)}{\sin(x/2)} \right)^2$$

for $x \notin 2\pi \mathbb{Z}$, and complete the proof of (3.4.7).

3.12. a) Let $P(\gamma) = \sum_{|n| \leq N} c_n e^{-\pi i n \gamma / \Omega}$ be a trigonometric polynomial, and let $F \in L^1(\mathbb{T}_{2\Omega})$. Compute $F * P$. Is $F * P$ a trigonometric polynomial?

b) Let $F \in L^1(\mathbb{T}_{2\Omega})$ and $G \in L^\infty(\mathbb{T}_{2\Omega})$. Prove that

$$\| F * G \|_{L^\infty(\mathbb{T}_{2\Omega})} \leq \| F \|_{L^1(\mathbb{T}_{2\Omega})} \| G \|_{L^\infty(\mathbb{T}_{2\Omega})}.$$

c) Let $F \in L^2(\mathbb{T}_{2\Omega})$ and $G \in L^1(\mathbb{T}_{2\Omega})$. Prove that

$$\| F * G \|_{L^2(\mathbb{T}_{2\Omega})} \leq \| F \|_{L^2(\mathbb{T}_{2\Omega})} \| G \|_{L^1(\mathbb{T}_{2\Omega})}.$$

3.13. a) Abel's *partial summation formula* is

$$\sum_{n=p}^{q} a_n b_n = \sum_{n=p}^{q-1} A_n (b_n - b_{n+1}) + A_q b_q - A_{p-1} b_p, \qquad \text{(E3.1)}$$

where $\{a_n, b_n : n = 0, 1, \ldots\} \subseteq \mathbb{C}$, $0 \leq p \leq q$, $A_{-1} \equiv 0$, and $A_n \equiv \sum_{j=0}^{n} a_j$. Prove (E3.1).

b) Prove that if $p, q \in \mathbb{Z}$, $p < q$, and $x \neq 2\pi k$ for any $k \in \mathbb{Z}$, then

$$\left| \sum_{n=p}^{q} e^{inx} \right| \leq \frac{1}{|\sin(x/2)|};$$

and so

$$\forall \gamma \neq 2\Omega k, \qquad \left| \sum_{n=p}^{q} e^{-\pi i n \gamma/\Omega} \right| \leq \frac{1}{|\sin(\pi\gamma/2\Omega)|}.$$

c) Let $f[p] \geq f[p+1] \geq \cdots \geq f[q] \geq 0$. Prove that

$$\forall \gamma \neq 2\Omega k, \qquad \left| \sum_{n=p}^{q} f[n] e^{-i\pi n\gamma/\Omega} \right| \leq \frac{f[p]}{|\sin(\pi\gamma/2\Omega)|}.$$

[Hint. Use the partial summation formula (E3.1).]

3.14. Let $\Omega > 0$ and let $\alpha \in (0, \Omega)$. Let $F = 1$ on $\mathbb{T}_{2\Omega}$, and define $F_{(\alpha)}$ on $\mathbb{T}_{2\Omega}$ by setting $F_{(\alpha)} = \mathbf{1}_{[-\alpha,\alpha)}$ on $[-\Omega, \Omega)$.

   a) Compute the Fourier coefficients of F and of $G = F - cF_{(\alpha)}$ for $c \in \mathbb{C}$.

   b) Does there exist $c \neq 0$ such that for all $\alpha \in (0, \Omega)$

$$\|F\|_{L^2(\mathbb{T}_{2\Omega})} = \|G\|_{L^2(\mathbb{T}_{2\Omega})}?$$

   c) Let $k \in \mathbb{N}$ and let $\alpha = \Omega/k$. For which values of n can we assert that $\hat{G}[n] = \hat{F}[n] = 0$.

   The point of part *a* is that even a small frequency perturbation of a signal causes almost all of the "temporal" data on $\mathbb{Z}$ to change. This is precisely the uncertainty principle phenomenon discussed in Remark 1.1.4. This lack of time *and* frequency localization in Fourier analysis can be circumvented to some extent in wavelet theory, e.g., Daubechies' book [Dau92].

3.15. Prove Theorem 3.3.7. The right side of (3.3.9) in Theorem 3.3.7 is defined for $G \in L^\infty(\mathbb{T}_{2\Omega})$. Why does Theorem 3.3.7 generally fail in this case?

3.16. Prove Theorem 3.4.4*b*. [Hint. Let $\epsilon > 0$; and let $G \in C(\mathbb{T}_{2\Omega})$ have the property that $\|F - G\|_{L^1(\mathbb{T}_{2\Omega})} < \epsilon/(2C)$, where $\|K_{(\lambda)}\|_{L^1(\mathbb{T}_{2\Omega})} \leq C, C \geq 1$. Then, by (3.4.1), you can show that

$$\|F - F * K_{(\lambda)}\|_{L^1(\mathbb{T}_{2\Omega})} < \frac{\epsilon}{2} + \|G - G * K_{(\lambda)}\|_{L^\infty(\mathbb{T}_{2\Omega})}$$
$$+ C\|G - F\|_{L^1(\mathbb{T}_{2\Omega})}.$$

The result follows from Theorem 3.4.4*a* and by taking a $\overline{\lim}$.]

3.17. In this excercise we shall approximate Fourier transforms on $\mathbb{R}$ by DFTs (Definition 3.8.1). Consider intervals $[a, b] \subseteq \mathbb{R}$ and $[\alpha, \beta] \subseteq \hat{\mathbb{R}}$, and denote their lengths by $T = b - a$ and $2\Omega = \beta - \alpha$. Suppose $2T\Omega \equiv N$ is a positive integer, and define

$$t_m = a + m\Delta t \quad \text{and} \quad \gamma_n = \alpha + n\Delta\gamma,$$

where $\Delta t \equiv 1/(2\Omega)$, $\Delta\gamma \equiv 1/T$, and $m, n = 0, 1, \ldots, N-1$. Let f be a continuous function on $[a, b]$, set $g[m] \equiv f(t_m)e^{-2\pi i m\alpha/(2\Omega)}$, and let G be the DFT of g. Show that

$$\forall n = 0, 1, \ldots, N-1, \qquad \int_a^b f(t)e^{-2\pi i t\gamma_n}\, dt \approx \frac{1}{2\Omega}e^{-2\pi i a\gamma_n} G[n].$$

[Hint. Use a Riemann sum approximation of the left side.]

We can effectively use this exercise to compute Fourier transforms, e.g., Exercises 3.18 and 3.19. The symbol "≈" means that the right side approximates the left side. This imprecise but meaningful statement can be quantified in terms of the Poisson Summation Formula (Theorem 3.10.8), e.g., [AG89], [BSS88], [BrHe95].

3.18. We computed the Fourier transform of the Gaussian $g(t) = (1/\sqrt{\pi})e^{-t^2}$ in Example 1.3.3.

a) Using the MATLAB *fft* function and the approach in Exercise 3.17, verify numerically that $\hat{g}(\gamma) = e^{-(\pi\gamma)^2}$. [Hint. To begin, let $[a, b] = [-32, 32]$, $[\alpha, \beta] = [-32, 32]$, $\Delta t = 1/64$, and $\Delta\gamma = 1/64$; and consider the vector

$$t = -32 : (1/64) : 32 - (1/64).]$$

b) With regard to PSF (Theorem 3.10.8), and in MATLAB terminology, compare *sum* $(f.*f)$ and *sum* $(fhat.*fhat)$, where $f \equiv \exp(-t.*t)$ and $fhat \equiv sqrt(pi) * \exp(-(pi*t).\hat{}2)$.

3.19. a) Using the MATLAB *fft* function and the approach in Exercise 3.17, graph the following Fourier transform pairs:

$$d_\lambda \quad \longleftrightarrow \quad \mathbf{1}_{[-\lambda/2\pi,\lambda/2\pi]},$$

$$w_\lambda \quad \longleftrightarrow \quad \max(1 - \tfrac{|2\pi\gamma|}{\lambda}, 0),$$

$$p_\lambda \quad \longleftrightarrow \quad e^{-2\pi|\gamma|/\lambda},$$

$$g_\lambda \quad \longleftrightarrow \quad e^{-(\pi\gamma/\lambda)^2},$$

where $\{d_\lambda\}$, $\{w_\lambda\}$, $\{p_\lambda\}$, and $\{g_\lambda\}$ are the Dirichlet, Fejér, Poisson, and Gauss kernels, respectively.

b) Graph the de la Vallée–Poussin kernel $\{v_\lambda\}$ that was defined in Exercise 1.43, cf., Exercise 1.50 and Example 3.5.3*b*.

c) For each of the pairs in part *a*, as well as the de la Vallée–Poussin kernel, compare the behavior of the function with its transform as $\lambda > 0$ increases. Compare the manner and speed with which each transform converges to 1 as $\lambda > 0$ increases.

3.20. The following is a recursive MATLAB program (m-file) that calculates the two-step FFT algorithm of Theorem 3.9.1:

```
function y = myfft(x)
n = length(x);
if  n == 1
    y = 1;
else
    xo = x(2 : 2 : n); xe = x(1 : 2 : n − 1);
    w = exp((−2*pi*i/n)*(0 : n − 1));
    xohat = myfft(xo);
    y = [xehat xehat] + w.*[xohat xohat];
end
```

a) Make an ascii file called *myfft.m* in some directory, say *c:\mymfiles*. Give the MATLAB command

$$path('c:\mymfiles', path);$$

b) Make a vector f of random numbers, e.g., set $f \equiv rand(1, 256)$. Compute the DFT of f by means of *myfft* as well as by the MATLAB *fft* function. Compare the results.

3.21. Let $f, g \in \ell^1(\mathbb{Z})$ have "compact support", i.e., there are integers A, B, C, D for which $f[n] = 0$ except for $A \leq n \leq B$ and $g[n] = 0$ except for $C \leq n \leq D$. Recall the definition of $f * g$ in Definition 3.5.1. Part *a* is a *fast convolution algorithm*.

a) Let $N \geq B - A + D - C + 1$, and let F and G be the DFTs of f and g on $\mathbb{Z}_N$, respectively. Prove that if $A + C \leq n \leq B + D$, then

$$f * g[n] = \mathcal{F}_N^{-1}(FG)[n - A - C],$$

and that $f * g[n] = 0$ elsewhere.

b) Prove that a direct implementation of the definition of convolution requires $(B - A + 1)(D - C + 1)$ multiplications.

c) Using the FFT algorithm of Section 3.9, show that the method of part *a* requires $3N \log_2 N + N$ multiplications.

3.22. The *z-transform* F of $f : \mathbb{Z} \longrightarrow \mathbb{C}$ is the function $F(f) : \mathbb{C} \longrightarrow \mathbb{C}$ defined by

$$F(f)(z) = \sum f[n]z^{-n}.$$

Of course, $F(f)$ may not exist for some $z \in \mathbb{C}$. Note that there is an intrinsic relationship between Fourier series and z-transforms; in fact, we have $\hat{f}(\gamma) = F(f)(e^{2\pi i \gamma}), \gamma \in \hat{\mathbb{R}}$.

Let $a, \omega \in \mathbb{C}$ and define $z_k = a\omega^k$ for $C \leq k \leq D$. Consider the sequences f, g, and h, where $f[n] = 0$ except for $A \leq n \leq B$, $h[n] = 0$ except for $C - B \leq n \leq D - A$, where it is defined as $h[n] = \omega^{n^2/2}$, and $g[n] = 0$ except for $A \leq n \leq B$, where it is defined as $g[n] = f[n]a^{-n}h[n]$. Prove that the z-transform F of f can be written as

$$F(f)(z_k) = \overline{h[k]}(g * h)[k].$$

This is the *chirp z-algorithm*.

3.23. a) Using MATLAB, implement the chirp z-algorithm of Exercise 3.22 to approximate $\int e^{-t^2} e^{-2\pi i t \gamma} dt$, cf., Example 2.10.5. [Hint. Approximate the Fourier transform at the points $\gamma \equiv \gamma_n \equiv n \Delta \gamma$, where $\Delta \gamma \equiv 0.01$ and $-100 \leq n \leq 100$. Use the values $f[m] \equiv e^{-(t_m)^2}$ of the Gaussian, where $t_m \equiv m \Delta t$, $\Delta t \equiv 0.01$, and $-300 \leq m \leq 300$. Show that the desired integral is approximately

$$F(f)(z_k)\Delta t,$$

where $f[m] = 0$ except for $-300 \leq m \leq 300$ and where $z_k \equiv (e^{-2\pi i \Delta t \Delta \gamma})^k$. Finally, invoke the fast convolution algorithm (with 1024 point FFTs) from Exercise 3.21 to perform the convolution used in the chirp z-algorithm of Exercise 3.22.]

b) Compare the number of multiplications required for the chirp z-transform in part *a* with the number of multiplications required in the computation of the same integral in Exercise 3.18. Note that the chirp z-transform calculation provides points γ_n on a finer mesh than the FFT approach of Exercise 3.18, and it requires only 65% of the number of multiplications.

3.24. Evaluate

$$\sum_{\substack{n\geq 1 \\ n \text{ odd}}} \frac{1}{n^2}.$$

[Hint. Consider Exercise 3.16.]

3.25. a) Let $F \in L^1(\mathbb{T})$ and $G \in L^\infty(\mathbb{T})$, and let $f = \{f[n]\}$ and $g = \{g[n]\}$ be their sequences of Fourier coefficients. Prove that

$$\lim_{n\to\infty} \int_{\mathbb{T}} F(\gamma)G(n\gamma)\,d\gamma = f[0]g[0].$$

The Riemann–Lebesgue Lemma (Theorem 3.1.5) is the special case of this result for $G(\gamma) = e^{2\pi i\gamma}$.

b) Let $F \in L^1(\mathbb{T})$ be real-valued. Using the notation from Example 3.1.4 and the result from part *a*, prove that

$$\lim_{k\to\infty}\sum_{n=1}^{\infty} \frac{b_{kn}}{n} = \lim_{k\to\infty}\sum_{n=1}^{\infty}(-1)^n \frac{a_{k(2n+1)}}{2n+1} = 0,$$

cf., [Lux62].

3.26. a) Let $F \in AC(\mathbb{T}_{2\Omega})$, and let $f = \{f[n]\}$ be the sequence of Fourier coefficients of F. Prove that $F' \in L^1(\mathbb{T}_{2\Omega})$, $(F')\check{\ }[n] = -(\pi in/\Omega)f[n]$ for each $n \in \mathbb{Z}$, and $\lim_{|n|\to\infty} nf[n] = 0$.

Thus, if $F \in C(\mathbb{T}_{2\Omega}) \cap BV(\mathbb{T}_{2\Omega})$ has Fourier coefficients $f[n]$, $n \in \mathbb{Z}$, and if

$$\overline{\lim_{|n|\to\infty}} |nf[n]| > 0,$$

then $F \notin AC(\mathbb{T}_{2\Omega})$.

b) Prove (3.5.3). [Hint. For $F \in AC(\mathbb{T}_{2\Omega})$, let $G = F' \in L^1(\mathbb{T}_{2\Omega})$ so that $\check{G}[0] = 0$ and $F(\gamma) = \int_0^\gamma G(\lambda)\,d\lambda + F(0)$ on $[0, 2\Omega]$. Apply part *a*, Hölder's Inequality, and Parseval's formula to the right side of (3.5.2).]

c) Let $G \in L^1(\mathbb{T}_{2\Omega})$ and suppose $\check{G}[0] = 0$. Prove

$$\int_0^{2\Omega} \int_0^\gamma G(\lambda)\,d\lambda d\gamma = -\int_0^{2\Omega} \gamma G(\gamma)\,d\gamma.$$

d) Let $F \in BV(\mathbb{T}_{2\Omega})$, and let $f = \{f[n]\}$ be the sequence of Fourier coefficients of F. Prove that

$$\exists M > 0 \text{ such that } \forall n \in \mathbb{Z}, \ |nf[n]| \leq M.$$

M can be taken as the *variation* of F, i.e., $M = \inf \left\{ \sum |F(\gamma_j) - (\gamma_{j-1})| \right\}$, where the infimum is taken over all finite partitions $\{\gamma_j\}$ of $[0, 2\Omega]$, cf., the definition of bounded variation in Definition 1.1.5.

Part *d* can be proved by a calculation for step functions and an approximation, but we also recommend an ingenious calculation due to M. Taibleson (*Fourier coefficients of functions of bounded variation*, Proc. Amer. Math. Soc. **18** (1967), 766), e.g., [Ben76, page 120].

3.27. Let $S(\gamma) = 1/(1 + \sin 2\pi\gamma)$. Then $1/S \geq 0$ on $\mathbb{T}$ has a zero at $\gamma = -\frac{1}{4} + n$. Prove that $S \notin L^1(\mathbb{T})$.

3.28. Let $P(x) = x^n + a_{n-1}x^{n-1} + \cdots + a_0$ be a polynomial of degree $n \geq 1$ with complex coefficients. By the Fundamental Theorem of Algebra there is $\alpha \in \mathbb{C}$ such that $P(\alpha) = 0$. The following facts are well known.

i) $P(x)$ has the unique representation
$$P(x) = (x - \alpha_1)(x - \alpha_2)\cdots(x - \alpha_n),$$
where $\{\alpha_1, \ldots, \alpha_n\} \subseteq \mathbb{C}$ is the set of zeros of P;

ii) The zeros and coefficients of P are related by the formulas
$$a_{n-1} = -\sum_{j=1}^{n} \alpha_j,$$
$$a_{n-2} = \sum_{i<j} \alpha_i \alpha_j,$$
$$a_{n-3} = -\sum_{i<j<k} \alpha_i \alpha_j \alpha_k,$$
$$\vdots$$
$$a_0 = \pm\alpha_1 \alpha_2 \cdots \alpha_n.$$

Fill in the details of the following brilliant persuasive rationale by Euler to establish that $\sum_{n=1}^{\infty} 1/n^2 = \pi^2/6$. The zeros of
$$P(x) = \frac{\sin x}{x} = 1 - \frac{x^2}{3!} + \frac{x^4}{5!} - \cdots$$
are $\{n\pi : n \in \mathbb{Z}\backslash\{0\}\}$. Arguing by analogy, Euler obtains
$$\frac{\sin x}{x} = \prod_{n=1}^{\infty}\left(1 - \frac{x^2}{(n\pi)^2}\right),$$
cf., Exercise 1.36; and, hence,
$$\frac{1}{3!} = \sum_{n=1}^{\infty}\frac{1}{(n\pi)^2},$$
by parts *i* and *ii*.

3.29. a) Prove that
$$\lim_{N\to\infty} \sum_{0<|n|\leq N} \frac{e^{-\pi i n\gamma/\Omega}}{n}$$
exists for each $\gamma \in (0, 2\Omega)$. [Hint. See Example 3.3.6*a*.]

b) Let $f[n] \geq f[n+1]$ for $n \geq 1$, and let $\lim_{n\to\infty} f[n] = 0$. Prove that

$$\sum_{n=1}^{\infty} f[n] e^{-\pi i n \gamma / \Omega}$$

exists for each $\gamma \in (0, 2\Omega)$, and that the convergence is uniform on any closed subset of $(0, 2\Omega)$.

c) Let $f[n] \geq f[n+1]$ for $n \geq 1$, and assume $nf[n] \leq C$. Prove that

$$\sup_{N} \left| \sum_{n=1}^{N} f[n] \sin \frac{\pi n \gamma}{\Omega} \right| < \infty,$$

and that the series converges to a continuous function on $(0, 2\Omega)$, cf., Exercise 3.43.

The relation between this exercise and pseudomeasures supported by arithmetic progressions is found in [Ben71, Section 4.3].

3.30. Prove Abel's original "Abelian theorem": *if $\sum_{n=1}^{\infty} a_n x^n$ converges on $[0, 1)$ and $\sum_{n=1}^{\infty} a_n = S$, then*

$$\lim_{x \to 1-} \sum_{n=1}^{\infty} a_n x^n = S. \qquad (E3.2)$$

[Hint. Use (E3.1).]

The first Tauberian Theorem, by Tauber in 1897, came as a response to the problem of finding some sort of converse to Abel's result. Tauber proved that *if $f(x) \equiv \sum_{n=1}^{\infty} a_n x^n$ converges on $[0, 1)$, $f(x-) \equiv S$, and $\lim_{n\to\infty} na_n = 0$, then $\sum_{n=1}^{\infty} a_n = S$*. The boundedness condition "$\lim_{n\to\infty} na_n = 0$" is the "Tauberian condition" required to effect the converse. This is a special case of the Tauberian condition "$\varphi \in L^{\infty}(\mathbb{R})$" used in Wiener's Tauberian Theorem (Theorem 2.9.12), e.g., [Ben75, Section 2.3].

3.31. Prove that

$$\forall \gamma \in (0, 2\pi), \qquad -\log\left(2\sin\frac{\gamma}{2}\right) = \sum_{n=1}^{\infty} \frac{\cos n\gamma}{n}. \qquad (E3.3)$$

[Hint. By a power series expansion we have

$$\operatorname{Re}\log\left(\frac{1}{1-z}\right) = \sum_{n=1}^{\infty} \frac{r^n \cos n\gamma}{n},$$

where $z = re^{i\gamma}$ and $r \in [0, 1)$. Compute

$$\lim_{r \to 1-} \sum_{n=1}^{\infty} \frac{r^n \cos n\gamma}{n} = -\log\left(2\sin\frac{\gamma}{2}\right)$$

on $(0, 2\pi)$. Since the right side of (E3.3) exists on $(0, 2\pi)$ by Exercise 3.29b, we can apply Exercise 3.30 to obtain (E3.3).]

3.32. Let $\Delta(ABC)$ be a triangle with vertices A, B, C; and let a and c be points on the segments BC and BA, respectively, so that

$$\angle(cCB) = \angle(cCA) \quad \text{and} \quad \angle(aAB) = \angle(aAC).$$

Assume $|Aa| = |Cc|$. Prove that $\Delta(ABC)$ is an isosceles triangle. [Hint. Use the law of sines.] This is a difficult exercise.

3.33. Prove that $\sum_{n=1}^{\infty} \cos(2\pi n\gamma)$ diverges for all $\gamma \in [0, 1)$, and that $\sum_{n=1}^{\infty} \sin(2\pi n\gamma)$ diverges for all $\gamma \in (0, 1)$.

3.34. Analogous to the definition of the Hilbert transform in Definition 2.5.11, we define the Hilbert transform of the sequence $f = \{f[n]\}$ as the sequence $\mathcal{H}f$, where

$$(\mathcal{H}f)[n] = \frac{1}{\pi} \sum_{m \in \mathbb{Z}, \, m \neq n} \frac{f[m]}{n - m}.$$

Prove that if $f \in \ell^2(\mathbb{Z}) \backslash \{0\}$ is real-valued, then

$$\|\mathcal{H}f\|_{\ell^2(\mathbb{Z})} < \|f\|_{\ell^2(\mathbb{Z})}.$$

This is the analogue on $\mathbb{Z}$ of Theorem 2.5.12a. [Hint. First show that if m, n are fixed and unequal and if $j \neq m, n$, then

$$\sum_j \frac{1}{(j-n)(j-m)} = \frac{2}{(m-n)^2}.$$

Use this fact to compute

$$\|\mathcal{H}f\|_{\ell^2(\mathbb{Z})}^2 = \sum_n f[n]^2 \sum_{j \neq n} \frac{1}{(j-n)^2} + \sum_n \sum_{m \neq n} f[n]f[m]\frac{2}{(m-n)^2} \quad \text{(E3.4)}$$

for f vanishing off some finite set. Since $2f[n]f[m] \leq f[n]^2 + f[m]^2$, we can invoke (3.3.8) to bound the right side of (E3.4) by $\pi^2 \sum f[n]^2$.] This clever, elementary proof is due to [Graf94], cf., [HLP52, pages 206ff., 212, 226, 235] for other proofs, [OS75, Chapter 7] for signal processing applications, and [Ivi85, pages 129ff.] for applications in analytic number theory.

3.35. Let $\{z_1, \ldots, z_n\} \subseteq \mathbb{C}$. Prove there is $S \subseteq \{1, \ldots, n\}$ such that

$$\sum_{j=1}^n |z_j| \leq 4\sqrt{2} \left| \sum_{j \in S} z_j \right|. \quad \text{(E3.5)}$$

[Hint. Begin by dividing $\mathbb{C}$ into four "diagonal" quadrants.] In particular, if $F(\gamma) = \sum f[n]e^{-\pi i t_n \gamma/\Omega}$, where $f = \{f[n]\}$ is a finite sequence and $\{t_n\} \subseteq \mathbb{R}$, then

$$\|F\|_{L^\infty(\hat{\mathbb{R}})} \leq \|f\|_{\ell^1(\mathbb{Z})} \leq 4\sqrt{2} \inf_{\gamma \in \hat{\mathbb{R}}} \left| \sum_{n \in S} f[n]e^{-\pi i t_n \gamma/\Omega} \right|$$

for some finite sequence $S \subseteq \mathbb{Z}$, cf., Definition 3.1.8.

Inequalities such as (E3.5) are used in measure theory and to prove versions of Schur's Lemma, e.g., [Ben76, Section 6.2]. The constant $4\sqrt{2}$ can be replaced by

π, which is best possible, e.g., [Ben76, page 172] for references to more advanced material.

3.36. Assume $\sum |c_n|^2 < \infty$ and

$$\lim_{N\to\infty} \sum_{|n|\le N} c_n e^{-2\pi i n y} = 0 \quad \text{a.e.}$$

Prove that $c_n = 0$ for all n, cf., the discussion in Section 3.2.6.

3.37. Let $0 \le a < b < 1$. Prove that $[a, b]$ is not a U-set. In fact, if $E \subseteq \mathbb{T}$ is measurable and $|E| > 0$, then E is not a U-set, cf., the discussion in Section 3.2.4.

3.38. a) Prove that

$$\left(\int_0^1 \frac{dt}{1+t^2}\right)^2 = \int_0^{\pi/4} \frac{\log(1+\cos 2\theta)}{\cos 2\theta} \, d\theta = \frac{3}{4}\int_0^1 \frac{\log(1+t)}{t}\, dt, \quad \text{(E3.6)}$$

e.g., [Ebe83].

b) We know the value of the left side of (E3.6). Expanding $t^{-1}\log(1+t)$ in a Maclaurin series, evaluate

$$\sum_{n=1}^{\infty} (-1)^{n+1} \frac{1}{n^2}.$$

c) Using part b, evaluate $\zeta(2)$ using the fact that

$$\zeta(2) - \sum_{n=1}^{\infty} (-1)^{n+1} \frac{1}{n^2} = \frac{1}{2}\zeta(2).$$

3.39. Prove that π is irrational, cf., [Ben77] for history and perspective on this material. [Hint. Assume $\pi^2 = a/b$, where $a, b \in \mathbb{N}$. Define $f_n(x) = x^n(1-x)^n/n!$ on $[0, 1]$ for each $n \ge 1$; and set

$$F_n(x) = b^n \{\pi^{2n} f_n(x) - \pi^{2n-2} f_n^{(2)}(x) + \pi^{2n-4} f_n^{(4)}(x) - \cdots$$
$$+ (-1)^j \pi^{2n-2j} f_n^{(2j)}(x) \cdots (-1)^n f_n^{(2n)}(x)\}.$$

Compute

$$\frac{d}{dx}(F'_n(x) \sin \pi x - \pi F_n(x) \cos \pi x) = \pi^2 a^n f_n(x) \sin \pi x,$$

and deduce that

$$\forall n \ge 1, \qquad \pi \int_0^1 f_n(x) \sin \pi x \, dx \in \mathbb{Z}.$$

Obtain a contradiction for large n since $0 < f_n(x) < 1/n!$ and $\sin \pi x > 0$ on $(0, 1)$.]

One can also show, more simply than for the case of π, that $e \notin \mathbb{Q}$. In fact, e and π are not only irrational, but are *transcendental*, i.e., they are not zeros of any polynomial P having rational coefficients. It is not known whether or not $\pi + e$ or π^e is irrational. On the other hand, e^π is transcendental.

3.40. A sequence $\{r_n : n \in \mathbb{N}\} \subseteq \mathbb{R}$ is *equidistributed modulo* 1 or, equivalently, *uniformly distributed modulo* 1 if the sequence $\{(r_n)\} \subseteq [0, 1)$ of fractional parts of the r_n are uniformly distributed in the sense that

$$\lim_{N \to \infty} \frac{\nu(I, N)}{N} = |I|$$

for every interval $I \subseteq (0, 1)$, where $\nu(I, N)$ is the number of elements from $\{(r_1), \ldots, (r_N)\}$ contained in I. $((r)$ is defined as $r - [r]$, where $[r]$ is the largest integer less than or equal to r.)

The Weyl Equidistribution Theorem (Remark 1.9.2) asserts that $\{r_n\}$ is equidistributed modulo 1 if and only if

$$\forall n \in \mathbb{Z}\setminus\{0\}, \qquad \lim_{N \to \infty} \frac{e^{2\pi i n r_1} + e^{2\pi i n r_2} + \cdots + e^{2\pi i n r_N}}{N} = 0,$$

e.g., [KK64], [Cha68, pages 84–90]. As a corollary, prove Kronecker's Theorem in one dimension: *if $r \in \mathbb{R}\setminus\mathbb{Q}$, then $\{(nr) : n \in \mathbb{N}\}$ is dense in $[0, 1)$.* [Hint. Use Exercise 3.13*b*.]

3.41. The d-dimensional version of Kronecker's Theorem was stated in Section 3.2.10. Prove that it is equivalent to the following assertion. *Let $\{\gamma_1, \ldots, \gamma_d\} \subseteq \mathbb{R}$ be linearly independent over the rationals, let $\gamma_0 = 0$, and let $c_0, c_1, \ldots, c_d \in \mathbb{C}$; then*

$$\sup_{t \in \mathbb{R}} \left| \sum_{j=0}^{d} c_j e^{-2\pi i t \gamma_j} \right| = \sum_{j=0}^{d} |c_j|.$$

3.42. Let $F \in L^1(\mathbb{T}_{2\Omega})$. Prove the following results.

a) If $\lim_{\lambda \to 0}(1/|\lambda|^2)\|\tau_\lambda F - F\|_{L^1(\mathbb{T}_{2\Omega})} = 0$, then $F = 0$ a.e.

b) If $\lim_{\lambda \to 0}(1/|\lambda|)\|\tau_\lambda F - F\|_{L^1(\mathbb{T}_{2\Omega})} = 0$, then F is a constant a.e.

c) If there exist $c, C > 0$ such that

$$\forall |\lambda| \le c, \qquad \|\tau_\lambda F - F\|_{L^1(\mathbb{T}_{2\Omega})} \le C|\lambda|,$$

then $F \in BV(\mathbb{T}_{2\Omega})$.

d) $\lim_{\lambda \to 0}\|\tau_\lambda F - F\|_{L^1(\mathbb{T}_{2\Omega})} = 0$.

Part *a* is a consequence of part *b*; part *b* follows from a refinement of FTC, e.g., [Ben76, pages 141–142] and Remark 1.7.9 on the Lebesgue set; part *c* is due to Hardy and Littlewood (1928), e.g., [Ben76, pages 124–126]; and part *d* is an elementary and fundamental fact from Appendix A. The converse of part *c* results from a straightforward calculation using the classical form of the Jordan Decomposition Theorem (Remark 1.7.4).

3.43. a) Let $\Omega > 0$, and let $f \in c_0(\mathbb{Z})$ have the property that $\{nf[n] : n \in \mathbb{N}\}$ decreases monotonically to 0 after a certain point. Prove that $\sum_{n=1}^{\infty} f[n] \sin(\pi n \gamma / \Omega)$ converges uniformly on $\hat{\mathbb{R}}$, cf., Example 3.3.4, Exercise 3.29, and [Zyg59, Volume I, pages 182–183].

b) Let $f[n] = 1/(n \log n)$, $n \geq 3$. By part a,

$$F(\gamma) = \sum_{n=3}^{\infty} \frac{1}{n \log n} \sin\left(\frac{\pi n \gamma}{\Omega}\right)$$

converges uniformly on $\hat{\mathbb{R}}$. Prove that $F \in C(\mathbb{T}_{2\Omega}) \backslash A(\mathbb{T}_{2\Omega})$, cf., Example 1.4.4 and Example 3.3.4 a.

3.44. Verify the inclusions and inequalities (3.1.8)–(3.1.11). Prove that the inclusions are proper, e.g., Exercise 3.43.

3.45. Use the Uniform Boundedness Principle (Theorem B.8), in a manner similar to Example 3.4.9c, to prove that there are continuous functions whose Fourier series diverge at a point, cf., the discussion in Section 3.2.8. This proof, due to Lebesgue (1905), is short and nonconstructive. [Hint. Define the linear functionals

$$\begin{aligned} L_N : C(\mathbb{T}) &\longrightarrow \quad \mathbb{C} \\ F &\longmapsto \quad S_N(F)(0). \end{aligned}$$

Let $F_N \in C(\mathbb{T})$ equal sgn D_N except on small intervals about the discontinuities of sgn D_N; further, construct F_N so that $|F_N| \leq 1$. Then

$$\|L_N\| \geq |L_N(F_N)| = \left| \int_{\mathbb{T}} D_N(\gamma) F_N(\gamma)\, d\gamma \right|,$$

and the right side is close to $\|D_N\|_{L^1(\mathbb{T})}$. Thus, $\{\|L_N\|\}$ is unbounded, and the result is obtained analogous to Example 3.4.9c.]

Fejér's *construction* in 1911 of continuous functions whose Fourier series diverge at a point makes *implicit* use of the Uniform Boundedness Principle. His proof provides divergence at much larger sets of measure 0 than single points, e.g., [Rog59, pages 75–77], [Zyg59, Volume I, Chapter VIII.1], cf., the theorem of Kahane and Katznelson quoted in Section 3.2.8. On the other hand, Fejér's type of example has unbounded partial sums. Using Riesz products, Zygmund (1948) constructed $F \in L^\infty(\mathbb{T})$ for which $\{S_N(F)\}$ is *uniformly bounded* and $S(F)$ diverges on uncountable dense sets of measure 0, e.g., [Zyg59, Volume I, page 302].

3.46. If $F \in L^1(\mathbb{T})$, define

$$M(F)(\gamma) = \sup_N \{|S_N(F)(\gamma)|\}.$$

The main part of Carleson's proof of the Lusin Conjecture (Section 3.2.8) is his theorem that

$$\exists C \text{ such that } \forall F \in L^2(\mathbb{T}), \ \|M(F)\|_{L^2(\mathbb{T})} \leq C \|F\|_{L^2(\mathbb{T})}. \tag{E3.7}$$

Using (E3.7) prove the Lusin Conjecture:

$$\forall F \in L^2(\mathbb{T}), \qquad \lim_{N \to \infty} S_N(F) = F \quad \text{a.e.}$$

With regard to (E3.7), compare Zygmund's result quoted in Exercise 3.45.

3.47. Let $F \in L^1(\mathbb{T})$ and $G \in L^\infty(\mathbb{T})$ have Fourier coefficients $f = \{f[n]\} \in A(\mathbb{Z})$ and $g = \{g[n]\} \in A'(\mathbb{Z})$, respectively. Compute $(FG)\check{}$ and $(F * G)\check{}$ in terms of f and g.

3.48. a) Let $F \in A(\mathbb{T})$ be nonnegative. With regard to Remark 3.6.5d on the Fejér–Riesz Theorem, verify whether or not there is a sequence $\{B_n\}$ of trigonometric polynomials on $\mathbb{T}$ for which

$$\lim_{n\to\infty} \|F - |B_n|^2\|_{A(\mathbb{T})} = 0,$$

cf., the Szegö Factorization Theorem stated in Section 3.7.

b) Let $F \in C(\mathbb{T})$, and suppose the sequence $f = \{f[n]\}$ of Fourier coefficients of F is nonnegative, i.e., $f[n] \geq 0$ for each n. Prove that $F \in A(\mathbb{T})$.

[Hint. f is a positive distribution of $\mathbb{Z}$ so that F is continuous and positive definite on $\mathbb{T}$.]

There are also several classical ways of proving part *b*.

3.49. A fundamental problem in harmonic analysis is to quantify properties such as support, smoothness, convergence, and decay of f and its approximants in terms of the behavior of $\hat{f}$; and many of our results can be put in this context, e.g., the Riemann–Lebesgue Lemma, Exercise 1.13, Exercise 3.26, and Example 3.5.3 where we proved that $C^1(\mathbb{T}) \subseteq A(\mathbb{T})$. In this regard, prove that if $1 \leq m < \infty$, then

$$\forall F \in C^m(\mathbb{T}), \qquad \lim_{N\to\infty} \|F - S_N(F)\|_{L^\infty(\mathbb{T})} = 0.$$

Estimate $\|F - S_N(F)\|_{L^\infty(\mathbb{T})}$ in terms of N and m.

3.50. Prove the assertions in Definition 3.1.9d.

3.51. a) Let $L \in \mathcal{L}(L^2(\mathbb{T}))$, i.e., $L : L^2(\mathbb{T}) \longrightarrow L^2(\mathbb{T})$ is a continuous linear function (Definition B.6). If $F \in L^2(\mathbb{T})$, prove that

$$L(F)(\gamma) = \sum (L_d \check{F})[n] e^{-2\pi i n \gamma}$$

where

$$(L_d \check{F})[n] = \sum_m \ell_{mn} \check{F}[m]$$

and

$$\ell_{mn} = \langle Le_{-m}, e_{-n} \rangle.$$

Thus, the action (operation) of L on $L^2(\mathbb{T})$ is *equivalent* to the action of the matrix $L_d \equiv \{\ell_{mn} : m, n \in \mathbb{Z}\}$ on $\ell^2(\mathbb{Z})$.

b) Now, further suppose that L is *translation invariant*, i.e., $L(\tau_\gamma F) = \tau_\gamma (LF)$ for all $\gamma \in \mathbb{T}$ and all $F \in L^2(\mathbb{T})$, e.g., Exercise 2.58. Prove that L_d is a diagonal matrix.

Parts *a* and *b* allow us to conclude that $\{e_{-m}\}$ *diagonalizes all continuous translation-invariant linear functions on* $L^2(\mathbb{T})$; and it is also clear that *each e_{-m} is an eigenfunction of the derivative operator*. This fact accounts for the classical success of Fourier analysis in dealing with linear partial differential equations, since it provides exactly the advantage of doing matrix calculations in a basis that diagonalizes the matrix. In more recent times, parts *a* and *b* are the starting point for the harmonic analysis of singular integral operators, e.g., the Calderón–Zygmund Theory [Ste70], as well as LTI systems in engineering, e.g., Definition 2.6.5.

3.52. Let $\{r_j : j = 0, \pm 1, \ldots, \pm N\} \subseteq \mathbb{C}$ satisfy the condition $r_j = \bar{r}_{-j}$ for each j; and define the $(N+1) \times (N+1)$ matrix $R = (r_{jk})$, where $j, k \geq 0$ and $r_{jk} \equiv r_{j-k}$. Also, define the functional L on the trigonometric polynomials $F(\gamma) = \sum_{n=-N}^{N} f[n]e^{-2\pi i n \gamma}$ by the rule

$$L(F) = \sum_{n=-N}^{N} f[n]\bar{r}_n.$$

   a) Prove that if $R \gg 0$ and $1 \leq j \leq N$, then $r_0 > |r_j|$.

   b) Prove that for each polynomial $G(\gamma) = \sum_{n=0}^{N} g[n]e^{-2\pi i n \gamma}$,

$$L(|G|^2) = \sum r_{j-k}\, g[k]\overline{g[j]}.$$

   c) We shall say that the functional L is *positive*, written $L \geq 0$, if $L(F) \geq 0$ and

$$L(F) = 0 \quad \text{implies} \quad F = 0$$

   for all nonnegative trigonometric polynomials $F(\gamma) = \sum_{n=-N}^{N} f[n]e^{-2\pi i n \gamma}$. Prove that $R \gg 0$ if and only if $L \geq 0$ cf., Exercise 2.52a for the analogous situation in terms of positive semidefinite matrices. [Hint. Part b is used in both directions, and the direction to prove $L \geq 0$ requires the Fejér–Riesz Theorem.]

3.53. Prove that $(\sin x)^{-2} < x^{-2} + 1$ on $(0, \pi/2]$. The proof is elementary, e.g., [Mon71, page 155–156]. This inequality should be compared with the inequality $(2/\pi)x \leq \sin x$ on $[0, \pi/2]$, which is usually used in the proof of Jordan's Inequality (Exercise 2.62).

   This inequality can be used to prove the following number-theoretic "sieve theorem" due to Bombieri, e.g., [Mon71, Chapters 2 and 3]. *Let $\delta > 0$ and $R \subseteq \mathbb{R}$ have the property that the distance of $r - s$ to any integer is greater than or equal to δ for all unequal $r, s \in R$; then*

$$\delta \sum_{\gamma \in R} |F(\gamma)|^2 \leq \left(\delta N + \frac{2}{\sqrt{3}} + 3\delta \right) \sum |c_n|^2$$

for all trigonometric polynomials

$$F(\gamma) = \sum_{n=M+1}^{M+N} c_n e^{-2\pi i n \gamma}.$$

   Because of the Parseval formula, it is interesting to note that if the elements of R are (relatively) equally spaced modulo 1, then $\delta \sum_{\gamma \in R} |F(\gamma)|^2$ is a Riemann sum approximating $\|F\|_{L^2(\mathbb{T})}^2$. The relation of such approximations to sampling theory is the subject of [Ben92b, pages 492–494].

3.54. a) Prove that if $W \in L^1(\mathbb{T})$ is positive on $\mathbb{T}$, then $\log W \in L^1(\mathbb{T})$.

   b) Prove that if $W \in L^1(\mathbb{T})$ is nonnegative, then $\log W \in L^1(\mathbb{T})$ if and only if

$$\int_{\mathbb{T}} \log W(\gamma)\, d\gamma > -\infty.$$

3.55. We discussed the Littlewood Flatness Problem in Remark 3.8.11*b*. In this regard, prove that if $F(\gamma) = \sum_{n=0}^{N-1} f[n]e^{-2\pi i n\gamma}$, then $|F|$ and $|f|$ cannot simultaneously take constant values on $\mathbb{T}$ and $\{0, 1, \ldots, N-1\}$, respectively.

3.56. Consider the following formal calculation for $f : \mathbb{R} \times \mathbb{R} \longrightarrow \mathbb{C}$, where $f(t, u) \equiv f(t-u)$:

$$\iint f(t, u)e^{-2\pi i(t\lambda+u\gamma)}\, dt\, du$$

$$= \int \left(\int f(v)e^{-2\pi i(u+v)\lambda}e^{-2\pi i u\gamma}\, dv \right) du \qquad (E3.8)$$

$$= \hat{f}(\lambda) \int e^{-2\pi i u(\lambda+\gamma)}\, du = \hat{f}(\lambda)\delta(\lambda+\gamma)$$

$$= \begin{cases} \hat{f}(\lambda), & \text{if } \lambda = -\gamma, \\ 0, & \text{if } \lambda \neq -\gamma. \end{cases}$$

a) Recalling the definition of a Toeplitz matrix $R = (r_{jk})$, viz., $r_{jk} = r_{j-k}$, prove that the two-dimensional DFT of a Toeplitz matrix is a diagonal matrix, cf., Exercise 3.51.

b) Provide the hypotheses and details to make (E3.8) into a theorem.

3.57. Let $\mu \in M(\mathbb{T})$ and assume $\lim_{n\to\infty} \check{\mu}[n] = c \in \mathbb{C}$. Prove that $\lim_{n\to-\infty} \check{\mu}[n] = c$. [Hint. By subtracting $c\delta$ from μ we can assume, without loss of generality, that $\lim_{n\to\infty} \check{\mu}[n] = 0$. By the Radon–Nikodym Theorem, let $\mu = F|\mu|$, where F is μ-measurable and $|F| \equiv 1$, e.g., [Ben76, Chapter 5]. If $G \equiv \overline{F}/F$, then

$$\overline{\check{\mu}[-n]} = \int_{\mathbb{T}} e^{2\pi i n\gamma} G(\gamma)\, d\mu(\gamma). \qquad (E3.9)$$

If G is a trigonometric polynomial the right side tends to 0 as $n \longrightarrow \infty$, and so the result is obtained by (E3.9). The result for $G = \overline{F}/F$ follows by the Weierstrass Approximation Theorem (Theorem 3.4.6 and Remark 3.4.8) and approximation properties of bounded μ-measurable functions (such as G) by continuous functions, cf., [Ben76, pages 201].]

3.58. Complete the proof of the Classical Sampling Theorem, which uses the PSF, i.e., provide the mathematical justification for the formal steps in *Proof 1* of Theorem 3.10.10.

3.59. Let $s(t) = d_{2\pi\Omega}(t) = (\sin 2\pi\Omega t)/(\pi t)$; and assume equation (3.10.21), i.e.,

$$f = T \sum f(nT)\tau_{nT}s,$$

is true for all $f \in PW_\Omega$. Prove that $2T\Omega \leq 1$, e.g., [Ben92b, page 450].

3.60. Let $F \in L^2(\mathbb{T})$, and let $f = \{f[n]\}$ be the sequence of Fourier coefficients of F. Assume

$$\sum_{n=1}^{\infty} \frac{1}{n^{1/2}} \left(\sum_{|k|=n}^{\infty} |f[k]|^2 \right)^{1/2} < \infty.$$

Prove that $F \in A(\mathbb{T})$.

      We have commented often on the difficulty of characterizing $A(\mathbb{T})$. The result of this exercise can be considered the first step in a fairly sophisticated classical line of thinking, e.g., [Ben75, pages 150–154].

3.61. As in Exercise 3.51, we shall say that $L \in \mathcal{L}(L^2(\mathbb{T}))$ is *translation invariant* if

$$\forall F \in L^2(\mathbb{T}) \text{ and } \forall \gamma \in \mathbb{T}, \quad L(\tau_\gamma F) = \tau_\gamma(LF),$$

cf., Definition 2.6.5. Prove that the translation-invariant continuous linear operators $L : L^2(\mathbb{T}) \longrightarrow L^2(\mathbb{T})$ are precisely of the form

$$\forall F \in L^2(\mathbb{T}), \qquad LF = G * F, \tag{E3.10}$$

where $G \in A'(\mathbb{T})$, i.e., $g \equiv \check{G} \in \ell^\infty(\mathbb{Z})$; and also show that $\|g\|_{\ell^\infty(\mathbb{Z})} = \|L\|$, where $\|L\|$ is defined in Definition B.6. [Hint. If L is defined by (E3.10), then it is easy to see that $L \in \mathcal{L}(L^2(\mathbb{T}))$ and that L is translation invariant; and it is elementary to prove the norm equality. The converse is a consequence of Exercise 3.51. In fact, if $L \in \mathcal{L}(L^2(\mathbb{T}))$ is translation invariant, then $(LF)\check{} = L_d f$, where L_d is the diagonal matrix of Exercise 3.51, which has entries $g[n] \equiv \langle Le_{-n}, e_{-n} \rangle$ on the diagonal. If $g \equiv \{g[n]\}$, then $\check{G} \equiv g \in \ell^\infty(\mathbb{Z})$ since $|g[n]| \leq \|L\|$, and $LF = G * F$ for this G.]

      In Example 2.6.6 we stated the L^1 analogue of the above result. In the L^1 setting, $A'(\mathbb{T})$ is replaced by $M(\mathbb{T})$. In $L^1(\mathbb{T})$, if L is also a projection, i.e., $L \circ L = L$ on $L^1(\mathbb{T})$, then the corresponding measure $\mu \in M(\mathbb{T})$ is an idempotent measure. Idempotent measures were discussed in Remark 3.10.13.

3.62. Let $T > 0$ and define $\mu = \sum e^{2\pi i n t/T}$ on $\mathcal{S}(\mathbb{R})$ by the rule

$$\forall f \in \mathcal{S}(\mathbb{R}), \qquad \mu(f) = \lim_{N \to \infty} \left(\sum_{|n| \leq N} e^{2\pi i n t/T} \right)(f(t)).$$

Recall from Chapter 2 that $g(f)$ denotes $\int g(t) f(t)\, dt$.

a) Prove that $\mu \in \mathcal{S}'(\mathbb{R})$.

b) Prove that $\mu \in M(\mathbb{T})$ and

$$T \sum \delta_{nT} = \sum e^{2\pi i n t/T}.$$

A

Real Analysis

$L^1(\mathbb{R})$ and the L^1-norm $\| \cdots \|_{L^1(\mathbb{R})}$ were introduced in Definition 1.1.1, and are a basic space and norm in real analysis, i.e., advanced calculus and real variables. This appendix lists results from real analysis, e.g., [AB66], [Apo57], [Ben76], [Rud66]. There are many excellent texts.

A.1 Definition. Convergence of Sequences

a. $\mathbb{N}$ is the set positive integers, $\mathbb{Z}$ is the set of integers, $\mathbb{R}$ is the set of real numbers, and $\mathbb{C}$ is the set of complex numbers.

b. A sequence of $\{c_n\} \subseteq \mathbb{C}$ *converges* to $c \in \mathbb{C}$ as $n \to \infty$, written $\lim_{n \to \infty} c_n = c$ if

$$\forall \epsilon > 0, \; \exists N = N(\epsilon), \quad \text{such that} \quad \forall n \geq N,$$
$$|c_n - c| < \epsilon. \tag{A.1}$$

The "statement" (A.1) is read as follows: for every $\epsilon > 0$, there is $N \in \mathbb{N}$ depending on ϵ, such that for any $n \geq N$ the inequality, $|c_n - c| < \epsilon$, is valid.

c. If $\{c_n\} \subseteq \mathbb{C}$ is a sequence and if a subsequence $\{c_{n_k}\} \subseteq \{c_n\}$ converges to $c \in \mathbb{C}$, then c is a *limit point* of $\{c_n\}$.

Let $\{c_n\} \subseteq \mathbb{R}$, and suppose there is $s \in \mathbb{R}$ with the properties that

$$\forall \epsilon > 0, \; \exists N = N(\epsilon), \quad \text{such that} \quad \forall n > N, \; c_n < s + \epsilon$$

and

$$\forall \epsilon > 0 \text{ and } \forall N, \; \exists n > N \quad \text{such that} \quad c_n > s - \epsilon.$$

Then s is the *limit superior* of $\{c_n\}$, and we write

$$s = \varlimsup_{n \to \infty} c_n.$$

The definition of limit superior means that for each $\epsilon > 0$, eventually (depending on ϵ) all terms of $\{c_n\}$ are less than $s + \epsilon$ and infinitely many terms of $\{c_n\}$ are greater than $s - \epsilon$. Intuitively, s is the "greatest" limit point of $\{c_n\}$. The *limit inferior* of $\{c_n\} \subseteq \mathbb{R}$ is defined as

$$\varliminf_{n\to\infty} c_n = - \varlimsup_{n\to\infty} (-c_n).$$

d.   If $\{a_n\}, \{b_n\} \subseteq \mathbb{R}$ are sequences, then it is not difficult to prove that

$$\varlimsup (a_n + b_n) \leq \varlimsup a_n + \varlimsup b_n.$$

A.2 Definition. Convergence of Functions

Let $\{f_n\}$ be a sequence of functions $f_n : X \to \mathbb{C}$, where $X \subseteq \mathbb{R}$.

a. $\{f_n\}$ *converges pointwise* to a function $f : X \to \mathbb{R}$ as $n \to \infty$, written

$$\forall t \in X, \qquad \lim_{n\to\infty} f_n(t) = f(t),$$

if

$$\forall t \in X \text{ and } \forall \epsilon > 0, \ \exists N = N(t, \epsilon) \in \mathbb{N} \quad \text{such that}$$

$$\forall n \geq N, \ |f_n(t) - f(t)| < \epsilon.$$

b. $\{f_n\}$ *converges uniformly* on X to a function $f : X \to \mathbb{R}$ as $n \to \infty$, written

$$\lim_{n\to\infty} f_n = f \quad \text{uniformly on} \quad X,$$

if

$$\forall \epsilon > 0, \ \exists N = N(\epsilon), \quad \text{such that} \quad \forall t \in X \text{ and } \forall n \geq N,$$

$$|f_n(t) - f(t)| < \epsilon.$$

In this case, N is independent of the variable t. This often allows us to *switch operations* legitimately. An example is the following result where the operations are integration and taking limits, cf., Theorem A.5 and [Ben76, Chapter 6] (where a characterization for the validity of such switching is given).

A.3 Theorem. $\lim_{n\to\infty} \int_a^b f_n(t)\,dt = \int_a^b f(t)\,dt$

Let $\{f_n\}$ be a sequence of continuous functions $f_n : [a, b] \to \mathbb{C}$, which converges uniformly on $[a, b]$ to a function $f : [a, b] \to \mathbb{C}$. Then f is continuous on $[a, b]$

and

$$\lim_{n\to\infty} \int_a^b f_n(t)\, dt = \int_a^b \left(\lim_{n\to\infty} f_n(t)\right) dt = \int_a^b f(t)\, dt.$$

A.4 Definition. Lebesgue Measure Zero

The *Lebesgue measure* of an interval with endpoints a and b, $b > a$, is defined as its length $b - a$. If $X = \bigcup_{j=1}^\infty (a_j, b_j)$, where $b_{j+1} \le a_j$, then the *Lebesgue measure* of X, denoted by $|X|$, is defined as $|X| = \sum_{j=1}^\infty (b_j - a_j)$. Unless there is possible confusion, we shall usually refer to the *measure* of X instead of the *Lebesgue measure* of X. This notion of measure extends naturally to many other sets $X \subseteq \mathbb{R}$, and we write $|X|$ to denote their measure.

A set $X \subseteq \mathbb{R}$ is a set of *measure* 0 if for each $\epsilon > 0$ there is a countable set $\{(a_j, b_j) \subseteq \mathbb{R} : j = 1, \ldots\}$ of intervals such that

$$X \subseteq \bigcup_{j=1}^\infty (a_j, b_j) \quad \text{and} \quad \Sigma(b_j - a_j) < \epsilon.$$

A property is valid *almost everywhere*, written *a.e.*, if it is true for all $t \in \mathbb{R}\backslash X$, where X is a set of measure 0. (Technically, an element of $f \in L^1(\mathbb{R})$ is a set of functions g that are equal a.e. and for which $\int |g(t)|\, dt < \infty$. In this book we shall think of elements in $L^1(\mathbb{R})$ as functions, and no problems will arise.)

The next result gives insight into the structure of $L^1(\mathbb{R})$. Recall the definition of $C_c^\infty(\mathbb{R})$ from Definition 2.2.1.

A.5 Theorem. Fundamental Properties of $L^1(\mathbb{R})$

Let $f \in L^1(\mathbb{R})$ and let $\epsilon > 0$.

a. *There is a function $g_\epsilon = \Sigma c_j \mathbf{1}_{[a_j, b_j)}$, $b_j \le a_{j+1}$, a finite sum, such that*

$$\|f - g_\epsilon\|_{L^1(\mathbb{R})} < \epsilon.$$

b. *There is a function $g_\epsilon \in C_c^\infty(\mathbb{R})$, such that*

$$\|f - g_\epsilon\|_{L^1(\mathbb{R})} < \epsilon.$$

c. *There is $\delta = \delta(\epsilon)$ such that if $|u| < \delta$, then*

$$\|f - \tau_u f\|_{L^1(\mathbb{R})} < \epsilon.$$

(Recall that $(\tau_u f)(t) = f(t - u)$ as a function of t.)

d. *If $\|f\|_{L^1(\mathbb{R})} = 0$, then f is the 0-element of $L^1(\mathbb{R})$, i.e., $f = 0$ a.e.*

A.6 Theorem. Fatou Lemma

Let $\{f_n\} \subseteq L^1(\mathbb{R})$ and $g \in L^1(\mathbb{R})$ be real-valued functions that have the property that for each n, $g \leq f_n$ a.e. If f is a function for which $\lim_{n \to \infty} f_n = f$ a.e., then $f \in L^1(\mathbb{R})$ and

$$\int f(t)\, dt \leq \varliminf_{n \to \infty} \int f_n(t)\, dt.$$

(Recall that $\int f(t)\, dt$ indicates integration over $\mathbb{R}$.)

A.7 Remark. Perspective on the Fatou Lemma

Fatou proved Theorem A.6 in 1906, and used it to prove the Parseval Theorem for $F \in L^2(\mathbb{T}_{2\Omega})$, e.g., [Haw70, pages 168–172]. Earlier, Lebesgue had proved the Parseval Theorem for $L^\infty(\mathbb{T}_{2\Omega}) \subseteq L^2(\mathbb{T}_{2\Omega})$. The Fatou Lemma can be used to prove the Beppo Levi Theorem, which was originally proved in 1906, as well as the general form of the Lebesgue Dominated Convergence Theorem (LDC), which Lebesgue published in 1908. Statements of these latter two results now follow.

A.8 Theorem. Beppo Levi Theorem

Let $\{f_n\} \subseteq L^1(\mathbb{R})$ and $g \in L^1(\mathbb{R})$ have the property that for each n, $g \leq f_n$ a.e. If $f_n \leq f_{n+1}$ a.e. for each n and $\lim_{n \to \infty} \int f_n(t)\, dt$ is finite, then $\lim_{n \to \infty} f_n \equiv f \in L^1(\mathbb{R})$ and

$$\int f(t)\, dt = \lim_{n \to \infty} \int f_n(t)\, dt,$$

cf., [Ben76, Theorem 3.7] for a relaxation of the monotonicity of $\{f_n\}$.

We shall use the notation *LDC* in referring to the following result.

A.9 Theorem. Lebesgue Dominated Convergence Theorem

Let $\{f_n\} \subseteq L^1(\mathbb{R})$, and let f be a function for which $\lim_{n \to \infty} f_n = f$ a.e. If $g \in L^1(\mathbb{R})$ has the property that

$$\forall n, \quad |f_n(t)| \leq g(t) \quad a.e. \quad on \quad \mathbb{R},$$

then $f \in L^1(\mathbb{R})$ and

$$\lim_{n \to \infty} \|f_n - f\|_{L^1(\mathbb{R})} = 0.$$

A.10 Definition. Measurable Functions and L^p-Spaces

a.　A complex-valued function f defined a.e. on $\mathbb{R}$ is *(Lebesgue) measurable* if there is a sequence $\{f_n\}$ of continuous functions on $\mathbb{R}$ for which $\lim_{n\to\infty} f_n = f$ a.e. The elements of $L^1(\mathbb{R})$ are measurable; whereas the function f, defined by $f(t) = 1/t$ a.e. on $\mathbb{R}$, is measurable but is not only not in $L^1(\mathbb{R})$ but is not in $L^1_{\text{loc}}(\mathbb{R})$. *All of our functions in the statements of all of our theorems are assumed to be measurable.* However, we are not interested in measurable functions per se; and a real knowledge of this concept is not required in this book even though it is fundamental in any systematic treatment of real analysis.

b.　Let $p \in [1, \infty)$. $L^p(\mathbb{R})$ is the space of functions $f : \mathbb{R} \to \mathbb{C}$, for which the L^p-norm of f,

$$\|f\|_{L^p(\mathbb{R})} = \left(\int |f(t)|^p \, dt \right)^{1/p}, \tag{A.2}$$

is finite. There are subtleties in this definition related to the meaning of Lebesgue integration and the parenthetical remark in Definition A.4. We shall not deal with such issues, and, as with measurable functions, they do not play a role in understanding the material in this book.

c.　If $p = \infty$ and the aforementioned subtleties are once again acknowledged and forgotten, then $L^\infty(\mathbb{R})$ is defined as the space of all functions that are bounded except possibly on a subset of measure zero. The L^∞-*norm* $\|f\|_{L^\infty(\mathbb{R})}$ of $f \in L^\infty(\mathbb{R})$ is intuitively the supremum of $|f|$ on $\mathbb{R}$. Technically,

$$\|f\|_{L^\infty(\mathbb{R})} = \inf\{M : |\{t \in \mathbb{R} : |f(t)| > M\}| = 0\}. \tag{A.3}$$

This complicated array of symbols reduces to

$$\|f\|_{L^\infty(\mathbb{R})} = \sup\{|f(t)| : t \in \mathbb{R}\}$$

in the case $f \in C_c^\infty(\mathbb{R})$.

d.　The L^p-spaces can be defined on subintervals of $\mathbb{R}$ as well as other sets X. In such cases, if $p < \infty$, then the integration over $\mathbb{R}$ in (A.2) is replaced by integration over X, with a similar adjustment in (A.3).

A.11 Example. Perspective on LDC

a.　In light of LDC, it is natural to ask if the hypothesis,

$$\lim_{n \to \infty} \|f_n\|_{L^1(\mathbb{R})} = 0,$$

allows us to conclude that $\lim_{n\to\infty} f_n = 0$ a.e., cf., [Ben76, Chapter 6]. The following example shows that the answer is "no". For each $n \in \mathbb{N}$ consider the unique representation $n = 2^j + k, 0 \le k < 2^j$; and for each n consider the interval $I_n = [k/2^j, (k+1)/2^j) \subseteq [0, 1)$. We define

$$f_n = \mathbf{1}_{I_n};$$

and note that $\{I_n\}$ keeps "sweeping over" $[0, 1)$ with finer and finer partitions, i.e., $I_1 = [0, 1), I_2 = [0, \frac{1}{2}), I_3 = [\frac{1}{2}, 1), I_4 = [0, \frac{1}{4}), I_5 = [\frac{1}{4}, \frac{1}{2})$, etc. Clearly, then, $\lim_{n\to\infty} f_n(t)$ does *not* exist for any $t \in [0, 1)$. On the other hand, if $p \in (0, \infty)$, then $\| f_n \|_{L^p(\mathbb{R})} = (1/2^j)^{1/p}$, and hence $\lim_{n\to\infty} \| f_n \|_{L^p(\mathbb{R})} = 0$ since $j = j(n) \to \infty$ as $n \to \infty$.

b. Generally, if $\lim_{n\to\infty} \| f_n \|_{L^\infty(\mathbb{R})} = 0$, then $\lim_{n\to\infty} f_n = 0$ a.e.

c. There is a reasonable sort of converse to LDC, which is due to F. Riesz: *if $\{f_n\}$ is a sequence of functions on $\mathbb{R}$ with the property that*

$$\forall \epsilon > 0, \quad \lim_{n\to\infty} |\{t \in \mathbb{R} : |f_n(t)| \ge \epsilon\}| = 0, \tag{A.4}$$

then there is a subsequence $\{f_{n_k}\} \subseteq \{f_n\}$ for which

$$\lim_{k\to\infty} f_{n_k} = 0 \quad a.e.$$

Note that (A.4) holds if $\lim_{n\to\infty} \| f_n \|_{L^1(\mathbb{R})} = 0$. If "$\lim f_n = 0$" in the sense of (A.4), we say that $\{f_n\}$ *converges in measure* to 0.

A.12   Theorem. Fubini Theorem (1907)

If $f(x, y)$ is integrable on the rectangle R defined by $a \le x \le b, c \le y \le d$, then the functions $x \mapsto f(x, y)$ and $y \mapsto f(x, y)$ are integrable for almost all values of y and x, respectively. Furthermore, the functions

$$y \longmapsto \int_a^b f(x, y)\, dx \quad \text{and} \quad x \longmapsto \int_c^d f(x, y)\, dy$$

are integrable and

$$\int_R f(x, y)\, dx dy = \int_c^d \left(\int_a^b f(x, y)\, dx \right) dy = \int_a^b \left(\int_c^d f(x, y)\, dy \right) dx,$$

where "$dxdy$" is so-called "product measure", e.g., [Rud66].

Tonelli's Theorem, which follows, can be viewed as a converse to Fubini's Theorem with the further hypothesis, $f \ge 0$. Equivalently, we can think of Tonelli's Theorem as giving conditions so that f is integrable, thereby allowing us to use Fubini's Theorem.

A.13 Theorem. Tonelli Theorem (1909)

Let f be nonnegative on the rectangle R defined in Theorem A.12. Then

$$\int_R f(x, y)\, dxdy = \int_c^d \left(\int_a^b f(x, y)\, dx \right) dy = \int_a^b \left(\int_c^d f(x, y)\, dy \right) dx.$$

These equations mean that if any one of the expressions is infinite, then the other two are infinite, and that if any one is finite, then the other two are finite and all three are equal.

Fubini's and Tonelli's Theorems can be stated quite generally on spaces of the form $X \times Y$ instead of R and for measures other than Lebesgue measure. For convenience we shall refer to one or the other collectively as the *Fubini–Tonelli Theorem*; and we shall not discuss *product measure* or go into any of the subtleties relating measurable functions on $X \times Y$ with their restrictions to X or Y. For $X = Y = \mathbb{R}$, or for even more general sets X and Y, the Fubini–Tonelli Theorem is as follows.

A.14 Theorem. Fubini–Tonelli Theorem

a. *(Tonelli). If $f : X \times Y \to \mathbb{C}$ and*

$$\int_X \left(\int_Y |f(x, y)|\, dy \right) dx < \infty, \tag{A.5}$$

then f is integrable, i.e., $f \in L^1(X \times Y)$.

b. *(Fubini). Let $f \in L^1(X \times Y)$, and define the function $f_x : Y \to \mathbb{C}$, resp., $f_y : X \to \mathbb{C}$, by*

$$f_x(y) = f(x, y), \quad resp., \quad f_y(x) = f(x, y).$$

Then $f_x \in L^1(Y)$ a.e. in x, $f_y \in L^1(X)$ a.e. in y, the functions defined a.e. by

$$\int_Y f_x(y)\, dy \quad and \quad \int_X f_y(x)\, dx$$

are in $L^1(X)$ and $L^1(Y)$, respectively, and

$$\int_{X \times Y} f(x, y)\, dxdy = \int_Y \int_X f_y(x)\, dxdy$$
$$= \int_X \int_Y f_x(y)\, dy\, dx. \tag{A.6}$$

c. If $f : X \times Y \to \mathbb{C}$ and (A.5) is valid, then each of the integrals in (A.6) is finite, and (A.6) is valid.

A.15 Theorem. Hölder Inequality

If $1 \le p \le \infty$ and $\frac{1}{p} + \frac{1}{q} = 1$, and if $f \in L^p(\mathbb{R})$ and $g \in L^q(\mathbb{R})$, then $fg \in L^1(\mathbb{R})$ and

$$\|fg\|_{L^1(\mathbb{R})} \le \|f\|_{L^p(\mathbb{R})} \|g\|_{L^q(\mathbb{R})}. \tag{A.7}$$

If $1 < p < \infty$, there is equality in (A.7) if and only if there are constants $A, B \ge 0$, not both 0, such that $A|f|^p = B|g|^q$ a.e.

A.16 Theorem. Minkowski Inequality

Let $p \ge 1$, and let f be a complex-valued function defined on $\mathbb{R} \times \mathbb{R}$. Then

$$\left(\int \left| \int f(t, u)\, dt \right|^p du \right)^{1/p} \le \int \left(\int |f(t, u)|^p\, du \right)^{1/p} dt, \tag{A.8}$$

i.e., the L^p-norm of a "sum" is less than or equal to the "sum" of the L^p-norms.

A.17 Remark. Consequences of Hölder's and Minkowski's Inequalities

a.   Minkowski's Inequality is true for many spaces X besides $\mathbb{R}$ and for measures, other than Lebesgue measure, on these spaces, cf., [HLP52, pages 146–150], [Zyg59, pages 18–19]. In particular, if

$$f_1, \ldots, f_n \in L^p(\mathbb{R}),$$

then

$$\left\| \sum_{j=1}^{n} f_j \right\|_{L^p(\mathbb{R})} \le \sum_{j=1}^{n} \|f_j\|_{L^p(\mathbb{R})} \tag{A.9}$$

for $1 \le p < \infty$; and thus $L^p(\mathbb{R})$ is a vector space. It is easy to see that $L^1(\mathbb{R})$ and $L^\infty(\mathbb{R})$ are vector spaces.

b.   Because of Hölder's Inequality, if $1 \le p \le r \le \infty$, then

$$\forall X \subseteq \mathbb{R}, \quad \text{for which } |X| < \infty, \qquad L^r(X) \subseteq L^p(X). \tag{A.10}$$

Using Fatou's Lemma and Minkowski's Inequality we can prove the following result for $1 \le p \le \infty$. The $p = \infty$ case is more elementary.

A.18 Theorem. Completeness of L^P

Let $1 \leq p \leq \infty$. $L^p(\mathbb{R})$ is complete, i.e., every Cauchy sequence in $L^p(\mathbb{R})$ converges in L^p-norm to an element of $L^p(\mathbb{R})$. (Recall that $\{f_n\} \subseteq L^p(\mathbb{R})$ is a Cauchy sequence *if*

$$\forall \epsilon > 0, \ \exists N \quad such \ that \quad \forall m, n \geq N, \ \|f_m - f_n\|_{L^p(\mathbb{R})} < \epsilon.)$$

We defined bounded variation in Definition 1.1.5. The following notion is a special case of both continuity and bounded variation.

A.19 Definition. Absolute Continuity

  a.   A function $F : [a, b] \to \mathbb{C}$ is *absolutely continuous* on $[a, b]$, written $F \in AC[a, b]$, if

$$\forall \epsilon > 0, \quad \exists \delta = \delta(\epsilon) \quad such \ that \quad \forall \{(x_j, y_j) \subseteq [a, b] : j = 1, \ldots, n\},$$

a finite disjoint family of intervals, we can conclude that

$$\sum_{j=1}^{n} (y_j - x_j) < \delta \quad implies \quad \sum_{j=1}^{n} |F(y_j) - F(x_j)| < \epsilon.$$

We define $AC(\mathbb{R})$ similarly.

  b.   A function $F : \mathbb{R} \to \mathbb{C}$ is *locally absolutely continuous*, written $F \in AC_{\mathrm{loc}}(\mathbb{R})$, if $F \in AC[a, b]$ for each interval $[a, b] \subseteq \mathbb{R}$.

A.20 Theorem. Fundamental Theorem of Calculus

  a.   *Let $f \in L^1[a, b]$ and let $r \in \mathbb{R}$. Define*

$$\forall t \in [a, b], \qquad F(t) = r + \int_a^t f(u) \, du$$

so that $F(a) = r$. Then $F \in AC[a, b]$ and

$$F' = f \quad a.e.$$

  b.   *A function $F : [a, b] \to \mathbb{C}$ is absolutely continuous on $[a, b]$ if and only if there is $f \in L^1[a, b]$ such that*

$$\forall t \in [a, b], \qquad F(t) - F(a) = \int_a^t f(u) \, du. \qquad (A.11)$$

A.21 Remark. Perspective on FTC

Theorem A.20 is denoted by FTC. Theorem A.20*a*, denoted by FTCI, asserts that *the derivative of the integral is the identity map.* Theorem A.20*b*, denoted by

FTCII, shows the importance of absolute continuity and asserts that *the integral of the derivative is the identity map.*

If $F \in AC[a, b]$, then F' exists a.e., $F' \in L^1[a, b]$, and (A.11) asserts that

$$F(t) - F(a) = \int_a^t F'(u) \, du.$$

A.22 Theorem. Integration by Parts

Let $f, g \in L^1[a, b]$, let $r, s \in \mathbb{C}$, and define the (absolutely continuous) functions

$$F(t) = r + \int_a^t f(u) \, du \quad and \quad G(t) = s + \int_a^t g(u) \, du$$

on $[a, b]$. Then

$$\int_a^b f(t) G(t) \, dt + \int_a^b g(t) F(t) \, dt = F(b) G(b) - F(a) G(a).$$

B

Functional Analysis

This appendix lists results from functional analysis that are used in the book. There are many excellent texts and expositions including [Die81], [GG81], [Hor66], [RN55], [Rud73], and [Tay58].

B.1 Definition. Compact Set

A set $S \subseteq \mathbb{R}$ is compact if, whenever

$$S \subseteq \bigcup_{\alpha} N_\alpha$$

for a collection of open intervals N_α, there is a finite subcollection $\{N_{\alpha_1}, \ldots, N_{\alpha_n}\}$ for which

$$S \subseteq \bigcup_{j=1}^{n} N_{\alpha_j}.$$

This definition generalizes to topological spaces. In the case of $\mathbb{R}$, compact sets $S \subseteq \mathbb{R}$ are precisely the closed bounded subsets of $\mathbb{R}$, e.g., [Apo57, Chapter 3].

B.2 Definition. Metric Space

a.   A *metric space* is a nonempty set M and a function $\rho : M \times M \to [0, \infty)$ satisfying the following properties:

$$\forall x, y \in M, \qquad \rho(x, y) \geq 0;$$
$$\forall x, y \in M, \qquad \rho(x, y) = 0 \quad \text{if and only if} \quad x = y;$$
$$\forall x, y \in M, \qquad \rho(x, y) = \rho(y, x);$$
$$\forall x, y, z \in M, \qquad \rho(x, z) \leq \rho(x, y) + \rho(y, z).$$

The last property is the *triangle inequality*, and the function ρ is a *metric*. An excellent reference for metric spaces is [Gle91].

b.   A sequence $\{x_n : n = 1, \ldots\}$, contained in a metric space M (with metric ρ), is a *Cauchy sequence* if

$$\forall \epsilon > 0, \ \exists N \quad \text{such that} \quad \forall m, n > N, \ \rho(x_m, x_n) < \epsilon.$$

If M is a metric space in which every Cauchy sequence $\{x_n\}$ converges to some element $x \in M$, i.e., $\lim_{n \to \infty} \rho(x_n, x) = 0$, then M is a *complete metric space*.

c.   Two metric spaces M_1 and M_2, with metrics ρ_1 and ρ_2, respectively, are *isometric* if there is a bijection $f : M_1 \to M_2$ such that

$$\forall x, y \in M_1, \qquad \rho_1(x, y) = \rho_2(f(x), f(y)).$$

In this case, f is an *isometry*. (Recall that a function $f : X_1 \to X_2$ is *bijective* if it is *injective* and *surjective*, i.e., if it is one-to-one and onto.)

d.   Let M be a metric space with metric ρ. A subset $V \subseteq M$ is *closed* if, whenever $\{x_n\} \subseteq V$ and $\lim_{n \to \infty} \rho(x_n, x) = 0$ for some $x \in M$, we can conclude that $x \in V$. The *closure* $\overline{X}$ of a subset $X \subseteq M$ is the set of all elements $x \in M$ for which there is a sequence $\{x_n\} \subseteq X$ such that $\lim_{n \to \infty} \rho(x_n, x) = 0$.

B.3 Definition. Banach Space

a.   A vector space $B \neq \{0\}$ over $\mathbb{C}$ is a *normed vector space* if there is a function $\| \cdots \| : B \longrightarrow [0, \infty)$ such that

$$\forall x \in B, \qquad\qquad\qquad \|x\| = 0 \quad \text{if and only if} \quad x = 0;$$

$$\forall x, y \in B, \qquad\qquad\qquad \|x + y\| \leq \|x\| + \|y\|;$$

$$\forall c \in \mathbb{C} \text{ and } \forall x \in B, \qquad \|cx\| = |c| \, \|x\|.$$

The function $\| \cdots \|$ is a *norm*.

b.   A normed vector space is a metric space with metric ρ defined by $\rho(x, y) = \|x - y\|$. A complete normed vector space is a *Banach space*. Let $1 \leq p \leq \infty$. $L^p(\mathbb{R})$, with L^p-norm defined in Definition A.10, is a Banach space. This is a restatement of Theorem A.18.

c.   Let B be a normed vector space, and let $\{x_n : n = 1, \ldots\} \subseteq B$. $\sum x_n$ *converges* to $x \in B$ if $\lim_{n \to \infty} \|x - \sum_{j=1}^{n} x_j\| = 0$; and $\sum x_n$ is *absolutely convergent* if $\sum \|x_n\| < \infty$. It is straightforward to prove that *a normed vector space is a Banach space if and only if every absolutely convergent series is convergent*, e.g., [Ben76, page 232].

B.4 Definition. Hilbert Space

a.   A *Hilbert space* $H \neq \{0\}$ is a Banach space for which there is a function $\langle \ldots, \ldots \rangle : H \times H \longrightarrow \mathbb{C}$ such that

$$\forall x, y \in H, \qquad\qquad \overline{\langle x, y \rangle} = \langle y, x \rangle;$$

$$\forall x, y, z \in H, \qquad\qquad \langle x + y, z \rangle = \langle x, z \rangle + \langle y, z \rangle;$$

$$\forall c \in \mathbb{C} \text{ and } \forall x, y \in H, \qquad \langle cx, y \rangle = c \langle x, y \rangle;$$

$$\forall x \in H, \qquad\qquad \|x\| = \langle x, x \rangle^{1/2},$$

where $\| \cdots \|$ is the Banach space norm. $\langle \ldots , \ldots \rangle$ is an *inner product*.

   b. $L^2(\mathbb{R})$ is a Hilbert space with inner product defined by $\langle f, g \rangle = \int f(t) \overline{g(t)} dt$, cf., [Ben76, Example I.2.4, page 234] for a structural converse of the form that every Hilbert space is some type of L^2 space. The integral is well defined by Hölder's Inequality (Theorem A.15).

   c. If H is a Hilbert space (actually, completeness is not required), then the properties of part *a* can be used to give *simple* proofs of the following facts:

$$|\langle x, y \rangle| \le \|x\| \, \|y\| \tag{B.1}$$

and

$$\|x + y\|^2 + \|x - y\|^2 = 2(\|x\|^2 + \|y\|^2). \tag{B.2}$$

(B.1) is the *Schwarz Inequality*, which in the case $H = L^2(\mathbb{R})$ is *Hölder's Inequality*. Of course, this does not mean there is a simple proof of Hölder's Inequality since Schwarz' Inequality *assumes* the existence of an inner product, and Hölder's Inequality *shows* the existence of an inner product for $H = L^2(\mathbb{R})$.

(B.2) is the *parallelogram law*. It can be shown that *a Banach space is a Hilbert space*, i.e., there is an inner product with the required properties, *if and only if the parallelogram law is valid*.

B.5 Theorem. Moore–Smith Theorem

Let M be a complete metric space with metric ρ, and let $\{x_{m,n} : m, n \in \mathbb{N}\}$ be given. Assume there are sequences $\{y_m\}, \{z_n\} \subseteq M$ such that

$$\lim_{n \to \infty} \rho(x_{m,n}, y_m) = 0 \quad \text{uniformly in } m$$

and

$$\forall n \in \mathbb{N}, \qquad \lim_{m \to \infty} \rho(x_{m,n}, z_n) = 0.$$

Then there is $x \in M$ such that the limits,

$$\lim_{m \to \infty} \lim_{n \to \infty} \rho(x_{m,n}, x), \ \lim_{n \to \infty} \lim_{m \to \infty} \rho(x_{m,n}, x), \ \text{and} \ \lim_{m,n \to \infty} \rho(x_{m,n}, x),$$

all exist and are equal to 0. *The last limit signifies that*

$$\forall \epsilon > 0, \quad \exists N > 0, \quad \text{such that} \quad \forall m, n > N,$$

$$\rho(x_{m,n}, x) < \epsilon.$$

The Moore–Smith Theorem is elementary to prove, e.g., [Ben76, pages 235–237], and should be compared with LDC and its refinements, e.g., [Ben76, Chapter 6].

B.6 Definition. Continuous Linear Functions

a. Let B_1 and B_2 be nonzero normed vector spaces with norms $\| \cdots \|_{B_1}$ and $\| \cdots \|_{B_2}$, respectively. The *norm* of a linear function $L : B_1 \to B_2$ is

$$\|L\| = \sup\{\|Lx\|_{B_2} : x \in B_1 \text{ and } \|x\|_{B_1} \le 1\}. \tag{B.3}$$

Thus, $\|L\|$ is the smallest constant $C \ge 0$ such that $\|Lx\|_{B_2} \le C\|x\|_{B_1}$ for all $x \in B_1$.

b. Clearly, *if* $\|L\| < \infty$, *then* L *is continuous*. The space of continuous linear functions $L : B_1 \to B_2$ is denoted by $\mathcal{L}(B_1, B_2)$. It is not difficult to show that *if* B_2 *is a Banach space, then* $\mathcal{L}(B_1, B_2)$ *is a Banach space, where* $L = c_1 L_1 + c_2 L_2$ *is defined by* $Lx = c_1 L_1(x) + c_2 L_2(x)$ *and* $\|L\|$ *is defined by* (B.3).

If $B = B_1 = B_2$ we denote $\mathcal{L}(B, B)$ by $\mathcal{L}(B)$. If $B = B_1$ and $B_2 = \mathbb{C}$ we denote $\mathcal{L}(B, \mathbb{C})$ by B'. B' is the space of *continuous linear functionals* on B and is called the *dual space* of B.

c. Let B_1 and B_2 be Banach spaces, with duals B_1' and B_2', and let $L \in \mathcal{L}(B_1, B_2)$. The *adjoint* of L is the linear function

$$\begin{aligned} L^* : B_2' &\longrightarrow B_1' \\ y &\longmapsto L^* y, \end{aligned}$$

where $(L^* y)(x)$ is defined to be $y(Lx)$ for all $x \in B_1$. It is easy to see that $L^* \in \mathcal{L}(B_2', B_1')$.

In the case of Hilbert spaces H_1 and H_2, we write

$$\forall x \in H_1 \text{ and } \forall y \in H_2, \qquad \langle Lx, y \rangle_{H_2} = \langle x, L^* y \rangle_{H_1} \tag{B.4}$$

to define the adjoint. In (B.4), we have used the fact that Hilbert spaces H are *reflexive*, i.e., $H' = H$, cf., Theorem B.14*b*.

Let M_1 and M_2 be metric spaces with metrics ρ_1 and ρ_2, respectively. A function $f : M_1 \to M_2$ is *uniformly continuous* if

$$\forall \epsilon > 0, \; \exists \delta > 0 \quad \text{such that} \quad \rho_1(x, y) < \delta \text{ implies } \rho_2(f(x), f(y)) < \epsilon.$$

B.7 Theorem. Uniformly Continuous Extensions

    a. *Let M_1 be a metric space, let M_2 be a complete metric space, and let V be a subset of M_1. Assume that $f : V \to M_2$ is uniformly continuous. Then f has a unique uniformly continuous extension to $\overline{V} \subseteq M_1$.*

    b. *Let B_1 and B_2 be Banach spaces, and let $L \in \mathcal{L}(B_1, B_2)$. Then L is uniformly continuous.*

    c. *Let B_1 and B_2 be Banach spaces, and let V be a subspace of B_1. If $L : V \to B_2$ is a continuous linear function, then L has a unique continuous linear extension to $\overline{V}$.*

The following result is also called the *Banach–Steinhaus Theorem*.

B.8 Theorem. Uniform Boundedness Principle

    a. *Let M be a complete metric space with metric ρ, and let $\mathcal{C}$ be a set of continuous functions $f : M \to \mathbb{C}$. Assume*

$$\forall x \in M, \ \exists C_x > 0 \quad \text{such that} \quad \forall f \in \mathcal{C}, \ |f(x)| \le C_x.$$

Then there is a constant $C > 0$ and a nonempty open ball U, i.e., $U = \{x \in M : \rho(x, x_0) < r\}$ for some $x_0 \in M$ and $r > 0$, such that

$$\forall x \in U \ \text{and} \ \forall f \in \mathcal{C}, \qquad |f(x)| \le C.$$

    b. *Let B_1 be a Banach space, let B_2 be a normed vector space, and let $\mathcal{C} \subseteq \mathcal{L}(B_1, B_2)$. Then one of the following is true.*

    i. *$\exists C$ such that $\forall L \in \mathcal{C}, \ \|L\| \le C$.*

    ii. *There is a nonempty set $V \subseteq B_1$ such that $\overline{V} = B_1$ and*

$$\forall x \in V, \qquad \sup_{L \in \mathcal{C}} \|Lx\| = \infty.$$

    (V is also the intersection of a countable family of open sets.)

    c. *Let B_1 be a Banach space, let B_2 be a normed vector space, and let $\{L_n\} \subseteq \mathcal{L}(B_1, B_2)$ have the property that*

$$\forall x \in B_1, \quad \exists Lx \in B_2 \quad \text{such that} \quad \lim_{n \to \infty} \|L_n x - Lx\| = 0.$$

Then $L \in \mathcal{L}(B_1, B_2)$.

B.9 Theorem. Open Mapping Theorem

Let B_1 and B_2 be Banach spaces, and let $L \in \mathcal{L}(B_1, B_2)$ be bijective. Then L^{-1} exists and $L^{-1} \in \mathcal{L}(B_2, B_1)$.

B.10 Remark. Baire Category Theorem

The Uniform Boundedness Principle and the Open Mapping Theorem both depend on completeness. In particular, they can be proved using the *Baire Category Theorem*, which asserts that *if M is a complete metric space, then the intersection of any countable family of dense open subsets of M is dense in M.*

Combining the Open Mapping Theorem with the Banach–Alaoglu Theorem [Ben76, pages 246–247], we have the following result.

B.11 Theorem. Injectivity, Surjectivity, and Continuity for L, L^*, L^{-1}, and $(L^*)^{-1}$

Let B_1 and B_2 be normed vector spaces with dual Banach spaces B_1' and B_2', and let $L \in \mathcal{L}(B_1, B_2)$.

a. $L^*(B_2') = B_1'$ *if and only if L^{-1} exists and $L^{-1} \in \mathcal{L}(L(B_1), B_1)$.*

b. *Let B_1 and B_2 be Banach spaces. $L(B_1) = B_2$ if and only if $(L^*)^{-1}$ exists and $(L^*)^{-1} \in \mathcal{L}(L^*(B_2'), B_2')$. Further, if L^{-1} exists, then it is in $\mathcal{L}(L(B_1), B_1)$.*

c. *Let B_1 and B_2 be Banach spaces, let L be injective, and assume $\overline{L(B_1)} = B_2$. The following are equivalent.*

    i. $L(B_1) = B_2$.

    ii. *There is $C > 0$ such that for all $y \in B_2'$, $\|y\|_{B_2'} \le C\|L^* y\|_{B_1'}$, i.e., L^* is an open mapping.*

    iii. $L^*(B_2') = B_1'$.

Let B_1 and B_2 be vector spaces, let $V \subseteq B_1$ be a subspace, and let $K : V \to B_2$ be a linear function. Using the Axiom of Choice, it can be shown that there is a linear function $L : B_1 \to B_2$ for which $Lx = Kx$ for all $x \in V$, e.g., [Tay58, pages 40–41]. If B_1 is a normed vector space, $B_2 = \mathbb{C}$, and K is continuous, then the following theorem provides a continuous norm-preserving extension $L : B_1 \to \mathbb{C}$ of K, e.g., [Ben76, pages 244–246]. Completeness is not required as it is for Theorems B.8 and B.9.

B.12 Theorem. Hahn–Banach Theorem

Let $V \subseteq B$ be a subspace of the normed vector space B. If $K : V \to \mathbb{C}$ is linear and continuous, then there is $L \in B'$ such that $L = K$ on V and $\|L\| = \|K\|$.

B.13 Corollary. Existence of Continuous Linear Functionals

Let B be a normed vector space.

a. *If $x, y \in B$ and $x \ne y$, then there is $L \in B'$ such that $Lx \ne Ly$.*

b. *Let V be a closed subspace of B and let $y \in B \backslash V$. There is $L \in B'$ such that $Ly = 1$ and $Lx = 0$ for all $x \in V$.*

PROOF. **a.** Let V be the subspace of B generated by $x - y$, and define a linear function K on V by the rule $K(c(x - y)) \equiv c\|x - y\|$ for $c \in \mathbb{C}$. Note that $\|K\| = 1$ so that K is continuous. Apply Theorem B.12 to obtain an extension $L \in B'$. Then $0 < \|x - y\| = L(x - y) = L(x) - L(y)$. This completes part a.

b. Let $d > 0$ have the property that $\|x - y\| \geq d$ for all $x \in V$, and let $V_y = \{x + cy : x \in V \text{ and } c \in \mathbb{C}\}$. Define the linear function K on V_y by the rule $K(x + cy) = c$. Thus, $K = 0$ on V and $K(y) = 1$. Also, K is continuous on V_y since $|K(x + cy)| = d^{-1}|c|d \leq d^{-1}|c|\|c^{-1}x + y\| = d^{-1}\|x + cy\|$. The result follows from Theorem B.12. ∎

B.14 Theorem. L^p-Duality Theorem

a. $L^1(\mathbb{R})' = L^\infty(\mathbb{R})$, where $g : L^1(\mathbb{R}) \to \mathbb{C}$ is well defined by $g(f) = \int g(t)f(t)\,dt$ for all $f \in L^1(\mathbb{R})$.

b. Let $1 < p < \infty$ and define p' by $\frac{1}{p} + \frac{1}{p'} = 1$. Then $L^p(\mathbb{R})' = L^{p'}(\mathbb{R})$, where $g : L^p(\mathbb{R}) \to \mathbb{C}$ is well defined by $g(f) = \int g(t)f(t)\,dt$ for all $f \in L^p(\mathbb{R})$. In particular, the Hilbert space $H = L^2(\mathbb{R})$ has the property that $L^2(\mathbb{R})' = L^2(\mathbb{R})$.

c. $L^\infty(\mathbb{R})'$ is the space of finitely additive bounded measures on the Borel algebra $\mathcal{B}(\mathbb{R})$, e.g., Chapter IV of N. Dunford and J. T. Schwartz, Linear Operators, Part I, John Wiley and Sons, New York, 1957. ($\mathcal{B}(\mathbb{R})$ was defined in Remark 2.7.4).

B.15 Remark. Riesz, Helly, and Hahn–Banach Theorems

a. F. Riesz introduced the L^p spaces, $p \neq 2$, in 1910 and proved the L^p-Duality Theorem (Theorem B.14b). Using a complicated argument with Lagrange multipliers, he then proved the following theorem. *Let $p > 1$, and let $\{f_n\} \subseteq L^p(\mathbb{R})$ and $\{c_n\} \subseteq \mathbb{C}$ be given sequences; then*

$$\exists f \in L^{p'}(\mathbb{R}) \quad \text{such that} \quad \forall n, \ c_n = \int f(t)f_n(t)\,dt \tag{B.5}$$

if and only if there is $C > 0$ such that for all finite sequences $\{\lambda_n\} \subseteq \mathbb{C}$,

$$\left| \sum \lambda_n c_n \right| \leq C \left\| \sum \lambda_n f_n \right\|_{L^p(\mathbb{R})}. \tag{B.6}$$

The proof that (B.5) implies (B.6) is clear. If $p = 2$ and $\{f_n\}$ is orthonormal then Riesz' Theorem is equivalent to the Riesz–Fischer Theorem (Remark 3.4.11c).

b. In 1912, the Austrian mathematician E. Helly (1884–1943) (who spent his last years in the USA as an actuary) gave an elementary proof that (B.6) is a sufficient condition for the validity of (B.5). In so doing, he proved a result that is *equivalent* to the Hahn–Banach Theorem [Die81].

c. The proof of (B.5) from (B.6) is a consequence of the Hahn–Banach Theorem as follows. Assume (B.6) and let $V = \text{span}\{f_n\} \subseteq L^p(\mathbb{R})$. (The *span* of a set X contained in a vector space B is the set of all finite linear combinations $x \equiv \sum \lambda_n x_n$, where $\lambda_n \in \mathbb{C}$ and $x_n \in X$.) Define $K : V \to \mathbb{C}$ by the rule that

$K(\sum \lambda_n f_n) = \sum \lambda_n c_n$ for any finite sequence $\{\lambda_n\} \subseteq \mathbb{C}$. Clearly, K is linear, and it is continuous on V by (B.6). Thus, by Theorem B.12, there is $L \in L^{p'}(\mathbb{R})$ for which $L = K$ on V. Hence, $c_n = f(f_n) = \int f(t) f_n(t) dt$ for each n, where we have denoted L by f.

B.16 Remark. Embeddings of Dual Spaces

a. Let $B_1 \subseteq B_2$, where B_1 and B_2 are normed vector spaces; and let $I : B_1 \to B_2$ be the identity mapping with adjoint $I^* : B_2' \to B_1'$. By definition,

$$\forall x \in B_1 \text{ and } \forall y \in B_2', \qquad (I^*y)(x) = y(x),$$

i.e., $I^*y = y$ on $B_1 \subseteq B_2$.

b. Assume I, and hence I^*, are continuous. Note that *if $y \in B_2'$, then $y|_{B_1}$, the restriction of y to B_1, is an element of B_1'.* To see this first note that since $B_1 \subseteq B_2$ and $y \in B_2'$, $y|_{B_1}$ is linear on B_1. $y|_{B_1}$ is also continuous on B_1 because of the continuity of I. In fact, since y is continuous on B_1 with the induced topology from B_2, it is continuous on B_1 with its given norm convergence because this latter topology is stronger (finer) than the B_2 criterion. (Continuity of a function for a given topology on its domain implies continuity for any stronger topology on that domain.)

c. To say that B_2' is *embedded* in B_1', in which case we write $B_2' \subseteq B_1'$, we mean that I^* is a continuous injection. This means that whenever $I^*y = 0 \in B_1'$, then $y = 0$, i.e., $y(x) = 0$ for all $x \in B_2$.

d. In the setting of parts *a* and *b*, we further assume $\overline{B}_1 = B_2$. Let $y \in B_2'$ have the property that $I^*y = 0 \in B_1'$. Suppose $x \in B_2$ and $\lim_{n \to \infty} \|x_n - x\|_{B_2} = 0$, where $\{x_n\} \subseteq B_1$. Then $\lim_{n \to \infty} y(x_n) = y(x)$ and $y(x_n) = (I^*y)(x_n) = 0$. Thus, $y(x) = 0$, and so $y \in B_2'$ is the 0-element. Hence, I^* is a continuous injection. I^* is also the identity function, i.e., for all $y \in B_2'$, $I^*y = y$ on a dense subspace of B_2.

e. We can summarize parts *a–d* by the following *Embedding Theorem: let B_1 and B_2 be normed vector spaces; if $B_1 \subseteq B_2$ in the sense that the identity map $I : B_1 \to B_2$ is continuous, and if $\overline{B}_1 = B_2$, then $B_2' \subseteq B_1'$.* Several embeddings are listed in Definition 2.4.5c and Example 2.4.6b, d, f. Some of these involve topological structures more general than Banach spaces. This generality does not give rise to any problems since the Embedding Theorem is true quite generally.

B.17 Example. $C_b(\mathbb{R})$, $M_b(\mathbb{R})$, and Duality

a. Let $C_b(\mathbb{R})$ be the Banach space of continuous bounded functions on $\mathbb{R}$ taken with the L^∞-norm $\| \cdots \|_{L^\infty(\mathbb{R})}$; and let $C_0(\mathbb{R})$ be the closed subspace of $C_b(\mathbb{R})$ whose elements f satisfy the condition that $\lim_{|t| \to \infty} f(t) = 0$. Recall from Theorem 2.7.3 (RRT) that $C_0(\mathbb{R})' = M_b(\mathbb{R})$. $C_b(\mathbb{R})$ is a closed subspace of $L^\infty(\mathbb{R})$, and $C_b(\mathbb{R})'$ is the space of finitely additive bounded regular measures on the Borel algebra $\mathcal{B}(\mathbb{R})$, cf., Theorem B.14c and the reference given there.

b. Since the continuous inclusion map $I : C_0(\mathbb{R}) \to C_b(\mathbb{R})$ is not dense, we *cannot* conclude that $C_b(\mathbb{R})' \subseteq M_b(\mathbb{R})$, as is apparent from the characterizations of $C_0(\mathbb{R})'$ and $C_b(\mathbb{R})'$ in part *a*, cf., Remark B.16.

c. The characterizations of $C_0(\mathbb{R})'$ and $C_b(\mathbb{R})'$ in part *a do* imply that

$$M_b(\mathbb{R}) \subseteq C_b(\mathbb{R})'. \tag{B.7}$$

In this regard, if $\mu \in C_0(\mathbb{R})'$, then μ extends to an element $\mu_e \in C_b(\mathbb{R})'$ by the Hahn–Banach Theorem. Of course, there is no a priori guarantee of a unique extension.

On the other hand, and without invoking the characterization of $C_b(\mathbb{R})'$ stated in part *a*, we can see the validity of (B.7) in the following way.

If $f \in C_b(\mathbb{R})$ we can choose $\{f_n\} \subseteq C_0(\mathbb{R})$ for which $\lim_{n\to\infty} f_n = f$ pointwise on $\mathbb{R}$ and $\sup_n \|f_n\|_{L^\infty(\mathbb{R})} = \|f\|_{L^\infty(\mathbb{R})} < \infty$. Then we apply a form of LDC for $L^1_\mu(\mathbb{R})$ which allows us to assert that $f \in L^1_{|\mu|}(\mathbb{R})$ and $\lim_{n\to\infty} \|f_n - f\|_{L^1_{|\mu|}(\mathbb{R})} = 0$, e.g., [Ben76, Chapter 3]. The integral $\mu(f)$ is well defined, i.e., it is independent of the sequence $\{f_n\} \subseteq C_0(\mathbb{R})$. Further, $\mu : C_b(\mathbb{R}) \to \mathbb{C}$ is linear. To prove the continuity of μ, let $f \in C_b(\mathbb{R})$, let $\epsilon > 0$, and choose $\{f_n\}$ as above. Then

$$\exists N > 0 \quad \text{such that} \quad \forall n \geq N, \ |\mu(f - f_n)| < \epsilon;$$

and so, for such n,

$$|\mu(f)| \leq \epsilon + |\mu(f_n)| \leq \epsilon + \|\mu\|\|f\|_{L^\infty(\mathbb{R})}.$$

This is true for all $\epsilon > 0$, and so $\mu \in C_b(\mathbb{R})'$. We designate μ so defined on $C_b(\mathbb{R})$ by μ^*.

The inclusion (B.7) is accomplished by the mapping $\mu \mapsto \mu^*$. The fact that many extensions μ_e of μ exist does not contradict (B.7). In fact, $\nu_e \equiv \mu_e - \mu^* \in C_b(\mathbb{R})'$ vanishes on $C_0(\mathbb{R})$; and if ν_e is not identically 0 on $C_b(\mathbb{R})$, then μ_e is not countably additive on $\mathcal{B}(\mathbb{R})$ and so it does not correspond to an element of $M_b(\mathbb{R})$.

C

Fourier Analysis Formulas

General Formulas

$$f \longleftrightarrow F.$$

$$F(\gamma) = \hat{f}(\gamma) = \int f(t)e^{-2\pi i t \gamma}\, dt, \qquad f(t) = \int F(\gamma)e^{2\pi i t \gamma}\, d\gamma.$$

$$f_\lambda(t) = \lambda f(\lambda t), \quad f_\lambda(t) \longleftrightarrow \frac{\lambda}{|\lambda|} F\left(\frac{\gamma}{\lambda}\right), \qquad \lambda \in \mathbb{R}\backslash\{0\}.$$

$$f \longleftrightarrow F \quad \text{implies} \quad F(t) \longleftrightarrow f(-\gamma).$$

$$\tau_u f(t) = f(t - u), \qquad (\tau_u f) \longleftrightarrow e^{-2\pi i u \gamma} F(\gamma), \qquad e^{2\pi i t \gamma_0} f(t) \longleftrightarrow F(\gamma - \gamma_0).$$

$$f_1 * f_2 \longleftrightarrow F_1 F_2, \qquad f_1 f_2 \longleftrightarrow F_1 * F_2.$$

$$\int f_1(t)f_2(t)\, dt = \int F_1(-\gamma)F_2(\gamma)\, d\gamma, \qquad \int f_1(t)\overline{f_2(t)}\, dt = \int F_1(\gamma)\overline{F_2(\gamma)}\, d\gamma,$$

$$\|f\|_{L^2(\mathbb{R})} = \|F\|_{L^2(\hat{\mathbb{R}})}.$$

$$f'(t) \longleftrightarrow 2\pi i\gamma F(\gamma), \qquad f^{(n)}(t) \longleftrightarrow (2\pi i\gamma)^n F(\gamma),$$

$$-2\pi i t f(t) \longleftrightarrow F'(\gamma), \qquad (-2\pi i t)^n f(t) \longleftrightarrow F^{(n)}(\gamma).$$

$$\int f(at - b)e^{-2\pi i t \gamma}\, dt = \frac{1}{a}e^{-2\pi i (b/a)\gamma} F\left(\frac{\gamma}{a}\right), \qquad a > 0.$$

Poisson Summation Formula

a. $\lambda \sum f(\lambda j) = \sum F\left(\frac{n}{\lambda}\right), \lambda > 0,$

b. $\left(\sum \delta_{nT}\right)^{\wedge} = \frac{1}{T}\sum \delta_{n/T}, T > 0.$

Special Kernels

$$\textbf{Dirichlet} \quad d(t) = \frac{\sin t}{\pi t}, \quad \textbf{Fejér} \quad w(t) = \frac{1}{2\pi}\left(\frac{\sin(t/2)}{t/2}\right)^2,$$

$$\textbf{Gauss} \quad g(t) = \frac{1}{\sqrt{\pi}}e^{-t^2}, \quad \textbf{Poisson} \quad p(t) = \frac{1}{\pi(1+t^2)},$$

$$\textbf{Triangle} \quad \Delta(t) = \max(1 - |t|, 0).$$

$$\int d_\lambda(t)\, dt = \int w_\lambda(t)\, dt = \int g_\lambda(t)\, dt = \int p_\lambda(t)\, dt = 1,$$

$$w_\lambda(t) = \frac{2\pi}{\lambda}d_{\lambda/2}(t)^2.$$

$$\mathbf{1}_{[-T,T)} \longleftrightarrow d_{2\pi T}, \qquad d_\lambda \longleftrightarrow \mathbf{1}_{[-\lambda/(2\pi),\lambda/(2\pi))},$$

$$T\Delta_{1/T} \longleftrightarrow w_{2\pi T}, \qquad w_\lambda \longleftrightarrow \max\left(1 - \frac{|2\pi\gamma|}{\lambda}, 0\right),$$

$$e^{-2\pi r|t|} \longleftrightarrow p_{1/r}, \qquad p_\lambda \longleftrightarrow e^{-2\pi|\gamma|/\lambda},$$

$$e^{-\pi r t^2} \longleftrightarrow g_{\sqrt{\pi/r}}, \qquad g_\lambda \longleftrightarrow e^{-(\pi\gamma/\lambda)^2},$$

$$T, \lambda, r > 0.$$

Heaviside Function H

a. $H(t) \longleftrightarrow \frac{1}{2\pi i}pv\left(\frac{1}{\gamma}\right) + \frac{1}{2}\delta(\gamma),$

b. $\frac{1}{2}\delta(t) - \frac{1}{2\pi i}pv\left(\frac{1}{t}\right) \longleftrightarrow H(\gamma),$

c. $H(t) - \frac{1}{2} \longleftrightarrow \frac{1}{2\pi i}pv\left(\frac{1}{\gamma}\right).$

D

Contributors to Fourier Analysis

Abel, Niels Henrik	1802–1829	Norway
Akhiezer (Achieser), Naum Ilich	1901–1980	Ukraine
Alembert, Jean Le Rond d'	1717–1783	France
Archimedes	c. 287–212B.C.	Greece (Syracuse, Sicily, Italy)
Arzelá, Cesare	1847–1912	Italy
Ascoli, Giulio	1843–1896	Italy
Babenko, Konstantin Ivanovich	1919–1987	Russia
Banach, Stefan	1892–1945	Poland
Bary, Nina Karlovna	1901–1962	Russia
Bernoulli, Daniel	1700–1782	Switzerland
Bernstein, Sergei Natanovich	1880–1968	Russia
Besicovitch, Abram Samoilovitch	1891–1970	Berdjansk, Russia, England
Bessel, Friedrich Wilhelm	1784–1846	Germany
Beurling, Arne Karl-August	1905–1986	Sweden

Obviously this is an incomplete list of past contributors to Fourier analysis.

With the complexity of history's chaotic paths through time, it is impossible to draw any serious conclusions from nationalistic labeling. Our listing of countries should be read in this spirit, viz., as a superficial organizational device. Further, many contributors made significant contributions in several countries, e.g., Erdélyi, Euler, Pólya, and Reiter. Others have a labyrinthine heritage, e.g., Cantor. Dirichlet, one of the protagonists of this book, succeeded Gauss at Göttingen; and he was, indeed, German born (at Düren between Aachen and Köln). On the other hand, his grandfather lived in Verviers, Belgium; and several generations of the Dirichlet family lived in the Province of Liège. In fact, the family name originated from the Belgian village of Richelette, not so far from Maastricht in The Netherlands.

Boas, Ralph Philip	1912–1992	USA
Bôcher, Maxime	1867–1918	USA
Bochner, Salomon	1899–1982	USA
Bois-Reymond, Paul David Gustav du	1831–1889	Germany
Bohr, Harald	1887–1951	Denmark
Bolzano, Bernhard	1781–1848	Czechoslovakia (Bohemia)
Bosanquet, Lancelot Stephen	1903–1984	England
Bromwich, Thomas John I'Anson	1875–1929	England
Cantor, Georg Ferdinand Ludwig Philip	1845–1918	Germany
Carleman, Torsten	1892–1949	Sweden
Carlini, Francesco	1783–1862	Italy
Carlson, Fritz David	1888–1952	Sweden
Carslaw, Horatio Scott	1870–1954	Scotland, Australia
Cauchy, Augustin Louis	1789–1857	France
Cesàro, Ernesto	1859–1906	Italy
Cooper, Jacob Lionel Bakst	1915–1979	South Africa, England
Cornu, Marie Alfred	1841–1902	France
Cramér, Harald	1893–1985	Sweden
De Moivre, Abraham	1667–1754	France, England
De Morgan, Augustus	1806–1871	India (Madura), England
Darboux, Jean Gaston	1842–1917	France
Dedekind, Julius Wilhelm Richard	1831–1916	Germany
Denjoy, Arnaud	1884–	France
Dini, Ulisse	1845–1918	Italy
Dirichlet, Johann Peter Gustav Lejeune (Lejeune-Dirichlet)	1805–1859	Germany
Doetsch, Gustav	1892–1977	Germany
Doss, Raouf	–1988	USA
Eberlein, William F.	1917–1986	USA
Erdélyi, Arthur	1908–1977	Hungary, Scotland, USA, Scotland
Euler, Leonhard	1707–1783	Switzerland, Russia

Fatou, Pierre Joseph Louis	1878–1929	France
Fejér, Lipót	1880–1959	Hungary
(b. Weisz. In Hungarian, "white" is "fehér")		
Fermat, Pierre de	1601–1665	France
Fischer, Ernst	1875–1954	Germany
Flett, Thomas Muirhead	1923–1976	England
Fourier, Jean Baptiste Joseph	1768–1830	France
Franklin, Philip	–1965	USA
Fréchet, Maurice René	1878–1973	France
Fredholm, Erik Ivar	1866–1927	Sweden
Fresnel, Augustin	1788–1827	France
Fubini, Guido	1879–1943	Italy
Gabor, Dennis	1900–1979	Hungary, England
Gauss, Carl Friedrich	1777–1855	Germany
Gibbs, Josiah Willard	1839–1903	USA
Glicksberg, Irving Leonard	1925–	USA
Green, George F.	1793–1841	England
Hankel, Hermann	1839–1873	Germany
Hardy, Godfrey Harold	1877–1947	England
Hausdorff, Felix	1868–1942	Germany
Heaviside, Oliver	1850–1925	England
Heine, Heinrich Eduard	1821–1881	Germany
Heisenberg, Werner	1901–1976	Germany
Herglotz, Gustav	1881–1953	Germany
Hermite, Charles	1822–1905	France
Herrero, Domingo A.	–1991	USA
Herz, Carl Samuel	1930–1995	USA, Canada
Hilbert, David	1862–1943	Germany
Hille, Einar	1894–1980	Sweden, USA
Hirschman, Jr., Isadore Isaac	1922–1990	USA
Hobson, Ernest William	1856–1933	England
Hölder, (Ludwig) Otto	1859–1937	Germany
Ingham, Albert Edward	1900–1967	England
Jackson, Dunham	1888–1946	USA
Jacobi, Karl Gustav Jacob	1804–1851	Germany
Jensen, Johan Ludvig William Valdemar	1859–1925	Denmark
Jordan, (Marie-Ennemond) Camille	1838–1922	France

Kac, Mark	1914–1984	Poland, USA
Karamata, Jean	–1967	Switzerland
Lord Kelvin (William Thomson)	1824–1907	Ireland, Scotland
Kennedy, Patrick Brendan	1929–1966	Ireland
Kirchhoff, Gustav Robert	1824–1887	Germany
Kluvánek, Igor	1931–1993	Slovakia, Australia
Kolmogorov, Andrei Nikolaevich	1903–1987	Russia
Köthe, Gottfried	1905–1989	Austria, Germany
Krein, Mark Grigorivich	1907–1989	Odessa (Ukraine)
Kronecker, Leopold	1823–1891	Germany
Kuzmin, Rodion Osieviz	1891–1949	Russia
Lagrange, Joseph-Louis (Lagrangia)	1736–1813	Italy (Torino), France
Landau, Edmund	1877–1938	Germany
Laplace, Pierre Simon de	1749–1827	France
Legendre, Adrien Marie	1752–1833	France
Lebesgue, Henri-Léon	1875–1941	France
Leeuw, Karel de	–1978	USA
Leibenzon, Leonid Samuilovich	1879–1951	Russia
Levi, Beppo	1875–1928	Italy
Levinson, Norman	1912–1975	USA
Lévy, Paul P.	1886–1971	France
Lewy, Hans	1904–1988	Germany, USA
Lindelöf, Ernst	1870–1946	Finland
Liouville, Joseph	1809–1882	France
Lipschitz, Rudolf Otto Sigismund	1832–1903	Germany
Littlewood, John Edensor	1885–1977	England
Loomis, Lynn Harold	1915–1994	USA
Lowdenslager, David B.	–1963	USA
Lozinskii, Sergei Mikhailovich	1914–1985	Russia
Lusin, Nikolai Nikolaevich	1883–1950	Russia
Maclaurin, Colin	1698–1746	Scotland
Mandelbrojt, Szolem	1899–1983	Poland, France
Marcinkiewicz, Josef	1910–1940	Poland
Mellin, Hjalnar	1854–1933	Finland
Mengoli, Pietro	1626–1686	Italy
Mensov, Dimitrii Egven'evich	1892–1988	Russia
Mertens, Franz	1840–1927	Austria
Michelson, Albert Abraham	1852–1931	USA

Minkowski, Hermann	1864–1909	Russia, Switzerland, Germany
Osgood, William Fogg	1864–1943	USA
Paley, Raymond Edward Alan Christopher	1907–1933	England
Parseval (des Chênes), Marc-Antoine	1755–1836	France
Peano, Giuseppe	1858–1932	Italy
Pichorides, Stylianos K.	1940–1992	Greece
Plancherel, Michel	1885–1967	Switzerland
Poisson, Siméon Denis	1781–1840	France
Pollard, Harry	1919–1985	USA (Boston)
Pollard, Samuel	1894–1945	China, England
Pólya, George (György)	1887–1985	Hungary, Switzerland, USA
Porcelli, Pasquale	–1972	USA
Pringsheim, Alfred	1850–1941	Germany, Switzerland
Pythagoras	c. 570–500B.C.	Greece
Rademacher, Hans	1892–1969	Germany, USA
Radon, Johann	1887–1956	Czechoslovakia
Rajchman, Aleksander	1890–1940	Poland
Lord Rayleigh (Strutt, John William)	1842–1919	England
Reiter, Hans Jakob	1921–1992	Austria, USA, England, The Netherlands, Austria
Riemann, Georg Friedrich Bernhard	1826–1866	Germany
Riesz, Frigyes	1880–1956	Hungary
Riesz, Marcel	1886–1969	Hungary, Sweden
Rivière, Nestor M.	1940–1978	Argentina, USA
Rogosinski, Werner W.	–1964	England
Rubio de Francia, José Luis	1949–1988	Spain
Salem, Raphaël	1898–1963	Greece, France
Schaeffer, A. C.	–1957	USA
Schmidt, Erhard	1876–1959	Germany
Schmidt, Robert		Germany (Kiel)
Schoenberg, Isaac J.	1903–1990	Romania, USA
Schrödinger, Edwin	1887–1961	Austria

Schuster, Sir Arthur R.	1851–1934	Germany, England
Smith, Henry John Stephen	1826–1883	England
Steinhaus, Hugo		
Stieltjes, Thomas Jan	1856–1894	The Netherlands
Stirling, James	1692–1770	Scotland
Stromberg, Karl Robert	1931–1994	USA
Szász, Otto	1884–1952	Hungary, USA
Szegö, Gábor	1895–1985	Hungary, USA
Szidon, S.		
Tamarkin, Jacob David	1888–1945	Russia, USA
Taylor, Brook	1685–1731	England
Thomson, William (Lord Kelvin)	1824–1907	Ireland, Scotland
Titchmarsh, Edward Charles	1899–1963	England
Tonelli, Leonida	1885–1946	Italy
Vallée-Poussin, Charles Jean de la	1866–1962	Belgium
van der Corput, Jan G.	1890–1975	The Netherlands
van Vleck, Edward Burr	1863–1943	USA
Viète, François	1540–1603	France
Vinogradov, Ivan Matveevich	1891–1983	Russia
Vitali, Giuseppe	1875–1932	Italy
Volterra, Vito	1860–1940	Italy
von Neumann, John (János Neumann)	1903–1957	Hungary, USA
Walsh, Joseph Leonard	1895–1973	USA
Weierstrass, Karl Theodor Wilhelm	1815–1897	Germany
Weiss, Mary	1930–1966	USA
Weyl, Hermann	1885–1955	Germany, USA
Widder, David Vernon	1899–1990	USA
Wiener, Norbert	1894–1964	USA
Wigner, Eugene Paul(Jenö Pál)	1902–1995	Hungary, USA
Wintner, Aurel	1903–1958	Hungary, USA
Young, Grace Chisholm	1868–1944	England
Young, William Henry	1863–1942	England
Zuckerman, H. S.	–1970	USA
Zygmund, Antoni	1900–1992	Poland, USA

Bibliography

[AB66]   E. Asplund and L. Bungart, *A First Course in Integration*, Holt, Rinehart and Winston, New York, 1966.

[ABe77]   W. Amrein and A. Berthier, On support properties of L^p-functions and their Fourier transforms, *J. Funct. Analysis*, 24 (1977), 258–267.

[AG89]   L. Auslander and F. A. Grünbaum, The Fourier transform and the discrete Fourier transform, *Inverse Problems*, 5 (1989), 149–164.

[AGR88]   A. K. Arora, S. K. Goel, and D. M. Rodriguez, Special integration techniques for trigonometric integrals, *Amer. Math. Monthly*, 95 (1988), 126–130.

[AKM80]   K. Anzai, S. Koizumi, and K. Matsuoka, On the Wiener formula of functions of two variables, *Tokyo J. Math.*, 3 (1980), 249–270.

[AMS73]   P. Antosik, J. Mikusiński, and R. Sikorski, *Theory of Distributions, the Sequential Approach*, Elsevier Scientific Publishing Company, Amsterdam, 1973.

[AN79]   E. Albrecht and M. Neumann, Automatische Stetigkeitseigenschaften einiger Klassen linearer Operatoren, *Math. Ann.*, 240 (1979), 251–280.

[AT79]   L. Auslander and R. Tolimieri, Is computing with the finite Fourier transform pure or applied mathematics?, *Bull. Amer. Math. Soc.*, 1 (1979), 874–897.

[AZ90]   J. Appell and P. Zabrejko, *Nonlinear Superposition Operators*, Cambridge University Press, London, 1990.

[Ach56]   N. I. Achieser, *Theory of Approximation*, F. Ungar Publishing Co., New York, 1956.

[Apo57]   T. Apostol, *Mathematical Analysis*, Addison-Wesley, Reading, MA, 1957.

[Ars66]   J. Arsac, *Fourier Transforms and the Theory of Distributions*, Prentice-Hall, Inc., Englewood Cliffs, NJ, 1966.

[Ash76]   J. M. Ash, editor, *Studies in Harmonic Analysis*, The Mathematical Association of America, Washington, D.C., 1976.

[BA83]   P. Burt and T. Adelson, The Laplacian pyramid as a compact image code, *IEEE Trans. Comm.*, 31 (1983), 532–540.

[BBE89]   J. J. Benedetto, G. Benke, and W. R. Evans, An n-dimensional Wiener–Plancherel formula, *Adv. in Appl. Math.*, 10 (1989), 457–487.

[BC49]   S. Bochner and K. Chandrasekharan, *Fourier Transforms*, Princeton University Press, Princeton, NJ, 1949.

[BCo95]   J. J. Benedetto and D. Colella, Wavelet analysis of spectrogram seizure chirps, *Proc. SPIE*, 2569 (1995), 512–521.

[BD81]   B. Brosowski and F. Deutsch, An elementary proof of the Stone–Weierstrass Theorem, *Proc. Amer. Math. Soc.*, 81 (1981), 89–92.

[BE81]   B. C. Berndt and R. J. Evans, The determination of Gauss sums, *Bull. Amer. Math. Soc.*, 5 (1981), 107–129.

[BF94]   J. J. Benedetto and M. W. Frazier, editors, *Wavelets: Mathematics and Applications*, CRC Press, Boca Raton, FL, 1994.

[BH92]   J. J. Benedetto and H. Heinig, Fourier transform inequalities with measure weights, *Adv. in Math.*, 96 (1992), 194–225.

[BHe93]   G. Benke and W. J. Hendricks, Estimates for large deviations in random trigonometric polynomials, *SIAM J. Math. Anal.*, 24 (1993), 1067–1085.

[BL94]   J. J. Benedetto and J. D. Lakey, The definition of the Fourier transform for weighted inequalities, *J. Funct. Anal.*, 120 (1994), 403–439.

[BM62]   A. Beurling and P. Malliavin, On Fourier transforms of measures with compact support, *Acta Math.*, 107 (1962), 291–309.

[BM67]   A. Beurling and P. Malliavin, On the closure of characters and the zeros of entire functions, *Acta Math.*, 118 (1967), 79–93.

[BMc66]   P. W. Berg and J. L. McGregor, *Elementary Partial Differential Equations*, Holden-Day, Inc., Oakland, CA, 1966.

[BMW91]   R. E. Blahut, W. Miller Jr., and C. H. Wilcox, *Radar and Sonar, Part I*, Springer-Verlag, New York, 1991.

[BNe74]   E. Beller and D. J. Newman, The minimum modulus of polynomials, *Proc. Amer. Math. Soc.*, 45 (1974), 463–465.

[BS82]   R. E. Berg and D. G. Stork, *The Physics of Sound*, Prentice-Hall, Inc., Englewood Cliffs, NJ, 1982.

[BSa94]   J. J. Benedetto and S. Saliani, Subband coding for sigmoidal non-linear operations, *Proc. SPIE*, 2242 (1994), 19–27.

[BSh66]   Z. I. Borevich and I. R. Shafarevich, *Number Theory*, Academic Press, New York, 1966.

[BrHe95]   W. L. Briggs and V. E. Henson, *The DFT, an Owner's Manual for the Discrete Fourier Transform*, Society for Industrial and Applied Mathematics, Philadelphia, PA, 1995.

[BSS88]   P. L. Butzer, W. Splettstösser, and R. L. Stens, The sampling theorem and linear prediction in signal analysis, *Jahresber. Deutsch. Math.-Verein.*, 90 (1988), 1–70.

[BSt86]   P. L. Butzer and E. L. Stark, "Riemann's Example" of a continuous nowhere differentiable function in the light of two letters (1865) of Christoffel to Prym, *Bull. Soc. Math. Belg. Sér A*, 38 (1986), 45–73.

[BT93]   J. J. Benedetto and A. Teolis, A wavelet auditory model and data compression, *Appl. Comput. Harmonic Anal.*, 1 (1993), 3–28.

[BTu59]   R. B. Blackman and J. W. Tukey, *The Measurement of Power Spectra*, Dover Publications, Inc., New York, 1959.

[BW94]   J. J. Benedetto and D. F. Walnut, Gabor frames for L^2 and related spaces, Chapter 3 in *Wavelets: Mathematics and Applications*, J. J. Benedetto and M. W. Frazier, editors, CRC Press, Boca Raton, FL, 1994.

[BZ96]   J. J. Benedetto and G. Zimmermann, Sampling operators and the Poisson Summation Formula, preprint.

[Bar78]   M. S. Bartlett, *An Introduction to Stochastic Processes*, 3rd edition, Cambridge University Press, London, 1978.

[Bary64]   N. K. Bary, *A Treatise on Trigonometric Series, Volumes* I and II, The MacMillan Company, New York, 1964 (1961).

[Bas84]   J. Bass, *Fonctions de Corrélation, Fonctions Pseudo- Aléatoires, et Applications*, Masson, Paris, 1984.

[Ben71]   J. J. Benedetto, *Harmonic Analysis on Totally Disconnected Sets*, Lecture Notes in Math. 202, Springer-Verlag, New York, 1971.

[Ben75]   J. J. Benedetto, *Spectral Synthesis*, Academic Press, New York, 1975.

[Ben76]   J. J. Benedetto, *Real Variable and Integration*, B. G. Teubner, Stuttgart, 1976.

[Ben77]   J. J. Benedetto, *A Mathematical Approach to Mathematics Appreciation. Lecture Notes 17*, Department of Mathematics, University of Maryland, College Park, MD, 1977.

[Ben80]   J. J. Benedetto, Fourier analysis of Riemann distributions and explicit formulas, *Math. Ann.*, 252 (1980), 141–164.

[Ben83]   J. J. Benedetto, Harmonic analysis and spectral estimation, *J. Math. Anal. and Appl.*, 91 (1983), 444–509.

[Ben91a]   J. J. Benedetto, The spherical Wiener–Plancherel formula and spectral estimation, *SIAM J. Math. Anal.*, 22 (1991), 1110–1130.

[Ben91b]   J. J. Benedetto, A multidimensional Wiener–Wintner theorem and spectrum estimation, *Trans. Amer. Math. Soc.*, 327 (1991), 833–852.

[Ben92a]   J. J. Benedetto, Stationary frames and spectral estimation, in *Probabilistic and Stochastic Methods in Analysis, with Applications*, J. S. Byrnes et al., editors, Kluwer Academic Publishers, Norwell, MA, 1992, 117–161.

[Ben92b]   J. J. Benedetto, Irregular sampling and frames, in *Wavelets: A Tutorial in Theory and Applications*, C. K. Chui, editor, Academic Press, New York, 1992.

[Bene84]   M. Benedicks, The support of functions and distributions with a spectral gap, *Math. Scand.*, 55 (1984), 285–309.

[Benk92]   G. Benke, On the minimum modulus of trigonometric polynomials, *Proc. Amer. Math. Soc.*, 114 (1992), 757–761.

[Ber87]   J.-P. Bertrandias et al., *Espaces de Marcinkiewicz, Corrélations, Mesures, Systèmes Dynamiques*, Masson, Paris, 1987.

[Beu89]   A. Beurling, *Collected Works*, *Volumes* 1 and 2, Birkhäuser, Boston, MA, 1989.

[Bir73]   G. Birkhoff, *A Source Book in Classical Analysis*, Harvard University Press, Cambridge, MA, 1973.

[Bom92]   E. Bombieri, Prime territory, *The Sciences*, Sept./Oct. (1992), 30–36.

[Bou65]   N. Bourbaki, *Intégration*, 2nd edition, Chapters 1–4, Hermann, Paris, 1965.

[Bra86]   R. Bracewell, *The Fourier Transform and Its Applications*, 2nd edition, McGraw-Hill Book Company, New York, 1986.

[Bre65]   H. J. Bremermann, *Distributions, Complex Variables, and Fourier Transforms*, Addison-Wesley Publishing Co., Inc., Reading, MA, 1965.

[Bri81]   D. R. Brillinger, *Time Series*, expanded edition, Holden-Day, Inc., San Francisco, CA, 1981.

[Bur79]   R. B. Burckel, *An Introduction to Classical Complex Analysis*, Birkhäuser-Verlag, Basel, 1979.

[Bur84]   R. B. Burckel, Bishop's Stone–Weierstrass theorem, *Amer. Math. Monthly*, 91 (1984), 22–32.

[But95]   P. L. Butzer, The Hausdorff–Young theorems of Fourier analysis and their impact, *J. Fourier Anal. Appl.*, 1 (1995), 113–130.

[CB78]   R. V. Churchill and J. W. Brown, *Fourier Series and Boundary Value Problems*, 3rd edition, McGraw-Hill Book Company, New York, 1978.

[CH53]   R. Courant and D. Hilbert, *Methods of Mathematical Physics*, Volume 1, John Wiley and Sons, New York, 1953 (1937).

[CT65]   J. W. Cooley and J. W. Tukey, An algorithm for the machine calculation of complex Fourier series, *Math. Comp.*, 19 (1965), 297–301.

[Cal64]   A. Calderón, Intermediate spaces and interpolation, the complex method, *Studia Math.*, 24 (1964), 113–190.

[Can55]   G. Cantor, *Contributions to the Founding of the Theory of Transfinite Numbers*, Dover Publications, Inc., New York, 1955 (1895, 1897).

[Car30]   H. S. Carslaw, *Fourier Series and Integrals*, 3rd edition, Dover Publications, Inc., 1930 (1906, 1921).

[Carl66]   L. Carleson, On convergence and growth of partial sums of Fourier series, *Acta Math.*, 116 (1966), 135–157.

[Carl80]   L. Carleson, Some analytic problem related to statistical mechanics, in *Euclidean Harmonic Analysis*, Lecture Notes in Math. 779, J. J. Benedetto, editor, Springer-Verlag, New York, 1980, 5–45.

[Cart91]   C. Carton-Lebrun, Continuity and inversion of the Hilbert transform of distributions, *J. Math. Anal. Appl.*, 161 (1991), 274–283.

[Cha68]   K. Chandrasekharan, *Introduction to Analytic Number Theory*, Springer-Verlag, New York, 1968.

[Cha70]   K. Chandrasekharan, *Arithmetical Functions*, Springer-Verlag, New York, 1970.

[Che80]   P. Chernoff, Pointwise convergence of Fourier series, *Amer. Math. Monthly*, 87 (1980), 399–400.

[Chi78]   D. G. Childers, editor, *Modern Spectrum Analysis*, IEEE Press, Piscataway, NJ, 1978.

[Col85]   J. F. Colombeau, *Elementary Introduction to New Generalized Functions*, North-Holland, Amsterdam, 1985.

[Con58]   F. W. Constant, *Theoretical Physics*, Addison-Wesley Publishing Company, Reading, MA, 1958.

[deJ64]   E. M. deJager, *Applications of Distributions in Mathematical Physics*, Mathematical Centre, Amsterdam, 1964.

[dHR93]   C. deBoor, K. Höllig, and S. Riemenschneider, *Box Splines*, Springer-Verlag, New York, 1993.

[DG79]   H. Dym and I. Gohberg, Extensions of matrix valued functions with rational polynomial inverses, *Integral Equations Operator Theory*, 2 (1979), 503–528.

[DM72]   H. Dym and H. P. McKean, *Fourier Series and Integrals*, Academic Press, New York, 1972.

[DM76]   H. Dym and H. McKean, *Gaussian Processes, Function Theory, and the Inverse Spectral Problem*, Academic Press, New York, 1976.

[DR71]   C. F. Dunkl and D. E. Ramirez, *Topics in Harmonic Analysis*, Appleton-Century-Crofts, New York, 1971.

[DV78]   P. Dierolf and J. Voigt, Convolution and S'-convolution of distributions, *Collect. Math.*, 29 (1978), 185–196.

[DVe90]   P. Duhamel and M. Vetterli, Fast Fourier transforms: A tutorial review and a state of the art, *Signal Process.*, 19 (1990), 259–299.

[Dau92]   I. Daubechies, *Ten Lectures on Wavelets*, CBMS-NSF Regional Conf. Ser. in Appl. Math. 61, Society for Industrial and Applied Mathematics., Philadelphia, PA, 1992.

[Daub79]   J. W. Dauben, *Georg Cantor: His Mathematics and Philosophy of the Infinite*, Harvard University Press, Cambridge, MA, 1979.

[Dav85]   B. Davies, *Integral Transforms and Their Applications*, 2nd edition, Springer-Verlag, New York, 1985.

[Die81]   J. Dieudonné, *History of Functional Analysis*, North-Holland, Amsterdam, 1981.

[Dir1829]   P. G. Lejeune-Dirichlet, Sur la convergence des séries trigonométriques que servent à représenter une fonction arbitraire entre des limites données, *J. Reine Angew. Math.*, 4 (1829), 157–169.

[Dir1837]   P. G. Lejeune-Dirichlet, Ueber die Darstellung ganz willkürlicher Functionen durch Sinus—und Cosinusreihen, *Rep. Phys.*, 1 (1837), 152–174.

[Don69]   W. Donoghue, *Distributions and Fourier Transforms*, Academic Press, New York, 1969.

[Dui91]   J. J. Duistemaat, Selfsimilarity of "Riemann's nondifferentiable function", *Nieuw Arch. Wisk.* (4) 9 (1991), 303–337.

[Ebe83]   W. F. Eberlein, Half-angles and Oldenburg's problem, preprint, 1983.

[Edw67]   R. E. Edwards, *Fourier Series*, Holt, Rinehart, and Winston, Inc., New York, 1967.

[Edwa74]   H. M. Edwards, *Riemann's Zeta Function*, Academic Press, New York, 1974.

[Erd62]   A. Erdélyi, *Operational Calculus and Generalized Functions*, Holt, Rinehart, and Winston, Inc., New York, 1962.

[FJW91]   M. W. Frazier, B. Jawerth, and G. Weiss, *Littewood-Paley Theory and the Study of Function Spaces*, CBMS Regional Conf. Ser. in Math. 79, American Mathematical Society, Providence, RI, 1991.

[Fef71]   C. Fefferman, On the convergence of Fourier series, *Bull. Amer. Math. Soc.*, 77 (1971), 744-745.

[Fef73]   C. Fefferman, Pointwise convergence of Fourier series, *Ann. of Math.* (2), 98 (1973), 551–571.

[Fel66]   W. Feller, *An Introduction to Probability Theory and Its Applications, Volume* 2, Wiley, New York, 1966.

[Fey86]   R. Feynman, *Surely You're Joking, Mr. Feynman*, Bantam Books, New York, 1986.

[Fou1822]   J. B. J. Fourier, *Théorie Analytique de la Chaleur*, 1822 (In *Oeuvres* I, with notes by G. Darboux).

[G-CRdeF85] J. García-Cuerva and J. L. Rubio de Francia, *Weighted Norm Inequalities*, Elsevier Science Publishers, Amsterdam, 1985.

[GG81]   I. Gohberg and S. Goldberg, *Basic Operator Theory*, Birkhäuser, Boston, MA, 1981.

[GL94]   I. Gohberg and H. J. Landau, Prediction and the inverse of Toeplitz matrices, *Internat. Ser. Numer. Math.*, 119 (1994), 219–229.

[GS58]   U. Grenander and G. Szegö, *Toeplitz Forms and their Applications*, University of California Press, Berkeley and Los Angeles, 1958.

[Gab93]   J.-P. Gabardo, Extensions of positive-definite distributions and maximum entropy, *Mem. Amer. Math. Soc.*, 102 (489), 1993.

[Gabo46]   D. Gabor, Theory of communication, *J. IEE*, 93 (1946), 429–457.

[Gar88]   W. Gardner, *Statistical Spectral Analysis*, Prentice-Hall, Englewood Cliffs, NJ, 1988.

[Gau66]   C. F. Gauss, *Disquisitiones Arithmeticae*, A. A. Clarke, S. J., translator, Yale University Press, New Haven, CT, 1966 (1801).

[Gib1893]   G. A. Gibson, On the history of the Fourier series, *Proc. Edinburgh Math. Soc.*, 11 (1893), 137–166.

[Gle91]   A. M. Gleason, *Fundamentals of Abstract Analysis*, Jones and Bartlett Publishers, Boston, MA, 1991 (1966).

[Glei92]   J. Gleick, *Genius—the Life and Science of Richard Feynman*, Vintage Books, New York, 1992.

[God72]   G. Godin, *The Analysis of Tides*, University of Toronto Press, Ontario, 1972.

[Gol61]   R. R. Goldberg, *Fourier Transforms*, Cambridge University Press, London, 1961.

[Gold59]   J. Golding, *Cubism: a History and Analysis 1907–1914*, 3rd edition, Belknap Press of Harvard University, Cambridge, MA, 1988 (1959).

[G-V92]   E. González-Velasco, Connections in mathematical analysis, *Amer. Math. Monthly*, 99 (1992), 427–441.

[Goo62]   I. J. Good, Analogues of Poisson's summation formula, *Amer. Math. Monthly*, 69 (1962), 258–266.

[Good68]   J. W. Goodman, *Introduction to Fourier Optics*, McGraw-Hill Book Company, New York, 1968.

[Gra84]   J. D. Gray, The shaping of the Riesz representation theorem: a chapter in the history of analysis, *Arch. Hist. Exact Sci.*, 31 (1984), 127–187.

[Graf94]   L. Grafakos, An elementary proof of the square summability of the discrete Hilbert transform, *Amer. Math. Monthly*, 101 (1994), 456–458.

[Grat72]   I. Grattan-Guinness, *Joseph Fourier 1768–1830*, in collaboration with J. R. Ravetz, The MIT Press, Cambridge, MA, 1972.

[Gus87]   K. E. Gustafson, *Partial Differential Equations and Hilbert Space Methods*, 2nd edition, John Wiley and Sons, New York, 1987.

[HH79]   E. Hewitt and R. Hewitt, The Gibbs–Wilbraham phenomenon: an episode in Fourier analysis, *Arch. Hist. Exact Sci.*, 21 (1979), 129–160.

[HJB84]   M. T. Heideman, D. H. Johnson, and C. S. Burrus, Gauss and the history of the fast Fourier transform, *IEEE ASSP Magazine*, October (1984), 14–21.

[HL95]   J. A. Hogan and J. D. Lakey, Extensions of the Heisenberg group by dilations and frames, *Appl. Comput. Harmonic Anal.*, 2 (1995), 174–199.

[HLP52]   G. H. Hardy, J. E. Littlewood, and G. Pólya, *Inequalities*, 2nd edition, Cambridge University Press, London, 1952 (first edition, 1934).

[HM89]   R. Hummel and R. Moniot, Reconstructions from zero crossings in scale space, *IEEE Trans. ASSP*, 37 (1989), 2111–2130.

[HR56]   G. H. Hardy and W. W. Rogosinski, *Fourier Series*, 3rd edition, Cambridge University Press, London, 1956 (1944, 1950).

[HS65]   E. Hewitt and K. Stromberg, *Real and Abstract Analysis*, Springer-Verlag, New York, 1965.

[HW65]   G. H. Hardy and E. M. Wright, *An Introduction to the Theory of Numbers*, 4th edition, Oxford University Press, London, 1965 (first edition, 1938).

[Har49]   G. H. Hardy, *Divergent Series*, Oxford University Press, London, 1949.

[Harr78]   F. Harris, On the use of windows for harmonic analysis, *Proc. IEEE*, 66 (1978), 51–83.

[Haw70]   T. Hawkins, *Lebesgue's Theory of Integration*, University of Wisconsin Press, Madison, WI, 1970.

[Hay85]   S. Haykin, editor, *Array Signal Processing*, Prentice-Hall, Englewood Cliffs, NJ, 1985.

[Hay94]   S. Haykin, *Neural Networks*, IEEE Press, Piscataway, NJ, 1994.

[Hea1894]   O. Heaviside, On operators in physical mathematics, *Proc. Royal Soc.*, 52 (1893), 504–529, and 54 (1894), 105–143.

[Hed80]   L. Hedberg, Spectral synthesis and stability in Sobolev spaces, in *Euclidean Harmonic Analysis*, Lecture Notes in Math. 779, J. J. Benedetto, editor, Springer-Verlag, New York, 1980.

[Hel64]   H. Helson, *Lectures on Invariant Subspaces*, Academic Press, New York, 1964.

[Hel83]   H. Helson, *Harmonic Analysis*, Addison-Wesley Publishing Company, Reading, MA, 1983.

[Her85]   D. Herrero, From my chest of examples of Fourier transforms, *Rev. Un. Mat. Argentina*, 32 (1985), 41–47.

[Herl93]   C. Herley, *Wavelets and Filter Banks*, Ph.D. dissertation, Columbia University, New York, 1993.

[Hig85]   J. R. Higgins, Five short stories about the cardinal series, *Bull. Amer. Math. Soc.*, 12 (1985), 45–89.

[Hil74]   E. Hille, *Analytic Function Theory*, *Volume* II, Chelsea Publishing Company, New York, 1974 (1962).

[Hob26]   E. W. Hobson, *The Theory of Functions of a Real Variable and the Theory of Fourier Series*, *Volume* II, 2nd edition, Cambridge University Press, 1926 (1907).

[Hof62]   K. Hoffman, *Banach Spaces of Analytic Functions*, Prentice-Hall, Inc., Englewood Cliffs, NJ, 1962.

[Hör83]   L. Hörmander, *The Analysis of Linear Partial Differential Operators*, *Volumes I* and *II*, Springer-Verlag, New York, 1983.

[Hor66]   J. Horváth, *Topological Vector Spaces*, Addison-Wesley Publishing Co., Reading, MA, 1966.

[IEEE69]   IEEE Trans. Audio Electroacoustics, 17 (2), June 1969, Special Issue on Fast Fourier Transforms.

[IEEE82]   *Proceedings IEEE*, 70 (9), September 1982, Special Issue on Spectral Estimation.

[Ivi85]   A. Ivić, *The Riemann Zeta Function*, John Wiley and Sons, New York, 1985.

[JN84]   N. S. Jayant and P. Noll, *Digital Coding of Waveforms*, Prentice-Hall, Inc., Englewood Cliffs, NJ, 1984.

[JNT75]   J. J. B. Jack, D. Noble, and R. W. Tsien, *Electric Current Flow in Excitable Cells*, Oxford University Press, London, 1975.

[Jac41]   D. Jackson, *Fourier Series and Orthogonal Polynomials*, Carus Monographs, Mathematical Association of America, Washington, D.C., 1941.

[Jaf94]   S. Jaffard, Wavelets and nonlinear analysis, Chapter 12 of *Wavelets: Mathematics and Applications*, J. J. Benedetto and M. W. Frazier, editors, CRC Press, Inc., Boca Raton, FL, 1994.

[Jan88]   A. J. E. M. Janssen, The Zak transform: a single transform for sampled time-continuous signals, *Philips J. Res.*, 43 (1988), 23–69.

[Jon82]   D. S. Jones, *The Theory of Generalized Functions*, Cambridge University Press, London, 1982.

[Joy86]   D. Joyner, *Distribution Theorems of L-Functions*, Longman Scientific and Technical, Essex, England, 1986.

[KJ55]   L. Kovasznay and H. Joseph, Image processing, *Proc. IRE*, 43 (1955), 560–570.

[KK64]   J. F. Koksma and L. Kuipers, editors, *Asymptotic Distribution Modulo 1*, P. Noordhoff, N.V., The Netherlands, 1964 (NUFFIC 1962).

[KS63]   J.-P. Kahane and R. Salem, *Ensembles Parfaits et Séries Trigonométriques*, Hermann, Paris, 1963.

[Kac59]   M. Kac, *Statistical Independence in Probability, Analysis, and Number Theory*, Carus Monograph 12, Mathematical Association of America, Washington, D.C., 1959.

[Kah70]   J.-P. Kahane, *Séries de Fourier Absolument Convergentes*, Springer-Verlag, New York, 1970.

[Kah80]   J.-P. Kahane, Sur les polynômes à coefficients unimodulaires, *Bull. London Math. Soc.*, 12 (1980), 321–342.

[Kahn67]   D. Kahn, *The Codebreakers*, The MacMillan Company, New York, 1967.

[Kat76]   Y. Katznelson, *An Introduction to Harmonic Analysis*, Dover Publications, New York, 1976 (1968).

[Ker90]   R. B. Kerby, *The Correlation Function and the Wiener-Wintner Theorem in Higher Dimensions*, Ph.D. dissertation, University of Maryland, College Park, MD, 1990.

[Kle56]   F. Klein, *The Icosahedron*, Dover Publications, New York, 1956 (1884, 1913).

[Kli72]   M. Kline, *Mathematical Thought from Ancient to Modern Times*, Oxford University Press, London, 1972.

[Kol41]   A. N. Kolmogorov, Stationary sequences in Hilbert space, G. Kallianpur, translator, *Bull. Moscow State Univ. Math.*, 2 (6) (1941).

[Koo64]   P. Koosis, Sur un théorème de Paul Cohen. *C. R. Acad. Sci. Paris Sér. I Math.*, 259 (1964), 1380–1382.

[Koo88]   P. Koosis, *The Logarithmic Integral I*, Cambridge University Press, London, 1988.

[Kör87]   T. W. Körner, Uniqueness for trigonometric series, *Ann. of Math.*, 126 (1987), 1–34.

[Kör88]   T. W. Körner, *Fourier Analysis*, Cambridge University Press, London, 1988.

[LP71]   L.-Å. Lindahl and F. Poulsen, *Thin Sets in Harmonic Analysis*, Marcel Dekker, Inc., New York, 1971.

[Lam66]   J. Lamperti, *Probability*, W. A. Benjamin, Inc., New York, 1966.

[Lam77]   J. Lamperti, *Stochastic Processes*, Springer-Verlag, New York, 1977.

[Lan87]   H. J. Landau, Maximum entropy and the moment problem. *Bull Amer. Math. Soc.*, 16 (1987), 47–77.

[Lar71]   R. Larsen, *An Introduction to the Theory of Multipliers*, Springer-Verlag, New York, 1971.

[Leb02]   H. Lebesgue, Intégral, longeur, aire, *Ann. Mat.*, 7 (1902), 231–359.

[Leb06]   H. Lebesgue, *Leçons sur les Séries Trigonométriques*, Gauthiers-Villars, Paris, 1906.

[Lee60]   Y. Lee, *Statistical Theory of Communication*, John Wiley, New York, 1960.

[Lem93]   P. G. Lemarié, Fonctions d'échelle pour les ondelettes de dimension n, preprint, 1993.

[Len72]   A. Lenard, The numerical range of a pair of projections, *J. Funct. Anal.*, 10 (1972), 410–423.

[Lev40]   N. Levinson, *Gap and Density Theorems*, Amer. Math. Soc. Colloq. Publ. 26, American Mathematical Society, Providence, RI, 1940.

[Lio86]   G. Lion, A simple proof of the Dirichlet-Jordan convergence test, *Amer. Math. Monthly*, 93 (1986), 281–282.

[Lit66]   J. E. Littlewood, On polynomials $\sum^n \pm z^m$, $\sum^n e^{\alpha_m i} z^m$, $z = e^{\theta i}$, *J. London Math. Soc.*, 41 (1966), 367–376.

[Lit68]   J. E. Littlewood, *Some Problems in Real and Complex Analysis*, D. C. Heath and Co., Lexington, MA, 1968.

[Loe66]   E. V. Loewenstein, The history and current status of Fourier transform spectroscopy, *Appl. Optics*, 5 (1966), 845–854.

[Log83]   B. F. Logan, Hilbert transform of a function having a bounded integral and a bounded derivative, *SIAM J. Math. Anal.*, 14 (1983), 247–248.

[Loj57]   S. Łojasiewicz, Sur la valeur et la limite d'une distribution en un point, *Studia Math.*, 16 (1957), 1–36.

[Lux62]   W. A. J. Luxemburg, A property of the Fourier coefficients of an integrable function, *Amer. Math. Monthly*, 69 (1962), 94–98.

[MG82]   J. D. Markel and H. H. Gray Jr., *Linear Prediction of Speech*, Springer-Verlag, New York, 1982.

[MP72]   J. H. McClellan and T. W. Parks, Eigenvalue and eigenvector decomposition of the discrete Fourier transform, *IEEE Trans. Audio Electroacoustics*, March 1972, 66–74.

[MPS81]   C. McGehee, L. Pigno, and B. Smith, Hardy's inequality and the L^1-norm of exponential sums, *Ann. of Math.* (2), 113 (1981), 613–618.

[MW57]   P. Masani and N. Wiener, The prediction theory of multivariate stochastic processes, I, *Acta Math.*, 98 (1957), 111–203.

[MZ93]   S. Mallat and Z. Zhang, Matching pursuits with time-frequency dictionaries, *IEEE Trans. Signal Processing*, 41 (1993), 3397–3415.

[Mac66]   D. H. Macmillan, *Tides*, C. R. Books, London, 1966.

[Mad92]   W. Madych, Some elementary properties of multiresolution analyses of $L^2(\mathbb{R}^n)$, in *Wavelets—A Tutorial in Theory and Applications*, C. Chui, editor, Academic Press, Boston, 1992, 259–294.

[Mak75]   J. Makhoul, Linear prediction: a tutorial review, *Proc. IEEE*, 63 (1975), 561–580.

[Mal79]   P. Malliavin, On the multiplier theorem for Fourier transforms of measures with compact support, *Ark. Mat.*, 17 (1979), 69–81.

[Mal82]   P. Malliavin, *Intégration et Probabilités, Analyse de Fourier et Analyse Spectrale*, Masson, Paris, 1982.

[Mall89]   S. Mallat, Multiresolution approximations and wavelet orthonormal bases of $L^2(\mathbb{R})$, *Trans. Amer. Math. Soc.*, 315 (1989), 69–87.

[Mas90]   P. Masani, *Norbert Wiener*, Birkäuser-Verlag, Boston, MA, 1990.

[Men68]   D. E. Mensov, Limits of indeterminancy in measure of trigonometric and orthogonal series, Trudy Mat. Inst. Steklov. 99 (1967); transl. in Proc. Steklov. Inst. Math. 99 (1968).

[Mey72]   Y. Meyer, *Algebraic Numbers and Harmonic Analysis*, North-Holland Publishing Company, Amsterdam, 1972.

[Mey81]   Y. Meyer, Multiplication of distributions, *Mathematical Analysis and Applications, Adv. in Math. Suppl. Stud.*, 7 (1981), 603–615.

[Mey90]   Y. Meyer, *Ondelettes et Opérateurs*, Hermann, Paris, 1990.

[Mey91]   Y. Meyer, editor, *Wavelets and Applications*, Springer-Verlag, New York, 1991.

[Mic62]   A. A. Michelson, *Studies in Optics*, University of Chicago Press, Chicago, IL, 1962.

[Mon71]   H. L. Montgomery, *Topics in Multiplicative Number Theory*, Lecture Notes in Math. 227, Springer-Verlag, New York, 1971.

[Mon94]   H. L. Montgomery, *Ten lectures on the Interface between Analytic Number Theory and Harmonic Analysis*, in CBMS Regional Conf. Ser. in Math. 84, American Mathematical Society, Providence, RI, 1994.

[Monn72]   A. F. Monna, The concept of function in the 19th and 20th centuries, in particular with regard to the discussions between Baire, Borel, and Lebesgue, *Arch. Hist. Exact Sci.*, 9 (1972), 57–84.

[Moz71]   C. J. Mozzochi, *On the Pointwise Convergence of Fourier Series*, Lecture Notes in Math. 199, Springer-Verlag, New York, 1971.

[Nah88]   P. J. Nahin, *Oliver Heaviside, Sage in Solitude*, IEEE Press, Piscataway, NJ, 1988.

[Ner71]   U. Neri, *Singular Integrals*, Lecture Notes in Math. 200, Springer-Verlag, New York, 1971.

[New95]   D. J. Newman, Finite type functions as limits of exponential sums, *J. Fourier Anal. Appl.*, Kahane Special Issue (1995), 479–483.

[Nyq28]   H. Nyquist, Certain topics in telegraph transmission theory, *AIEE Trans.*, 47 (1928), 617–644.

[OS75]   A. V. Oppenheim and R. Schafer, *Digital Signal Processing*, Prentice-Hall, Inc., Englewood Cliffs, NJ, 1975.

[OW83]   A. V. Oppenheim and A. S. Willsky, *Signals and Systems*, Prentice-Hall, Inc., Englewood Cliffs, NJ, 1983.

[Obe92]   M. Oberguggenberger, *Multiplication of Distributions and Applications to Partial Differential Equations*, Pitman Res. Notes in Math. Ser. 259, Longman Scientific and Technical, Harlow, 1992.

[Olv74]   F. W. J. Olver, *Asymptotics and Special Functions*, Academic Press, New York, 1974.

[PS76]   G. Pólya and G. Szegö, *Problems and Theorems in Analysis*, Volume II, 4th edition, C. E. Billigheimer, translator, Springer-Verlag, New York, 1976.

[PW33]   R. E. A. C. Paley and N. Wiener, Notes on the theory and applications of Fourier transforms I, II, *Trans. Amer. Math. Soc.*, 35 (1933), 348–355.

[PW34]   R. E. A. C. Paley and N. Wiener, *Fourier Transforms in the Complex Domain*, Amer. Math. Soc. Colloq. Publ. 19, American Mathematical Society, Providence, RI, 1934.

[Pap66]   A. Papoulis, Error analysis in sampling theory, *Proc. IEEE*, 54 (1966), 947–955.

[Pap77]   A. Papoulis, *Signal Analysis*, McGraw-Hill Book Company, New York, 1977.

[Per52]   S. Perlis, *Theory of Matrices*, Addison-Wesley, Reading, MA, 1952.

[Pie83]   J. R. Pierce, *The Science of Musical Sound*, Scientific American Books, Inc., New York, 1983.

[Pla10]   M. Plancherel, Contribution à l'étude de la réprésentation d'une fonction arbitraire par des intégrales défines, *Rend. Circ. Mat. Palermo* (2), 30 (1910), 289–335.

[Pla15]   M. Plancherel, Sur la convergence et la sommabilité par les moyennes de $\lim_{x=\infty} \int_a^x f(x) \cos(xy)\, dx$, *Math. Ann.*, 76 (1915), 315–326.

[Pla25]   M. Plancherel, Le développement de la théorie des séries trigonométriques dans le dernier quart de siècle, *Enseign. Math.*, 24 (1924–1925), 19–58.

[Pól54]   G. Pólya, *Induction and Analogy in Mathematics*. Princeton University Press, Princeton, NJ, 1954.

[Pou84]   M. Pourahmadi, Taylor expansion of $\exp(\sum_{k=0}^{\infty} a_k z^k)$ and some applications, *Amer. Math. Monthly*, 91 (1984), 303–307.

[Pric85]   J. F. Price, editor, *Fourier Techniques and Applications*, Plenum Press, New York, 1985.

[Pri81]   M. B. Priestley, *Spectral Analysis and Time Series*, Volume 1, Academic Press, New York, 1981.

[Ptá72]   V. Pták, Un théorème de factorisation, *C. R. Acad. Sci. Paris Sér. I Math.*, 275 (1972), 1297–1299.

[QS96]   H. Queffelec and B. Saffari, On Bernstein's inequality and Kahane's ultraflat polynomials, *J. Fourier Anal. Appl.*, 2 (1996).

[RL55]   M. Riesz and A. Livingston, A short proof of a classical theorem in the theory of Fourier integrals, *Amer. Math. Monthly*, 62 (1955), 434–437.

[RN55]   F. Riesz and B. Sz.-Nagy, *Functional Analysis*, 2nd edition, F. Ungar Publishing Co., New York, 1955.

[RR72]   L. R. Rabiner and C. M. Rader, editors , *Digital Signal Processing*, IEEE Press, New York, 1972.

[Red77]   R. M. Redheffer, Completeness of sets of complex exponentials, *Adv. in Math.*, 24 (1977), 1–62.

[Rei68]   H. Reiter, *Classical Harmonic Analysis and Locally Compact Groups*, Oxford University Press, London, 1968.

[Ric54]   S. O. Rice, Mathematical analysis of random noise, in *Noise and Stochastic Processes*, N. Wax, editor, Bell System Tech. J. 23 and 24, Dover Publications, Inc., New York, 1954.

[Rie1873] B. Riemann, Sur la possibilité de représenter une fonction par une série trigonométrique, *Bull. Sci. Math. Astron.*, 5 (1873), 20–48 and 79–96 (1854).

[Ries14]   F. Riesz, Démonstration nouvelle d'un théorème concernant les opérations fonctionnelles linéaires, *Ann. Sci. École Norm. Sup.*, 31 (1914), 9–14.

[Ries49]   F. Riesz, L'évolution de la notion d'intégrale depuis Lebesgue, *Ann. Inst. Fourier* (Grenoble), 1 (1949), 29–42.

[Rih85]   A. W. Rihaczek, *Principles of High-Resolution Radar*, Peninsula Publishing, Los Altos, CA, 1985.

[Rog59]   W. Rogosinski, *Fourier Series*, Chelsea Publishing Co., New York, 1959 (1930).

[Ros91]   J.-P. Rosay, A very elementary proof of the Malgrange-Ehrenpreis theorem, *Amer. Math. Monthly*, 98 (1991), 518–523.

[Rud62]   W. Rudin, *Fourier Analysis on Groups*, John Wiley and Sons, New York, 1962.

[Rud63]   W. Rudin, The extension problem for positive definite functions, *Illinois J. Math.*, 7 (1963), 532–539.

[Rud66]   W. Rudin, *Real and Complex Analysis*, McGraw-Hill Book Company, New York, 1966.

[Rud73]   W. Rudin, *Functional Analysis*, McGraw-Hill Book Company, New York, 1973.

[SB65]   P. Sconzo and J. J. Benedetto, The orbit of an equatorial satellite, Technical Report 0232-G, IBM Federal Systems Division, Cambridge, MA, 1965.

[SI64]   R. Shiraishi and M. Itano, On the multiplicative products of distributions, *J. Sci. Hiroshima Univ.*, 28 (1964), 223–235.

[SW71]   E. M. Stein and G. Weiss, *An Introduction to Fourier Analysis on Euclidean Spaces*, Princeton University Press, Princeton, NJ, 1971.

[Sal63]   R. Salem, *Algebraic Numbers and Fourier Analysis*, D.C. Heath, Boston, MA, 1963.

[Sal67]   R. Salem, *Oeuvres Mathématiques*, Hermann, Paris, 1967.

[Sch61]   L. Schwartz, *Méthodes Mathématiques pour les Sciences Physiques*, with the assistance of D. Huet, Hermann, Paris, 1961.

[Sch63]   L. Schwartz, Some applications of the theory of distributions, in *Lectures on Modern Mathematics*, Volume I, T. L. Saaty, editor, John Wiley and Sons, New York, 1963.

[Sch66]   L. Schwartz, *Théorie des Distributions*, Hermann, Paris 1966 (1950, 1951).

[Scha85]   L. Scharlau, *Quadratic and Hermitian Forms*, Springer-Verlag, New York, 1985.

[Sche60]   S. A. Schelkunoff, A mathematical theory of linear arrays, *Bell System Tech. J.*, 39 (1960), 80–107.

[Schi68]   L. I. Schiff, *Quantum Mechanics*, McGraw-Hill Book Company, 3rd edition, New York, 1968 (1955).

[Scho73]   I. J. Schoenberg, *Cardinal Spline Interpolation*, CBMS-NSF Regional Conf. Ser. in Appl. Math. 12, Society for Industrial and Applied Mathematics, Philadelphia, PA, 1973.

[Sea87]   G. F. C. Searle, *Oliver Heaviside, the Man*, C.A.M. Publishing, St. Albans, England, 1987.

[Sha58]   D. Shanks, Two theorems of Gauss, *Pacific J. Math.*, 8 (1958), 609–612.

[Ste70]   E. M. Stein, *Singular Integrals and Differentiability Properties of Functions*, Princeton University Press, Princeton, NJ, 1970.

[Ste93]   E. M. Stein, *Harmonic Analysis: Real-Variable Methods, Orthogonality, and Oscillatory Integrals*, with the assistance of T. S. Murphy, Princeton University Press, Princeton, NJ, 1993.

[Stei76]   B. D. Steinberg, *Principles of Aperture and Array System Design*, John Wiley and Sons, New York, 1976.

[Str88]   G. Strang, *Linear Algebra and its Applications*, Harcourt Brace Jovanovich, New York, 1988.

[Swe96]   D. Sweet, *Fundamental Principles of Ordinary Differential Equations*, CRC Press, Inc., Boca Raton, FL, 1996.

[TAL89]   R. Tolimieri, M. An, and C. Lu, *Algorithms for Discrete Fourier Transform and Convolution*, Springer-Verlag, New York, 1989.

[Tay58]   A. E. Taylor, *Introduction to Functional Analysis*, John Wiley and Sons, Inc., New York, 1958.

[Tit51]   E. C. Titchmarsh, *The Theory of the Riemann Zeta-Function*, Oxford University Press, London, 1951.

[Tit67]   E. C. Titchmarsh, *Introduction to the Theory of Fourier Integrals*, 2nd edition, Oxford University Press, London, 1967.

[Vai93]   P. P. Vaidyanathan, *Multirate Systems and Filter Banks*, Prentice-Hall, Englewood Cliffs, NJ, 1993.

[vN55]   J. von Neumann, *Mathematical Foundations of Quantum Mechanics*, Princeton University Press, Princeton, NJ, 1955 (1932, 1949).

[Walk91]   J. Walker, *Fast Fourier Transforms*, CRC Press, Boca Raton, FL, 1991.

[Wei65]   H. F. Weinberger, *A First Course in Partial Differential Equations*, John Wiley and Sons, New York, 1965.

[Wey50a]   H. Weyl, *The Theory of Groups and Quantum Mechanics*, Dover Publications, Inc., New York, 1950 (1928, 1930).

[Wey50b]   H. Weyl, Ramifications, old and new, of the eigenvalue problem, *Bull. Amer. Math. Soc.*, 56 (1950), 115–139.

[Wid41]   D. V. Widder, *The Laplace Transform*, Princeton University Press, Princeton, NJ, 1941.

[Wie33]   N. Wiener, *The Fourier Integral and Certain of Its Applications*, Cambridge University Press, London, 1933.

[Wie48]   N. Wiener, *Cybernetics*, The MIT Press, Cambridge, MA, 1948.

[Wie49]   N. Wiener, *Time Series*, The MIT Press, Cambridge, MA, 1949.

[Wie81]   N. Wiener, *Collected Works*, P. Masani, editor, The MIT Press, Cambridge, MA, 1981 (1976, 1979).

[Wik65]   I. Wik, Extrapolation of absolutely convergent Fourier series by identically zero, *Ark. Mat.*, 6 (1965), 65–76.

[Yos84]   K. Yosida, *Operational Calculus*, Springer-Verlag, New York, 1984.

[You80]   R. M. Young, *An Introduction to Nonharmonic Fourier Series*, Academic Press, New York, 1980.

[Zem87]   A. H. Zemanian, *Generalized Integral Transformations*, Dover Publications, Inc., New York, 1987.

[Zyg59]   A. Zygmund, *Trigonometric Series, Volumes I and II*, 2nd edition, Cambridge University Press, London, 1959.

Index of Notation

τ_t, 9, 97

$\Omega - BL$, 11

$\varnothing$, xiv, 171
$\equiv$, xiv
$\mathbf{1}_x$, xiv

$\int g(t)\,dt$, 1
$\int f(t)\,d\mu(t)$, 82
$\int_{\mathbb{T}_{2\Omega}} F(\gamma)\,d\gamma$, 156

$*$, 22, 96
$\tilde{\ }$, 51, 65, 100
$\longleftrightarrow$, 2, 54, 154, 225
$\left(\frac{p}{q}\right)$ (Legendre symbol), 237

$|X|$ (Lebesgue measure of X), 279
X^c (complement of X), xiv
$P \gg 0$ (P is positive definite), 114
$\|f\|_{L^p(\mathbb{R})}$ ($L^p(\mathbb{R})$ norm of f), 281
f_λ (dilation of f), 9
f_s (sampling of f), 250
$f[n]$ (Fourier coefficient of f), 153
$\overset{\circ}{f}_N$ (N-periodization of f), 248
$\overset{\circ}{f}_T$ (T-periodization of f), 254
$\hat{f}$ (Fourier transform of f), 2, 54, 89, 92, 154, 225
$\check{f}$ (inverse Fourier transform of f), 2, 54, 154, 225, 250
T' (distributional derivative of T), 80
T_g (distribution corresponding to function g), 78

Index of Names

Abel, 165, 192, 261, 267
Adelson, 45
Airy, 138
Archimedes, 110, 125
Artin, 239
Atal, 217
Auslander, L., 231

Bary, 173, 174, 175
Bass, J., 132
Beltrami, 172
Bernoulli, D., 167, 168
Bernoulli, J., 183
Bernstein, 173
Berry, 117
Bertrand, 172
Bertrandias, 49, 132
Betti, 172
Beurling, 91, 123, 167, 222, 223, 224, 251
Bôcher, 47
Bochner, 115, 117
Bohl, 49
Bombieri, 151, 273
Bonnet, 170
Boole, 104
Borel, 171
Braque, xii
Bremmer, 104
Brown, 173, 239
Burg, 205
Burkhardt, 166
Burrus, 239
Burt, 45
Butzer, 248

Cantor, 171, 172, 173
Carleman, 216, 222
Carleson, 175, 243, 247, 271
Carlini, 239
Carslaw, 166

Carson, 104
Catherine de Medici, 66
Cauchy, 169
Chernoff, P., 160
Chover, 204
Cicero, 125
Clairaut, 168
Cohen, P., 199, 259
Coifman, 247
Colombeau, 101
Cooley, 239
Cramér, 132

d'Alembert, 167, 168
de la Vallée-Poussin, 173
deLeeuw, 174
du Bois–Reymond, 175, 180, 188
Darboux, 168
Daubechies, 262
Dedekind, 239
Dicke, 165
Dieudonné, 184
Dini, 169, 171, 172
Dirac, 72
Dirichlet, 38, 160, 169, 170, 176, 177, 235, 237
Doetsch, 104
Doss, 117, 204
Dym, 204

Ehrenpreis, 105
Eichler, 239
Einstein, 126, 165
Euler, 165, 167, 168, 183, 232, 237, 266

Fagnano, 165
Fant, 217
Fantappié, 251
Fatou, 172, 175, 280
Fefferman, C., 247

Index

absolutely continuous, 285
absolutely convergent Fourier series, 162
adjoint, 290
algebraic number, 172
aliasing, 49, 258
all-pole model, 217
all-pole model method of spectral estimation, 220
almost everywhere, 279
amplitude, 7
analytic functionals, 251
antenna array, 238
approximate identity, 24, 186
apsidal line shift, 165
arithmetic progressions, 267
asymptotically unbiased estimator, 123
Atlantic submarine cable problem, 104
autocorrelation, 214
autoregressive (AR) model, 217
autoregressive moving average(ARMA) model, 217
Axiom of Choice, 292

Baire Category Theorem, 292
Banach algebra, 148, 199
Banach space, 288
Banach–Alaoglu Theorem, 355
Banach–SteinhausTheorem, 291
bandlimited, 11
bandpass, 109
barometric variations, 239
Beppo Levi Theorem, 280
Bessel Inequality, 194
Betti number, 172
Beurling Theorem, 224
Beurling–Malliavin Theorem, 222
bijective, 288
bilateral Laplace transform, 92, 146
binary expansion, 246

bit reversal, 244, 246
Blaschke products, 175
Bonnet Theorem, 34
Borel algebra, 112
bounded set, 75
bounded measures, 82
bounded positive measures, 83
bounded quadratic means, 126
bounded Radon measures, 82
bounded variation, 3
butterfly, 245

cable equation, 68
Calderón–Zygmund Theory, 272
Cantor function, 20, 67, 87
Cantor measure, 87, 147
Cantor set with ratio of dissection α, 174
Cantor–Lebesgue Lemma, 172
Carleson Theorem, 175, 247
carrier frequency, 10
carrier wave, 10
Cauchy principal value, 1
Cauchy sequence, 288
causal signal, 28, 108
Central Limit Theorem, 30
character, 226
charge, 83
chirp z-algorithm, 264
chirp signal, 133
chirp transform algorithm, 139
circle group, 155
Classical Sampling Theorem, 183, 256
Classical Uncertainty Principle Inequality, 125, 204
closed set, 75, 288
closed ideal, 148
closed translation invariant subspace, 131
closure problem, 223
closure theorems, 222
Cohen Factorization Theorem, 199